Guidebook: Toxic Substances Control Act

Volume II

Editor

George Dominguez

President
Springborn Regulatory Services, Inc.
Enfield, Connecticut

 **CRC Press**
Taylor & Francis Group
Boca Raton London New York

CRC Press is an imprint of the
Taylor & Francis Group, an **informa** business

CRC Press
Taylor & Francis Group
6000 Broken Sound Parkway NW, Suite 300
Boca Raton, FL 33487-2742

Reissued 2019 by CRC Press

© 1983 by Taylor & Francis Group, LLC
CRC Press is an imprint of Taylor & Francis Group, an Informa business

No claim to original U.S. Government works

A Library of Congress record exists under LC control number:

Publisher's Note
The publisher has gone to great lengths to ensure the quality of this reprint but points out that some imperfections in the original copies may be apparent.

Disclaimer
The publisher has made every effort to trace copyright holders and welcomes correspondence from those they have been unable to contact.

ISBN 13: 978-0-367-26303-4 (hbk)
ISBN 13: 978-0-367-26330-0 (pbk)
ISBN 13: 978-0-429-29265-1 (ebk)

Visit the Taylor & Francis Web site at http://www.taylorandfrancis.com and the
CRC Press Web site at http://www.crcpress.com

ACKNOWLEDGMENTS

In reviewing the enormous effort that went into the creation of this *Guidebook,* I can only once again complement those on my Editorial Board and our contributors without whose continuous aid and cooperation this book would not have been possible. In appreciation of their efforts, I would like to repeat the thanks given to them and to our CRC Press colleagues in the "Acknowledgments" made in what is now Volume I of this *Guidebook.*

As Editor, my task was more than eased by the expertise and resourcefulness of my Advisory Board. My personal thanks goes to each and every one of them for their invaluable assistance in reviewing and commenting on this publication from its conceptual inception to final publication. Obviously, this board, regardless of skill or knowledge, could accomplish little or nothing were it not for the actual articles that represent the substance of the guidebook. I know that I speak for the entire board when I express my thanks, which is hardly sufficient for the actual work involved, to each of our authors who not only took the time from their busy schedules to prepare their manuscripts but, more importantly, were willing to share their knowledge with us in the preparation of this publication. And last, but not least, we must express our gratitude to the publisher and its internal Editorial Staff, especially Benita Budd, Senior Editor, CRC Uniscience, who certainly eased the Editor's burden by doing all of the hard work and leaving him the glory. May she and the others at CRC Press know that their hard work does not go unnoticed or unappreciated.

George S. Dominguez
Editor
Wilton, Connecticut
1981

EDITOR

George S. Dominguez, M.B.A., is President of Springborn Regulatory Services, Inc., Enfield, Connecticut.

Mr. Dominguez received his B.S. and M.B.A. from Kentucky Christian University, Ashland. He is also a graduate of the Manhattan Medical and Dental Assistant School and the U.S. Army Intelligence School.

Mr. Dominguez is a member of the Board of Governors of the Synthetic Organic Chemical Manufacturers Association. He is a member of the Government Relations Committee and Chairman of the Chemical Regulations Advisory Committee of the Manufacturing Chemists Association. He is also a member of the Ecology Steering Committee and Chairman of the Toxic Substances Task Force of the National Association of Manufacturers.

Mr. Dominguez has had numerous publications and is the author of *Product Management; Marketing in a Shortage Economy; How to be a Successful Product Manager;* and *Business, Government, and the Public Interest* and *Government Relations.*

Mr. Dominguez was formerly Director, Government Regulations, Safety, Health, and Ecology Department, CIBA-GEIGY Corporation, Ardsley, New York. He has also taught and lectured widely, was a member of the faculty of New York University, Alamance Technical Institute, and has been a frequent guest lecturer at the New York University Graduate School of Education.

CONTRIBUTORS

George S. Dominguez
President
Springborn Regulatory Services, Inc.
Enfield, Connecticut

Irving L. Fuller, Jr.
Director, International Chemical Affairs
Office of Pesticides and Toxic Substances
U.S. Environmental Protection Agency
Washington, D.C.

Charles Ganz, Ph.D.
President
EN-CAS Analytical Laboratories
Winston-Salem, North Carolina

Richard deC. Hinds, Esq.
Partner
Cleary, Gottlieb, Steen, & Hamilton
Washington, D.C.

Peter Barton Hutt, Esq.
Partner
Covington & Burling
Washington, D.C.

Breck Milroy
Staff Advisor
International Chemical Affairs Staff
Office of Pesticides and Toxic Substances
U.S. Environmental Protection Agency
Washington, D.C.

Robert Polack, Esq.
Vice President
Reilly Tar and Chemical Corporation
Indianapolis, Indiana

Leon Starr, Ph.D.
Director
Environmental, Health, and Safety Affairs
Celanese Corporation
New York, New York

Robert M. Sussman, Esq.
Partner
Covington & Burling
Washington, D.C.

Carl W. Umland
Environmental Health Coordinator
Exxon Chemical Americas
Houston, Texas

TABLE OF CONTENTS

Chapter 1

INTRODUCTION

George Dominguez

TABLE OF CONTENTS

I. GENERAL

The first *Guidebook: Toxic Substances Control Act,* which by virtue of this publication has now become Volume I of this set, was published in July 1977, only 7 months after the Toxic Substances Control Act (TSCA) became effective. Since then, there have been a considerable number of more-or-less significant developments, not the least of which is a new Administration and a series of important new key Environmental Protection Agency personnel operating in a considerably reorganized, restaffed, and rebudgeted agency. While TSCA implementation may not have proceeded as far as some had been predicting in 1977, there nevertheless have been a considerable number of important regulatory, policy, enforcement, economic, scientific, and technical developments. All of these have unquestionably combined to create one undeniable effect — the operation of the chemical industry has, as predicted, alterated appreciably and will undoubtedly alter still further as many elements of the law not yet fully implemented, or initiated, begin to take effect in the years to come.

As we look back on these first few years of TSCA implementation, it is apparent that there have been many important developments that should be noted. Among the more significant of these developments that we can identify are the following:

1. EPA has completed the initial TSCA inventory
2. Section 8(e) Substantial Risk Notification requirements are fully operational
3. Premanufacturing notice (PMN) provisions are in effect and new chemicals are subject to PMN requirements
4. Significant New Use Rules (SNURs) are being actively considered and in a few cases actually proposed
5. Various PCB regulations are in place
6. TSCA inspections are taking place
7. Testing and other "guidelines" have been proposed
8. Various implementation "policies" and "strategies" have been formulated and issued

In short, there have been a number of developments which in themselves require updating and, more important, there is a need for a preview of what future TSCA implementation may hold in store for industry and the public. It is because of these and other developments of importance that this second volume of the *Guidebook: Toxic Substances Control Act* was undertaken and produced.

II. SCOPE AND COVERAGE

In this volume we have attempted not only to review major developments but also to examine their direct and indirect effects and implications within the context of future EPA TSCA activities. We are particularly fortunate in having had the contributions of a series of outstanding experts. They have provided a balance between both the theoretical and the practical — and, as in the past, the emphasis of this volume of the "Guidebook" is on the practical. In this context, special note should be taken of the outstanding chapters on PMNs, economic impacts, testing, and that very important subject, confidentiality. However, calling special attention to these chapters does not in any way imply any less importance of the other contributions which go to making this updating volume not only comprehensive but also inclusive of the major developments and trends in TSCA that are important to all those who are affected by this complex and comprehensive statute. Also, because they have become so important, I have included a chapter on international development in the "toxic" area (Chapter 9).

III. MAJOR REGULATORY DEVELOPMENTS

One area of special practical importance is the actual status of various regulations. As a special feature of this *Guidebook* we have prepared a series of comprehensive tables which will be found in the appendix that lists all of the major TSCA regulatory developments October 1977 through January 1981.

To add to the utility of these tables, they have been:

1.   Presented in chronological order
2.   Separated into regulations vs. rules
3.   Presented by topic, e.g., Testing, Confidentiality, etc.
4.   Indexed by title, subject, Federal Register (FR) reference, and keyed to the chapters
      in Volumes I and II of the *Guidebook* that have relevance to the subject matter involved

This unique series of regulatory tables permits the reader not only to follow the evolution of TSCA regulatory development but also quickly and conveniently to locate actual or proposed regulations by subject. These tables also provide the original FR reference citations so that those interested can also locate the full text originals. Moreover, by referencing them to specific chapters in both volumes of the *Guidebook,* it is possible to determine what regulations are related to topics covered in the two volumes of the *Guidebook* without the necessity of having direct references in the text, while at the same time having relevant descriptive and discussional material of both volumes in one reference source. We think that this unique approach will provide considerable additional insight and detailed TSCA information in a readily accessible format particularly useful to all of those involved with TSCA whether from the legal, financial, scientific, technical, compliance, or other perspectives.

IV. HEALTH AND ENVIRONMENTAL EFFECTS TESTING

While we had originally planned to include separate chapters on health and environmental effects testing, developments in the health effects testing area have been so rapid and diverse that rather than attempt to cover them here the reader is referred to the Testing Guidelines developed and released by the Organization for Economic Comperation and Development (OECD) in Paris. These represent a set of guidelines which have been adopted by the 24 member nation of the OECD which does include the U.S. In addition to these specific testing guidelines, the OECD has also been successful in reaching a decision on the application of Guidelines for Good Laboratory Practices (GLPs) (which are also available in full text from OECD in Paris) and the principle of the mutual acceptance of test data. This latter involves the acceptability of data from one OECD member nation by another, provided that the data were developed pursuant to the OECD Testing Guidelines themselves (or their equivalent) and based on using the OECD GLPs (or their equivalent).

On the other hand, developments have been far slower in the area of environmental effects testing, and therefore we have included a very comprehensive chapter on this topic of growing importance. While testing protocols in the environmental area are far from developed, this chapter provides excellent insights into the subject and the overall direction and thrust being taken.

V. USE AND OBJECTIVES OF THE *GUIDEBOOK*

When the first volume of this *Guidebook* was published, several specific uses and objectives were initiated. While circumstances have unquestionably changed, the utility and objectives of this overall publication and of Volume II in particular remain unchanged. The following points that were made in our introduction to Volume I are therefore just as relevant today to Volume II as they were then and bear repetition.

The overall objectives of this publication are

1.   A clear yet concise summary of the law itself, useful to both lawyers and nonlawyers
2.   An exhaustive yet understandable analysis of the law from both the legal and business
      viewpoints
3.   Practical recommendations for company compliance, planning activity, and testing
4.   An updating program of important rules and regulations, if justified by the demand
      for this work

From the viewpoint of utility, Volumes I and II individually or collectively are valuable for determining:

1.   Probable effects on chemical industry operations
2.   Actions, costs, alternatives, and compliance programs

3. Recommendations for organizational preparation
4. Review of major rules and regulations (as they are proposed and promulgated)
5. Evaluation of the effects of this law on all affected sectors of the chemical industry (including users)

VI. WHAT THE FUTURE HOLDS IN STORE

While the Reagan Administration, with its emphasis on regulatory relief changes in EPA and federal regulatory agency budget cuts and long-term commitment to a new "Federalism" (decentralization of federal activity and a return to state implementation), will have its impact on the future of TSCA and its implementation, it is also fair to say that this statute and the basic principles that it embodies are here to stay. With this in mind, then it is important that we understand the basic objectives and purposes of TSCA and that we keep abreast of the latest developments in its implementation. Hopefully, this updated volume of the TSCA *Guidebook* will provide not only a readable guide to these developments but also an important central reference work of present and future utility.

APPENDIX

Topical Index

Subject	Title	FR Ref.	Vol. I Relevancy	Vol. II Relevancy
Asbestos	Friable Asbestos-Containing Materials in Schools; Proposed Identification and Notification	45FR 61966	IV, VII, VIII, IX	III
Chloroflurocarbons	Ozone-Depleting Chlorofluorocarbons; Proposed Production Restriction	45FR 66726	IV, VII, VIII, IX	I, III, X
Citizen suit	Prior Notice of Citizen Suit; Proposed Procedural Regulation	44FR 43148	VII	I, II
Confidential business information	Confidential Business Information		IV, VII	II, III, V, VI, VII, X
	TSCA Confidential Business Information Security Manual		IV, VII	II, III, V, VI, XII, X
Data reimbursement	Data Reimbursement under Sections 4 and 5 Of The Toxic Substances Control Act	44FR 54284	IV, VII	III, V, VIII, IX
Dioxin	Tetrachlorodibenzo-*P*-Dioxin; Prohibition or Disposal	45FR 15592	VII, IX	III
Fully halogenated chlorofluoroalkanes	Fully Halogenated Chlorofluoroalkanes	43FR 11318	VII, IX	III
Fully halogenated chlorofluoroalkanes	Notification of Export for Polychlorinated Biphenyls and Fully Halogenated Chlorofluoroalkanes	44FR 56856	VII, IX	III
	Reporting Requirements for Manufacturer and Processors of Fully Halogenated Chlorofluoroalkanes and Fully Halogenated Chlorofluoroalkanes; Recodification	45FR 43721	VII, IX	III
Good laboratory practices	Proposed Health Effects Test Standards for Toxic Substances Control Act; Proposed Good Laboratory Practice Standards for Health Effects	44FR 44054	IV, VII, VIII	VIII, IX, X
Health and safety studies	Health and Safety Data Reporting, Submission of Lists and Copies of Health And Safety Studies	44FR 77470	IV, VII, VIII	V, VIII, IX
Imports/exports	Chemical Imports and Exports; Notification of Export	45FR 82844	VII, IX	III, X
Penalties	TSCA Civil Penalty Policy		IV, VII, IX	I, III
Polybrominated biphenyls	Submission of Notice of Manufacture or Importation of Polybrominated Biphenyls (PBBs) and TRIS, Final Rule	45FR 70728	VIII	III, V, VIII
Polychlorinated biphenyls	Polychlorinated Biphenyls	43FR 7150	IV, VII, IX	I
Polychlorinated biphenyls	Polychlorinated Biphenyls Manufacturing, Processing, Distribution in Commerce and Use Bans	43FR 24802	IV, VII, IX	I, III
	Polychlorinated Biphenyls Manufacturing, Processing, Distribution in Commerce and Use Prohibitions	44FR 31514	IV, VII, IX	I, III
	Notification of Export for Polychlorinated Biphenyls and Fully Halogenated Chlorofluoroalkanes	44FR 56856	IV, VII, IX	I, III
	Polychlorinated Biphenyls Manufacturing, Processing, Distribution in Commerce and Use Prohibitions; Proposed Restric-	45FR 30989	IV, VII, IX	I, III

Topical Index

Topical Index

Subject	Title	FR Ref.	Vol. I Relevancy	Vol. II Relevancy
Testing	Proposed Environmental Standards; and Proposed Good Laboratory Practice Standards for Physical, Chemicals, Persistence, and Ecological Effects Testing	45FR 77332	VII, VIII	III, IV, V, VI, VIII, IX, X

Chronological Index: Rule or Regulation

Chronology	Title	Subject	FR Ref.	Vol. I Relevancy	Vol. II Relevancy
12/2/77	Procedures for Rulemaking under Section 6 of the Toxic Substances Control Act	Section 6	42FR 61259	IV, XI	I, III
2/17/78	Polychlorinated Biphenyls (PCBs) Disposal and Marking	PCB's	43FR 7150	II	I, III
3/17/78	Fully Halogenated Chlorofluoroalkanes	Same	43FR 11318		I, III
6/7/78	Polychlorinated Biphenyls (PCBs) Manufacturing, Processing, Distribution in Commerce, and Use Bans	PCB's	43FR 24802	II	I, III
1/10/79	Premanufacture Notification Requirements and Review Procedures	PMNs	44FR 2242	VI, VII, VIII	III, V
3/23/79	Confidential Business Information	Same	44FR 17673	VII	VII
5/9/79	Proposed Health Effects Test Standards for Toxic Substances Control Act Test Rules	Testing	44FR 27334	IV, VII, VIII	III, IV, IX, X
5/31/79	Polychlorinated Biphenyls (PCBs) Manufacturing, Processing, Distribution in Commerce, and Use Prohibitions	PCB's	44FR 31514	II, IX	I, III, X
7/26/79	Proposed Health Effects Test Standards for Toxic Substances Control Act; Proposed Good Laboratory Practice Standards for Health Effects	Testing and GLPs	44FR 44054	VIII	VIII, IX
9/18/79	Data Reimbursement under Section 4 and 5 of the Toxic Substances Control Act	Data reimbursement	44FR 54284	VII, IX	III, VII, IX
10/2/79	Notification of Export for Polychlorinated Biphenyls and Fully Halogenated Chlorofluoroalkanes	PCBs and halogenchlorofluoralkanes	44FR 56856	II, VII, IX	X
10/16/79	Reproposal of Premanufacture Notice Forum and Provisions of Rules	PMNs	44FR 59764	II, VII, IX	III, V
12/31/79	Health and Safety Data Reporting; Submission of Lists and Copies of Health and Safety Studies	Health and safety studies	44FR 77470	VII, VIII, IX	III, IV, IX
3/11/80	Tetrachlorodibenzo-*P*-Dioxin; Prohibition of Disposal	Dioxin	45FR 15592	VII, IX	I, III

Chronological Index: Rule or Regulation

Chronology	Title	Subject	FR Ref.	Vol. I Relevancy	Vol. II Relevancy
5/9/80	Polychlorinated Biphenyls (PCBs) Manufacturing, Processing, Distribution in Commerce, and Use Prohibitions, Proposed Restrictions on Use at Agricultural Pesticide and Fertilizer Facilities	PCBs	45FR 30989	II, VII, IX	III
6/25/80	Prior Notice of Citizen Suit; Proposed Procedural Regulation	Citizen suit	45FR 43148	II, IX	I
6/30/80	Reporting Requirements for Manufacturers and Processors of Fully Halogenated Chlorofluoroalkanes and Fully Halogenated Chlorofluoroalkanes; Recodification	Fully halogenated chlorofluoroalkanes	45FR 43721	II, VII, IX	III
7/11/80	Toxic Substances Control Act; Records of Allegations of Significant Adverse Reactions to Health or the Environment	Records and reports	45FR 47008	II, VII, IX	III
8/15/80	Premanufacture Review Program; Proposed Processor Requirements	PMNs	45FR 54642	II, VII	III, V, VII, VIII, IX
9/17/80	Friable Asbestos-Containing Materials in Schools; Proposed Identification and Notification	Asbestos	45FR 61966	VII, IX	III
10/7/80	Ozone-Depleting Chlorofluorocarbons; Proposed Production Restriction	Chlorofluorocarbons	45FR 66726	VII, IX	III
10/24/80	Submission of Notice of Manufacture or Importation of Polybrominated Biphenyls (PBBs) and TRIS, Final Rules	PBBs and TRIS	45FR 70728	VII, IX	III, V
11/21/80	Proposed Environmental Standards; and Proposed Good Laboratory Practice Standards for Physical, Chemicals, Persistence, and Ecological Effects Testing	Testing	45FR 77332	VIII	III, V, VIII
12/16/80	Chemical Imports and Exports; Notification of Export	Imports/exports	45FR 82844	VII, IX	III, X

Chronological Index to Notices

Chronology	Title	Subject	FR Ref.	Vol. I Relevancy	Vol. II Relevancy
10/5/77	TSCA Interagency Testing Committee	Testing	42FR 55026	VII, VIII	IV
3/16/78	Notification of Substantial Risk under Section 8(e)	Substantial risk	43FR 11110	VII	IV
4/19/78	Second Report of the Interagency Testing Committee; Receipt and Request for Comments	Testing	43FR 16684	VII, VIII	IV

Chronological Index to Notices

Chronology	Title	Subject	FR Ref.	Vol. I Relevancy	Vol. II Relevancy
10/30/78	Third Report of the Inter-agency Testing Committee	Testing	43FR 50630	VII, VIII	IV
5/15/79	Toxic Substances Control; Premanufacturing Notification Requirements and Review Procedures; Statement of Interim Policy	PMNs	44FR 28564	IV, VII	V
6/1/79	Fourth Report of the Inter-agency Testing Committee	Testing	44FR 31866	VII, VIII	IV
12/7/79	Fifth Report of the Inter-agency Testing Committee	Testing	44FR 70664	VII, VIII	IV
3/10/80	TSCA Civil Penalty Policy	Penalties		IV, VII	I, II
4/24/80	PCB Penalty Policy	PCB		II, IV, VII	I, II
5/28/80	Sixth Report of the Inter-agency Testing Committee	Testing	45FR 35897	VII, VIII	IV
6/11/80	TSCA Confidential Business Information Security Manual	Confidential business information		IV, VII	VII
7/18/80	Acrylamide: Response to the Interagency Testing Committee	Testing	45FR 48510	VII, VIII	IV
1980	Enforcement Facts and Strategy Premanufacture Notification	PMNs		IV, VII	V
11/7/80	Toxic Substances Premanufacture and Review Procedures; Statement of Revised Interim Policy	PMNs	45FR 74378	IV, VII	V
1/27/81	Agency Policy to Premanufacture Testing of New Chemical Substances and Announcement of Rescheduled Meeting and Extension of Comment on Certain Environmental Test Standards	PMNs	46FR 8986	IV, VII	V

Chapter 2

JUDICIAL CONSTRUCTION OF THE TOXIC SUBSTANCES CONTROL ACT

Richard deC. Hinds

TABLE OF CONTENTS

I. INTRODUCTION

Most significant new federal statutes in the health and environmental area generate a considerable body of litigation, particularly in the early stages of implementation. The Toxic Substances Control Act (TSCA) which was enacted on October 1, 1976, is proving no exception. The issues litigated through July 1982 involve the obligation of the Environmental Protection Agency (EPA) to respond to testing recommendations of an interagency committee, EPA's regulation of polychlorinated biphenyls, the scope of EPA's authority to require submission of health and safety studies, the adequacy of EPA's procedures to maintain the secrecy of confidential information, and EPA's authority to use contractors to conduct investigations.

II. TESTING

In the most significant case litigated under TSCA, *National Resources Defense Council* v. *Costle*,[1] NRDC sought a declaratory judgment that EPA had failed to comply with its statutory duty either to initiate rulemaking proceedings to require testing of the chemicals designated for priority consideration by the TSCA Interagency Testing Committee (TITC) or explain why it had failed to do so.

Under Section 4(a) of TSCA,[2] when there are insufficient data to predict the health and environmental effects of a chemical substance, EPA must publish a rule requiring testing if it finds (1) that the manufacturing, processing, distribution, use, or disposal of the chemical may present an unreasonable risk or (2) there is or may be a substantial human or environmental exposure to the chemical.

In order to assist EPA, Section 4(e) established the TITC to make recommendations to EPA concerning the chemicals which should be given priority consideration. After EPA receives the TITC list of priority chemicals, it has 12 months either to initiate a rulemaking proceeding for testing under Section 4(a), or to publish in the *Federal Register* its reason for not doing so.

On October 1, 1977, the TITC recommended that ten chemical substances or categories of chemical substances be considered by EPA for rulemaking under Section 4(a).[3] In its response to the TITC's recommendation,[4] EPA indicated that it was not initiating Section 4(a) rulemaking proceedings for the recommended chemicals at that time. It stated that it would be inappropriate to take such action then because further evaluation of ongoing, planned, and completed studies was necessary before EPA could determine whether a Section 4(a) test rule was needed. In addition, EPA stated it had not promulgated standards for conducting tests under Section 4(a), and it believed that development of such standards was necessary before it could issue a test rule for specific chemicals.[5]

On April 10, 1978, the TITC added eight additional chemical substances and categories of chemical substances to its priority list for EPA action.[6] Once again, EPA's response was that it would not initiate a rulemaking at that time because it had not yet completed its review and evaluation of available information regarding the recommended chemicals.[7]

The major contention of NRDC in its suit was that the EPA *Federal Register* responses were not legally cognizable under the Act. NRDC claimed EPA must either initiate testing proceedings on a TITC-recommended chemical, or specifically determine that the chemical did not meet the criteria for testing under Section 4. Since EPA did neither, NRDC claimed EPA had failed to meet its statutory obligations.

The Court ruled that EPA's published responses to the TITC recommendations were insufficient to comply with the Congressional intent and mandate of the Act. The Court rejected EPA's argument that its progress reports concerning the development of its testing program satisfied its obligations under TSCA. According to the Court a progress report, providing generalizations on EPA's past and present difficulties, did not fulfill the statutory requirement that EPA provide reasons for not initiating rulemaking proceedings with respect to the TITC's recommendation. The Court found that EPA was in violation of section 4 and ordered it to submit to the Court its proposed plans for compliance.

---

[1]   No. 79-2411 (S.D. N.Y. February 4, 1980).

[2]   15 U.S.C. § 2603.

[3]   42 *Fed. Reg.* 55026 *et. seq.* (October 12, 1977).

[4]   43 *Fed. Reg.* 50134 (October 26, 1978).

[5]   EPA has since decided not to develop such standards before issuance of test rules. 47 *Feb. Reg.* 13012 (March 26, 82).

[6]   43 *Fed. Reg.* 16684 (April 19, 1978).

[7]   44 *Fed. Reg.* 28095 (May 14, 1979).

EPA thereafter submitted plans to eliminate the backlog of TITC recommendations. The agency has been responding to recent recommendations by issuance of a proposed rule or, in a few instances, an advance notice of proposed rulemaking within the statutory period, or a decision not to issue a test rule based on voluntary testing undertaken by industry.

III. PCBs

EPA's regulations defining the scope of the statutory ban on polychlorinated biphenyls ("PCBs") have been attacked by both industry and environmentalists. Section 6(e) of TSCA[8] provides that after July 1, 1979, no person may manufacture, process, or distribute PCBs in commerce and no person may use PCBs in other than a "totally enclosed manner", or any other manner which EPA determines will not present an unreasonable risk of injury to health or the environment. The term "PCBs" is not defined in Section 6(e).

Pursuant to Section 6(e), EPA promulgated regulations which prohibit the manufacture, processing, and distribution of PCBs and control their labeling and disposal.[9] The regulations define PCBs to mean "any chemical substance that is limited to the biphenyl molecule that has been cholorinated to varying degrees or any combination of substances which contain such substance."[9a]

An environmental group challenged EPA's determination that certain commercial uses of PCBs accounting for about 99 percent of all PCBs to be found in the United States were "totally enclosed", its decision to limit the applicability of the statutory ban to PCBs present at concentrations greater than 50 parts per million and its decision to authorize the continued use of 11 non-totally enclosed uses of PCBs. The D.C. Circuit Court agreed with the first two contentions and set those portions of the regulations aside. It sustained the Agency's regulations permitting the continued use and servicing of certain transformers and electromagnets containing PCBs as supported by substantial evidence. *Environmental Defense Fund v. Environmental Protection Agency.*[9b]

With respect to the 50-ppm regulatory cutoff, EDF argued that the statute required EPA to regulate all manufacture and use of PCBs. While noting that a literal reading of the statute supported this argument, the Court stated that the legislative history of Section 6(e) demonstrated that Congress did not intend to regulate ambient sources of PCBs. However, the Court ruled that EPA's regulatory cutoff could not be defended on that ground because it not only exempted ambient sources of PCBs under 50 ppm, it also exempted point sources of contamination under 50 ppm, including the incidental manufacture of PCBs as an impurity or byproduct of the manufacture of another substance. The court ruled that EPA had failed to carry its "heavy burden" of establishing that such a cutoff was justified either by administrative necessity or by the *de minimus* doctrine. Noting that the record contained evidence that PCBs were toxic in concentrations well below 50 ppm and more importantly that any exposure to PCBs could have adverse effects for the environment or humans, the Court stated that EPA had failed to demonstrate either that it was an administrative impossibility to regulate PCB concentrations under 50 ppm or that 50 ppm was such a low level that "the burdens of regulation yield a gain of trivial or no value." The Court strongly recommended that upon remand EPA rely upon the use authorization and exemption provision in Subsections 6(e)(2)(B) and 6(e)(3)(B) of the Act which require EPA to make a finding of no "unreasonable risk of injury to health or the environment", and, in the case of exemptions, good faith efforts to find substitutes.

The Court also set aside as unsupported by substantial evidence EPA's determinations that several PCB uses, including nonrailroad transformers, capacitors, and electromagnets, were "totally enclosed uses" and therefore exempt from regulation under the Act. The Court noted that EPA's regulations contained no procedures for inspection or self-reporting of leaks and that the administrative record failed to support EPA's finding that PCBs would not enter the environment from such uses, which included most uses of PCBs in the United States.

Because the Court's opinion in this case had the effect of banning many industrial activities, the Court agreed to stay its mandate during the period necessary for EPA to conduct a rulemaking. Thus, the 50-ppm regulatory cutoff remains in effect until EPA issues a new PCB rule (46 *Fed. Reg.* 27615, May 20, 1981).[10]

---

[8]   15 U.S.C. § 2605(e).

[9]   44 *Fed. Reg.* 31514 (May 31, 1979).

[9a]   40 C.F.R. § 761.2.

[9b]   636 F. 2d 1267 (D.C. Cir. 1980).

[10]   EPA has recently asked for comments on such a proposed new rule. 47 *Fed. Reg.* 24976 (June 8, 1982).

In the second case involving EPA's PCB regulations, *Dow Chemical Co.* v. *Costle*,[11] Dow claimed that EPA exceeded its statutory authority under Section 6(e) of TSCA[12] when it promulgated rules defining the scope of the statutory ban on manufacture of PCBs. Dow, in the manufacture of two of its products, produced an impurity that contains amounts of monochlorobiphenyl ("MCB"). It was Dow's contention that although MCBs were a form of chlorinated biphenyl, the fact that it had only a single chlorine atom on one of the two rings of the biphenyl molecule should remove it from the statutory prohibitions against PCBs. Dow argued that the prefix "poly" usually meant "two or more", and therefore an MCB cannot be considered a PCB. In addition Dow argued that the legislative history of Section 6(e) indicated that the ban on PCBs was premised on the fact that PCBs were extraordinarily resistant to degradation and thus were becoming widespread throughout the environment. Dow asserted that MCBs are biodegradable, do not accumulate in the environment, and therefore were not meant to be treated as PCBs. Consequently, Dow sought a declaratory judgment that Section 6(e) and the EPA regulation promulgated thereunder did not apply to PCBs, and an injunction to prevent EPA from enforcing its regulation against MCBs.

EPA moved to dismiss the action on the ground that Section 19(a)(1)(A) of TSCA[13] provides for exclusive jurisdiction in the courts of appeals over attempts to secure judicial review of a regulation promulgated pursuant to Section 6(e). Dow argued that while TSCA requires cases challenging the validity of regulations to be brought in the courts of appeals, jurisdiction to hear cases, such as Dow's, involving questions of application or interpretation of regulations was left with the district courts.

The Court disagreed with Dow's construction of the statute and held it did not have jurisdiction. It then went on to rule that even if Dow's reading of the statute were adopted it would still lack jurisdiction because, at the time the regulation was published, it was clear that it applied to MCBs. By questioning the authority of EPA to issue the regulations as opposed to an unexpected interpretation or application, Dow was thus challenging the validity of the regulation. Such a challenge to EPA's authority to regulate MCBs under Section 6(e) had to be made by means of a timely petition for review in a court of appeals. Consequently, the court held it did not have jurisdiction to review the regulations and dismissed the case.

This case is important both because it upholds the EPA position that the PCB regulations apply to MCBs and because it makes clear that pre-enforcement challenges to most EPA regulations must be brought in the courts of appeal within 60 days of issuance as provided in Section 19(a).

IV. HEALTH AND SAFETY STUDIES

In another case also brought by Dow, the Third Circuit rejected Dow's claims that EPA had exceeded the statutory authority granted to it by Section 8(d) of TSCA[14] when it issued a regulation requiring the submission of health and safety studies for certain chemical substances which had been designated by the TITC for testing.[15] *Dow Chemical Co.* v. *EPA*.[16]

Section 8(d) of TSCA authorizes EPA to require any person who manufactures or processes "for commercial purposes" or who distributes in commerce a chemical substance or mixture, to submit to EPA a list of health and safety studies which they have conducted, know to exist, or which are reasonably ascertainable by them. EPA is also authorized to request a listed study from any person in possession of it.

Under the regulation EPA issued pursuant to this section, all persons who manufactured, processed, or distributed in commerce one or more of the substances or categories of substances listed had to submit a list of all health and safety studies relating to each of the substances that were (a) in their possession; (b) conducted or initiated by them; or (c) contained or referenced in letters or memoranda in their files. In addition, copies of any completed health and safety studies in their possession had to be submitted, regardless of whether the substance was manufactured, processed, or distributed by them. Likewise, any person who actually possessed a study included in a list submitted to EPA had to submit a copy of the study, if requested, even if the person was not required to submit a list in the first place.

[11]   484 F. Supp. 101 (D. Dela. 1980).
[12]   15 U.S.C. § 2605(e).
[13]   15 U.S.C. § 2618(a)(1)(A).
[14]   15 U.S.C. § 2607(d).
[15]   43 Fed. Reg. 30984 (July 18, 1978).
[16]   605 F.2d 673 (3d Cir. 1979).

The regulation also defined "manufacture or process" to mean manufacture or process for a commercial purpose, which covers (a) for distribution in commerce, including test marketing, (b) use by the manufacturer or processor, including use as an intermediate, and (c) research and development.

Dow argued that EPA exceeded its authority when it included chemicals manufactured solely for purposes of research and development within the definition of chemicals manufactured for "commercial purposes". Dow claimed that the manufacture of small quantities of a chemical that it does not propose to offer in commerce, but that it will use solely for purposes of product research and development, is not manufacture "for commercial purposes". Therefore, EPA did not have the authority under Section 8(d) to include health and safety studies in such chemicals in the regulation.

The Court found that the language in the statute itself defeated Dow's interpretation. Other sections of TSCA applying to chemicals "manufactured for commercial purposes" had express exemptions for chemicals manufactured solely for research and development.[17] The Court reasoned that those provisions would be meaningless, unless Congress intended that chemicals manufactured for commercial purposes included chemicals manufactured solely for research and development.

The Court agreed with EPA that any activity engaged in for profit is commercial activity, and that a company, such as Dow, which engaged in product development in the hope of making profits, and not out of pure scientific interests, did so with a purpose that was commercial in nature. According to EPA, the "for commercial purpose" language was used to distinguish profit-making enterprises from nonprofit institutions such as a university or foundation.

Dow also challenged the authority of EPA under Section 8(d)(2) to seek copies of health and safety studies from persons not subject to the listing requirement of Section 8(d)(1). EPA contended that while the listing requirement in Section 8(d)(1) applies only to chemicals that are actually manufactured, processed, or distributed by the reporting company, there is no such limitation under Section 8(d)(2). According to EPA, Section 8(d)(2) thus allowed it to obtain copies of studies on designated chemicals in the possession of any company, even if the company did not manufacture, process, or distribute the chemical and therefore would not have to list it under Section 8(d)(1). Dow argued that a company is required to submit only copies of those studies that it is required to list under Section 8(d)(1).

The Court found the legislative history could be read to support either the Dow or EPA interpretation of Section 8(d)(2). However, the Court held that the language of Section 8(d)(2) appeared to comport more closely with EPA's interpretation. Although Section 8(d)(1) is expressly limited to manufacturers, processors, or distributors, Section 8(d)(2) applies to "*any person* who has possession of a study." The Court therefore concluded that EPA's regulation requiring copies of health and safety studies from persons not subject to the listing requirements of Section 8(d)(1) was within its delegated authority under Section 8(d).

Apart from clarifying the broad reach of Section 8(d) this *Dow* case is also significant because it dealt with the ability of a government agency to unilaterally deprive a court from reviewing its actions. After Dow initiated its lawsuit, EPA, apparently in response to the suit, withdrew the regulation at issue[18] and argued to the Court that the case was therefore moot.

EPA stated it withdrew the rule because Dow's complaint raised questions concerning EPA's compliance with the procedural requirements of the Administrative Procedure Act.[19] EPA added that a new Section 8(d) rule would be promulgated to correct any alleged procedural violations, and that the new rule would take precisely the same substantive position as the withdrawn rule.

The Court concluded that the case was not moot, finding that the parties were adverse, the dispute real and concrete, and that EPA had no intention of changing its substantive position. In addition, the court recognized the danger of allowing agencies to avoid judicial review, whenever they chose, by simply withdrawing the challenged rule.

V. CONFIDENTIALITY

Preserving the confidentiality of trade secrets is of great concern to many manufacturers because TSCA requires them to provide EPA with product data which may have considerable proprietary value. In *Polaroid Corp.* v. *Costle*,[20] Polaroid sought a preliminary injunction against an EPA order requiring Polaroid to

---

[17] Sections 5 and 8 of TSCA. 15 U.S.C. §§ 2604, 2607.
[18] 44 *Fed. Reg.* 6099 (January 31, 1979).
[19] 5 U.S.C. § 551 *et. seq.*
[20] 11 E.R.C. 2134 (D. Mass. 1978).

supply the identity of certain proprietary chemicals to EPA. Pursuant to the inventory reporting regulations,[21] EPA had required manufacturers to disclose the identity of their commercial chemicals in order to compile the Chemical Substance Inventory required by Section 8(b) of TSCA.[22] Polaroid refused to give the required information on certain chemicals because of its belief that the information constituted a confidential trade secret and that EPA's reporting regulations lacked sufficient procedures to protect the confidentiality of trade secrets.

The Court found that to grant the preliminary injunction sought it would have to determine that EPA's existing confidentiality regulation and interim procedures were inadequate. The Court concluded it did not have the jurisdiction to grant the specific relief requested by Polaroid because Section 19(a) of TSCA[23] provides that only a court of appeals could review EPA regulations promulgated under Section 8 of TSCA.

However, the Court did not dismiss the case, finding that it had jurisdiction to consider Polaroid's allegation that the statutory exceptions to confidentiality in Section 14[24] violated the Fifth Amendment. Polaroid argued that it would be deprived of property without due process of law if its trade secrets were released under authority of Section 14 which permits disclosure without notice to government employees, EPA contractors, Congress, and in certain other circumstances. The Court granted a preliminary injunction against such disclosure of Polaroid's confidential trade secrets pursuant to Section 14, pending further hearings and orders of the Court.

Before the Court could decide the constitutional issues, EPA promulgated new regulations relating to confidential business information[25] which contained sufficient safeguards against disclosures to satisfy Polaroid, and it withdrew its suit.

While the *Polaroid* case thus does not set any precedent, it does indicate a judicial willingness to consider claims that the provisions of Section 14 regarding trade secrets do not pass muster under the Constitution. EPA's procedures for handling confidential information, which contain provisions for notice prior to any disclosure under Section 14, may have resolved certain problems, but others remain. For example, Section 14(b)[9] as construed by EPA in its proposed premanufacture notification regulations,[26] requires the disclosure of health and safety studies submitted to EPA, including the specific chemical identity of the chemical that is the subject of the study, except in narrowly defined situations. Since the chemical identity of a new chemical is often a highly sensitive trade secret, there may well be further litigation concerning EPA's handling of confidential business information.

VI. INSPECTIONS

Concerns over loss of confidentiality of trade secrets have also given rise to a challenge to EPA's authority to use private consultants for enforcement purposes. Section 11(a) of TSCA authorizes the EPA Administrator "and any duly designated representative of the Administrator" to conduct inspections of chemical facilities to ensure compliance with the Act. In *Aluminum Company of America, et al.* v. *DuBois* (No. C80-1178V, June 11, 1981, W.D. Wash.), appeal pending (9th Cir.), Plaintiffs Alcoa and a subsidiary sued for a declaratory judgment that TSCA does not authorize EPA to use EPA contractors as "designated representatives of the Administrator" and for an injunction against inspections by non-EPA employed personnel. Alcoa stated that it was concerned that private consultants would observe highly confidential technological processes during an inspection of its plant and seek thereafter to become advisors to competitors or to market valuable information learned during the inspection. The Court, in a brief opinion, ruled that EPA could use non-Agency employees to conduct investigations under Section 11. The Court based its decision on a literal reading of the phrase "representative of the Administrator" as well as the legislative history of the Act.

---

[21]   42 *Fed. Reg.* 64572 (December 23, 1977).
[22]   15 U.S.C. § 2607(b).
[23]   15 U.S.C. § 2618(a).
[24]   15 U.S.C. § 2613.
[25]   43 *Fed. Reg.* 39997 (September 8, 1978).
[26]   Proposed Regulation § 720.43, 44 *Fed. Reg.* 2276 (January 10, 1979).

Chapter 3

ECONOMIC IMPACTS OF THE TOXIC SUBSTANCES CONTROL ACT
(PUBLIC LAW 94—469)

Leon Starr

TABLE OF CONTENTS

I. INTRODUCTION

The Toxic Substances Control Act (TSCA) was passed on October 11, 1976, and became effective on January 1, 1977. Congress intended the Act to provide the vehicle for data development with respect to the health and environmental effects of chemical substances in articles of commerce in the U.S. Furthermore, this Act was to protect human health and the environment and regulate commerce by requiring testing of, and necessary use restrictions of, certain chemical substances which may present unreasonable risks.

The Toxic Substance Control Act mentions economic factors in Sections 2(a), 3, 2(c), 4(b)1, 4(c)3, 4(c)4, 5(h), 6(c)1, 6(c)3, 6(c)4, 10(c), 10(d), 16(a)2, 24(a), and 25(a) (Ref. 1). Some of the factors relate to penalties, compensation, or indemnification, as well as specific economic effects as a result of rule making.

Congress recognized the very broad scope of the Act and the potential for serious economic dislocation which could result from the implementation of TSCA. The following abstracts from the policy section of the Act, Section 2, delineate some of these Congressional concerns.

> (2) adequate authority should exist to regulate chemical substances and mixtures which present an unreasonable risk of injury to health or the environment, and to take action with respect to chemical substances and mixtures which are imminent hazards; and
>
> (3) authority over chemical substances and mixtures should be exercised in such a manner as not to impede unduly or create unnecessary economic barriers to technological innovation while fulfilling the primary purpose of this chapter to assure that such innovation and commerce in such chemical substances and mixtures do not present an unreasonable risk of injury to health or the environment. 15 U.S.C. § 2601(b) (2), (3).

The further intent of Congress is shown below:

> (c) *Intent of Congress.* — It is the intent of Congress that the Administrator shall carry out this chapter in a reasonable and prudent manner, and that *the Administrator shall consider the environmental, economic, and social impact* of any action the Administrator takes or proposes to take under this chapter. 15 U.S.C. § 2601(c).

Many of the key rules and regulations under various sections of the Act have not been finalized. For example, the final rules under Sections 4, 5, 6, and 8, which are expected to result in the most significant economic impacts, are expected in 1982. This chapter will cover economic factors and studies of TSCA which have been published since the last Toxic Substances Guide, present more detail on those sections which are expected to cause significant economic impact, and describe some upcoming economic studies and some potential downstream economic effects resulting from the Act.

II. GENERAL ECONOMIC ANALYSES AND CONSIDERATIONS

There have been a number of references (Ref. 2—10) covering the overall impact of regulations on U.S. industry and the significant cost impact of regulation on industry has been recognized by the Federal Government. This recognition led to actions such as Executive Orders 12044 and 12291 which require a regulatory analysis on specific rules, and regulations which have a cost impact above $100 million. This thrust for more efficient and cost-effective regulations has also led to the establishment of the old Interagency Regulatory Liaison Group (IRLG), the Regulatory Council, the proposed Regulatory Reform Act, the establishment of the regulatory reform task force under Vice President Bush, and the recognized need for increased industry/regulatory agency cooperation.

M. Weidenbaum's work (Ref. 5) discusses the effect of regulation and attempts to estimate the impact of increasing government regulation on business. He estimated that in 1979 to 1981 the total costs of regulation, due to regulatory agencies such as the Environmental Protection Agency (EPA), the administration of the Occupational Safety and Health Act (OSHA) and of industry commissions would increase to over $100 billion from a 1974 level of $65 billion. The increase in costs for administration of the various government agencies was expected to more than double from $2.2 billion in 1974 to over $4.8 billion in 1979. The increase in funding for the EPA office dealing with the administration of TSCA is seen in Tables 1 and 2. Table 1 describes the Congressional action on funding for fiscal years 1978 through 1980. Table 2 indicates the funding breakdown for the years 1978 and 1979. Also, the large increases in approved personnel for 1978 to 1979 are included. The largest rates of increase were expected in the area of safety, health, energy, and environmental regulations. The estimates and conclusions of Weidenbaum's work have

Table 1

CONGRESSIONAL AUTHORIZATION FOR
TSCA
($, MILLIONS)[a]

	FY 1978	FY 1979	FY 1980	FY 1981[b]
EPA request	28.005	56.732	103.316	
House action	30.705	57.007	98.316	
Senate action	28.005	56.732	98.316	
Final action		56.732	98.316	94.104

[a] Including abatement and control, R & D, enforcement, Regional
— Excluding Management Account, EPA, Office of Industry
Assistance, Washington, D.C., 1979.

[b] *Chemical Regulation Reporter*, Bureau of National Affairs,
Washington, D.C., 1982, 1282.

Table 2

TOXIC SUBSTANCES — PRESIDENT'S BUDGET[a]

	Budget Authority ($, Millions)		Positions[b]	
	1978	1979	1978	1979
Abatement and control	22.9	41.6	221	428
(Hazard identification and assessment)	17.9	33.0	183	315
(Chemical Control)	5.0	8.6	38	113
Enforcement	2.3	4.6	48	85
R & D	3.6	10.5	45	60
Subtotal	28.8	56.7	314	573
Management	0.4	1.3	8	40
TOTAL	29.2	58.0	322	613

[a] Office of Industry Assistance, Environmental Protection Agency,
Washington, D.C., 1979.

[b] Including regional and HQ.

been questioned in recent reports (Ref. 6). The methodology on cost collection and extrapolation were critiqued and objections were raised that only the cost side of the cost/benefit ratio was addressed. While these are valid points to be considered when evaluating the overall impact of regulations, there does not seem to be any real question in government, industry, or the public that a significant cost impact of regulations does exist.

It is also recognized that regulations have provided and will continue to provide real benefits to our society. However, while costs are easily quantified, the rationale, methodologies, and techniques for determining and evaluating benefits of regulatory actions are still in the formative stage. Improved and accepted methodologies for risk/benefit and cost/benefit analysis for evaluating regulatory impact are necessary, especially when these concepts may be used for defining actions deriving from regulatory reform legislation. This chapter will not attempt to deal with the question of cost/benefit analysis. References 11 through 16 can be used for further reading on this subject.

A 1979 study on the cost impact of regulations was performed by A. Anderson & Company for the Business Round Table (Ref. 7). This study summarized the direct incremental costs incurred by 48 companies due to compliance with 6 regulatory agencies. These were the EPA, OSHA, Federal Trade Commission (FTC), Equal Employment Opportunity (EEO), Employee Retirement Income Security Act (ERISA), and Department of Energy (DOE). The companies responding to the study represented the mining, food products, wood products, chemicals, oil and gas, rubber, transportation, electrical, communications, wholesale trade and banking industries.

This study showed that the incremental compliance costs due to regulations promulgated by the agencies described above was over $2.6 billion. Of the 48 participating companies, those in the manufacturing sector were impacted most heavily, accounting for $2.3 billion of incremental costs out of the $2.6 billion total.

The study did not attempt to address the benefits of regulation or secondary economic effects, such as loss of productivity, inflation, international competitiveness, or investment disincentive.

One area that was highlighted was the impact of these incremental compliance costs on research and development activities. The participating companies identified $75 million of incremental R & D costs necessary for compliance with EPA regulations. These 1977 costs for all companies were about 1% of their total R & D budgets. However, for those companies in chemical or allied products businesses, 5% of their R & D expenditures were allocated to compliance.

The Dow Chemical Company evaluated the impact of federal regulation on its business operations (Ref. 18). Total company expenditures for regulatory activities had risen from $170 million in 1975 to over $268 million in 1977. Dow categorized regulatory compliance costs as: (1) appropriate costs; (2) questionable costs and (3) excessive costs. The Dow analysis indicated that the excessive cost category is increasing the most rapidly. This cost category had jumped from $50 million in 1975 to $150 million in 1977, a 200% increase. One may argue the methodology used by Dow to characterize costs, but the evaluation does indicate the significant increase in overall regulatory costs for just one chemical company since 1975. Furthermore, the full cost impacts of regulation under the Toxic Substances Control Act are yet to be realized.

The annual report of the Council of Environmental Quality describes a number of economic issues arising from the Toxic Substances Control Act (Ref. 19).

Prior to the passage of TSCA, there were a number of estimates of the economic impact on the U.S. chemical industry (Ref. 20). The EPA estimated the cost to industry to be between $80 to $140 million. The Chemical Manufacturers Association (CMA, formerly MCA) estimates were between $360 million and $1.3 billion, and a private estimate by Dow Chemical Company was $2 billion. The General Accounting Office reviewed these studies at the request of the Congress and estimated that the annual cost to the chemical industry would be between $100 to $200 million. Costs to other segments, such as downstream industries that use chemicals, have not been estimated. These costs would be highly dependent on how specific chemicals are regulated, as well as what actual TSCA implementation costs are incurred by the chemical industry.

The most detailed study of the economic impact of TSCA made to this writing has been the Foster D. Snell study (Ref. 21). This study was made prior to the enactment of TSCA and included a number of assumptions on the form full implementation of TSCA would take. This study assumed that regulation under the Act would require significant testing and reporting elements, including pre-market screening of new substances and for new applications of existing substances. Furthermore, the Act would introduce other compliance requirements, as well, and provide wide ranging discretionary powers to the EPA Administrator. The Foster D. Snell analysis of TSCA implementation was done on two levels, one broad, and the other selective interpretation by the Administrator. Also, they set two boundary conditions for industry: innovation maintained, in which R & D costs would increase to cover the increased complexity of the product research, development, and commercialization process; and R & D costs not increasing with testing and other TSCA compliance costs being part of the R & D budget, with the result being displacement of innovation.

The study findings showed that approximately 11.5 thousand companies would be burdened by some form of TSCA compliance activity. These would be 1000 in the basic chemical industry, 7400 in the chemical process industry, and about 3000 in the allied products industries. The annual sales range of these companies would be from about $5 million to over $1 billion. The increase in R & D expenditures needed to maintain current levels of effort, exclusive of TSCA compliance expenditures, ranged from 10 to 30%. In the low level of testing hypothesis, costs for animal and environmental testing were about $30,000 per chemical. To maintain innovation at this level of testing, large companies would have to increase R & D budgets by 10%. Smaller companies would be more heavily impacted and the study predicted a 75 to 80% reduction in new product introduction by smaller (below $50 million sales) companies. Larger companies would predict a corresponding 10 to 15% decrease. The extensive testing scenario called for very complex and detailed oncogenicity testing. In 1975, the cost estimate for a single detailed study was between $100 to $500 thousand. Recent (1981) projections, described under the Section 4 discussions, range between $1.5 to $2.0 milllion. In this case, large companies would have to increase R & D spending by up to 25% to maintain innovation. Small companies would again require larger percentage increases. The percent decrease in new product introduction would be 25% and 80 to 90% for large and small companies, respectively.

The Snell report also estimated major direct cost elements for a number of cases. For maintenance of innovation with comprehensive premarket screening of all major new substances and applications and selective screening of minor ones, the cost impact would be $600 million. The total annual costs increase to $1.3 billion when other costs, such as administration, testing of existing chemicals, and costs associated with selective use limitations and bans are added to this number. For selective premarket screening and testing, with innovation maintenance, the cost impact was $678 million, including the other cost elements above. If one assumes that new product innovation will be displaced, then the annual cost impacts of TSCA were estimated to range from $358 to $600 million.

The Snell report also attempted to describe some noneconomic and nonquantifiable effects of TSCA. For example, the report estimated an adverse effect on 1985 employment of 20,000 to 80,000 jobs in the maintenance and displacement of innovation scenarios, respectively. Also, due to the transition of the chemical industry to a highly regulated industry, it was predicted that the industry would become more concentrated, due to failure of small specialty chemical firms, movement of some R & D and commercial development overseas, and a trend to lower risk, less new product R & D. Furthermore, there would arise discontinuities in sales and marketing of some chemicals identified as hazardous, but before full scientific evaluation of test data had occurred.

It should be emphasized that the Foster D. Snell study was performed before TSCA was passed in the form we know it today. Now, 4 years after the passage of the Act, many of the rules and regulations which are expected to have the most significant economic impacts are yet to be fully promulgated. It is also important to note that some aspects of TSCA were modified after the Snell report, that EPA and industry are cooperating in attempts to make regulations reasonable and workable, and that the concerns about regulatory excesses in terms of economic impact and negative impacts on innovation have been raised by both Federal and industrial institutions.

Hopefully, all this will tend to have a moderating effect on the full impact of TSCA, while the objective of protecting the health and environment from substantial, significant, and unreasonable risks may still be realized. Furthermore, EPA and industry have proposed and begun a number of economic studies to obtain a current evaluation of the impact of TSCA. A number of current EPA-sponsored economic impact studies are shown in Table 3. Also, the CMA is sponsoring a comprehensive study of the economic impacts of TSCA. This study will be in two phases, a 1-year pilot study to collect past- and in-year costs followed by a 2- to 3-year study to continue to update cost impact on the chemical industry and the U.S. economy. The CMA study, through the use of outside consultants, will also attempt to estimate the benefits which derive from TSCA.

As we face the full implementation of TSCA, there are a number of actions business organizations should take to insure they are prepared for meeting their responsibilities under this Act. Many companies have established committees and systems to inventory chemicals, establish chemical handling procedures, train employees, and implement effective management information systems (Ref. 2, 4). Also, these committees help to assess the cost elements which make up the overall economic impact of TSCA. Some of the direct cost elements which should be considered when assessing the overall impact are testing costs, record keeping, personnel involved with TSCA activities, participation in the regulatory process by involvement with trade associations, agencies and Congress, and general administrative costs. These organizations that business establishes to insure that its activities and products are in compliance with TSCA are also necessary to protect upper management from the civil and criminal penalties under TSCA. Senior executives are the responsible parties according to EPA and will be included in enforcement proceedings under the Act.

As an example, the organization chart shown in Figure 1 is a composite example of how some companies are organized to comply with TSCA. Many large companies have organizations similar to that shown in Figure 1. Many small companies cannot afford extensive organizations, but usually have a high-level executive who has TSCA responsibilities. It is expected that implementation of TSCA will have a significant impact on companies that have fewer personnel and dollar resources available to address the requirements of the Act. The EPA has been informed by the Small Business Administration of a decision to assist small businesses adversely affected by TSCA (Ref. 3). Under Section 7(h) of the Small Business Act, companies could be eligible for loans to assist their compliance activities. This was an administrative decision of SBA

Table 3
EPA ECONOMIC IMPACT STUDIES CONTRACT LIST[a]

Firm	RFP or Contract number	Study	Expiration of period of performance
Arthur D. Little, Inc. (ADL)	68-01-4717	Level of effort (LOE)	9/30/80
Arthur Young	68-01-3930	LOE	10/3/79
American Management Systems (AMS)	68-01-5050	Indemnification study	1/31/80
Rand	68-01-3882	CFCs	9/30/79 — 6 Month extension requested in August
	WA 78-B277	Risk/benefit methodologies	
Mathtech	68-01-5859	TSCA economic analysis methodology	9/26/81
ICF/DPRA/MIT	68-01-5878	LOE for economic analysis of regulatory options	8/24/81
	WA 79-B047	Retrospective CFCs analysis	
Putnam, Hayes & Bartlett	68-01-5943	LOE for section 6 economic support	9/26/81
A. T. Kearney	68-01-5924	Economic analysis of hazard warning regulation	9/10/80
Mathtech	68-01-5864	LOE for testing rules — economic analysis	9/10/81
	WA 79-B058	Lab availability study	
	WA 79-B059	LOE for pre-manufacturing review support	
Research Triangle Institute	68-01-5818	Economic analysis of asbestos	1/12/80
	WA 79-B078	Economic analysis of TSCA	

[a]   Private communication, Office of Industry Assistance, Environmental Protection Agency, Washington, D.C., 1979.

since TSCA, unlike the Clean Air Act and the Federal Water Pollution Control Act, did not have small business loan provisions written into the statute. It is suggested that businesses which want additional information on the above contact the local field office of SBA or the Industry Assistance Office of EPA.

III. ANALYSIS OF TSCA SECTIONS

Figure 2 depicts a cash flow curve for a typical chemical product. The solid line shows the negative cash flow generated by a project as it moves through the research (R), plant construction (C), and initial phases of commercialization. The curve becomes positive and remains so during the mature phase of the product life cycle. The depths of the negative valley and the positive peak or shoulder after the crossover point will depend on the size of the project, the R & D investment, capital costs, and sales and market demand. The dotted line shown in this figure has been drawn to indicate the possible effect of regulation on the product cash flow curve. The dotted line shows the displacement in cash flow which may be due to increased R & D costs for toxicological and environmental testing, reimbursements, restrictions of sales to particular markets, delays due to additional testing, and reduced demand due to premature hazard classification of the specific product. This figure was not developed for any particular product or product line. However, this type of model may be useful for the reader to use when estimating regulatory impact on his particular business.

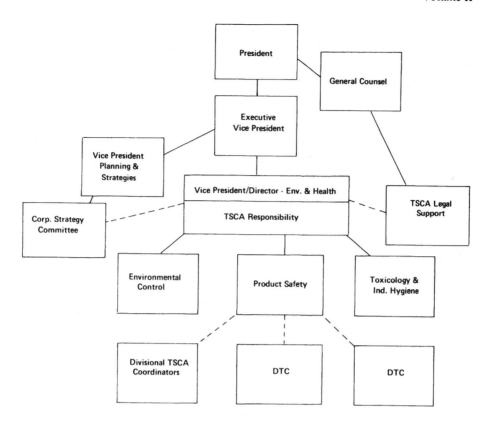

FIGURE 1. Location of TSCA responsibility in a company organization.

TOXIC SUBSTANCES CONTROL ACT (PL-94-469)

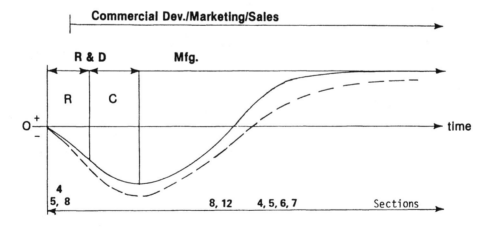

FIGURE 2. Regulatory/cash flow curve.

We will use this figure in an attempt to show what periods of the life of a product are most affected by particular TSCA sections. One could suggest that all sections of TSCA should be considered when initiating any research program. However, we will attempt to discuss those sections which we believe could result in the most significant economic impact at particular phases of the cash flow curve.

A. Section 4 — Testing of Chemical Substances and Mixtures

Under this section of the statute, the EPA Administrator could require testing if he finds that chemicals manufactured for distribution in commerce, their use in processing, and/or their disposal may present an unreasonable risk or injury to the health or the environment. He may also require testing if he finds a lack of sufficient data with which to make such a determination. The EPA will publish comprehensive testing rules covering health and environmental effects, including carcinogenesis, mutagenesis, teratogenesis, behavioral disorders and acute toxicity. They will also publish guidelines for epidemiology studies, hierarchical in-vivo and in-vitro tests, and good laboratory practices. The EPA has published for comment, health effects test standards for acute and subchronic health affects, chronic health effects and good laboratory practices (Ref. 22, 23). However, during 1982, these proposed test standards, which would have been unduly restrictive, have been withdrawn as standards and will be published as informal guidelines. Extensive comments were submitted by industry groups, toxicology laboratories and other interested parties. Final rules and dispositions are being issued at the present time.

Part 4(e) of this section provided for the establishment of a committee, the Interagency Testing Committee (ITC), whose responsibility is to determine priority lists of chemicals for review by the Administrator. The committee is made up of eight members from the following organizations:

1. One member appointed by the Administrator of EPA
2. One member appointed by the Secretary of Labor from OSHA
3. One member appointed by the Chairman of the Council on Environmental Quality
4. One member appointed by the Director of National Institute for Occupational Safety (NIOSH)
5. One member appointed by the Director of the National Institute of Environmental Health Sciences (NIEHS)
6. One member appointed by the Director of the National Cancer Institute (NCI)
7. One member appointed by the Director of the National Science Foundation (NSF)
8. One member appointed by the Secretary of Commerce

The Administrator has the discretion to add chemicals to the priority lists which have not been identified by the ITC. The Act states that the total number of substances and mixtures on the priority list should not, at any time, exceed 50. However, the ITC has published 10 lists (Ref. 24—33) which include specific chemicals and categories of chemicals. The inclusion of categories causes the actual number of priority chemicals identified for possible further testing to exceed 50. The EPA has recently begun to define the target chemicals in the categories for which they will require health and safety reporting (see discussions on 8[d]), and in subsequent testing rules. The chemicals and categories submitted by the ITC are shown in Tables 4 through 9. The administrator is required, under TSCA, to make a rule on testing of the compounds on the list within 12 months of the date when they are first designated by the ITC or publish in the *Federal Register* his reasons why testing is not required.

The process of establishing testing rules is complex and requires obtaining a detailed review of existing health effects data, in-depth scientific evaluation of the proposed general test rules, and good laboratory practices, and then deciding what test data should be requested. This involved process has resulted in the delays, beyond the 12 month requirement, by EPA, and the delays are reasonable according to most governmental and private sector scientists. However, the National Resources Defense Council sued EPA on this issue. They were granted a summary judgment in January, 1980 (Ref. 34) which requires EPA to follow a 1-year schedule for rule development. To date, EPA has not issued a final rule on any substance or category designated by ITC. The reasons for this are as follows:

1. Negotiated testing between EPA and consortiums of manufacturers or single manufacturers have obviated the need for formal rulemaking. EPA, however, still retains the authority to require or mandate testing via rulemaking should voluntary negotiated testing programs prove inadequate.
2. Several substances are either no longer commercially produced or are produced in such small volumes as to pose no real serious threat to health or the environment.
3. Final rules have not been promulgated as EPA became aware of on-going industry testing that would satisfy EPA's concerns.
4. Many ITC recommended substances reviewed by EPA for health and environmental testing were withdrawn from further rulemaking. EPA has advised in these situations that sufficient data already exist or that ITC concerns were not warranted, based upon more current use and exposure data.

Table 4
SUMMARY OF TESTING RECOMMENDATIONS BY THE TSCA
INTERAGENCY TESTING COMMITTEE

First Report[a]

Substance or category	Carcino-genicity	Muta-genicity	Terato-genicity	Other chronic effects	Environmental effects	Epidemiological study
Alkyl epoxides	X	X	X	X	X	X
Alkyl phthalates					X	
Chlorinated benzines, (Mono- and Di-)	X	X	X	X	X	X
Chlorinated paraffins	X	X	X	X	X	
Chloromethane	X	X	X	X		
Creosols	X	X	X	X	X	
Hexachloro-1,3-butadiene					X	
Nitrobenzene	X	X			X	
Toluene	X		X	X		X
Xylenes		X	X			X

[a]   Source: Reference 24.

Table 5
SUMMARY OF TESTING RECOMMENDATIONS BY THE TSCA
INTERAGENCY TESTING COMMITTEE

Substance or category	Carcino-genicity	Muta-genicity	Terato-genicity	Other toxic effects	Environmental effects	Epidemiological study
			Second Report[a]			
Acrylamide	X	X	X		X	X
Aryl phosphates	X	X	X	X	X	X
Chlorinated naphthalenes	X	X	X	X	X	
Dichloromethane	X	X	X	X	X	X
Halogenated alkyl epoxides	X	X	X	X	X	X
Polychlorinated terphenyls	X	X	X	X	X	
Pyridine	X	X	X	X	X	X
1,1,1,-trichloroethane	X	X	X	X	X	X
			Third Report[b]			
Chlorinated Benzenes, Tri-, Tetra- and Penta-	X	X	X	X	X	X
1,2-Dichloropropane	X	X	X	X	X	X
Glycidol and its derivatives	X	X	X	X		X

[a]   Source: Reference 25.
[b]   Source: Reference 26.

Table 6
SUMMARY OF TESTING RECOMMENDATIONS BY THE TSCA INTERAGENCY TESTING COMMITTEE

Fourth Report[a]

Substance or category	Types of Testing Recommended					
	Carcino-genicity	Muta-genicity	Terato-genicity	Other toxic effects	Environmental effects	Epidemiological study
Acetonitrile	X	X	X	X		X
Aniline and chloro-, bromo-, and/or nitro-anilines	X	X	X	X	X	X
Antimony	X	X	X	X	X	X
Antimony sulfide	X	X	X	X	X	X
Antimony trioxide	X	X	X	X	X	X
Cyclohexanone	X	X	X	X	X	X
Hexachlorocyclopentadiene	X	X	X	X	X	
Isophorone	X	X	X	X		X
Mesityl oxide	X	X	X	X		X
Methyl ethyl ketone				X		X
Methyl isobutyl ketone		X	X	X		X
4,4'-Methylenedi-analine	X	X	X	X	X	X

[a]   Source: Reference 27.

Table 7
SUMMARY OF TESTING RECOMMENDATIONS BY THE TSCA INTERAGENCY TESTING COMMITTEE

Fifth Report[a]

Substance or category	Type of Testing Recommended					
	Carcino-genicity	Muta-genicity	Terato-genicity	Other chronic effects	Environmental effects	Epidemiological study
Benzidine-based dyes					X	
o-Diamidine-based dyes				X	X	
o-Toludine-based dyes				X	X	
Hydroquinone	X		X		X	X
Quinone	X		X		X	

[a]   Source: Reference 28.

One can realize from the above that this section of TSCA can have a significant direct cost impact on the chemical industry. The need for additional testing may be required by EPA during the mature phase of the product life cycle. It is assumed that the request for testing would be justified, based on a preponderance of available scientific evidence pointing to a need for additional data to support a finding of unreasonable risk to health or environment. If the additional testing is required, it is in the best interest of those companies which are required to test to pool their testing information or form an ad-hoc group to perform the testing. The reasons for this will be described below in more detail. It is possible that products with marginal returns could not support additional testing and could be discontinued. The more likely case, with existing products, would be that testing would occur and attempts would be made to pass through these costs. A more negative impact may occur in companies which are performing R & D on new processes for existing products subject to a Section 4 rule, or which may be researching a new product line based on the product in question. In these cases, the cost of testing, or the costs of sharing the testing costs must be factored into the cash flow of the project.

Table 8
SUMMARY OF TESTING RECOMMENDATIONS BY THE TSCA
INTERAGENCY TESTING COMMITTEE[a]

Substance or category	Types of Testing Recommended					
	Carcino-genicity	Muta-genicity	Terato-genicity	Other chronic effects	Environmental effects	Epidemiological study
Sixth Report						
Phenylenediamines	X	X	X	X	X	X
Seventh Report						
Alkyltin compounds	X	X	X	X	X	X
Benzyl butyl phthalate	X			X	X	
Butyl glycolyl butyl phthalate		X		X	X	
Fluoroalkalenes	X	X	X	X		
Eighth Report						
2-Chlorotoluene	X	X	X	X	X	
Diethylenetriamine			X	X		
Hexachloroethane					X	

[a]   Source: References 29, 30, and 32.

Table 9
SUMMARY OF TESTING RECOMMENDATIONS BY THE TSCA
INTERAGENCY TESTING COMMITTEE[a]

Substance or category	Types of Testing Recommended					
	Carcino-genicity	Muta-genicity	Terato-genicity	Other chronic effects	Environmental effects	Epidemiological study
Ninth Report						
Chlorendic Acid					X	
4-Chlorobenzo-trifluoride				X	X	
Tris(2-chloroethyl)-phosphite			X	X	X	
Tenth Report						
Biphenyl					X	
Ethyltoluene		X		X	X	
Formamide	X		X	X		
1,2,4-Trimethylbenzene			X	X	X	

[a]   Source: References 31 and 33.

As mentioned above, companies which manufacture or process a chemical for which a Section 4 rule has been proposed or issued will probably cooperate to develop the required test data. The driving force behind this observation is that EPA will allow companies to share the cost of testing required by a Section 4 rule. EPA is encouraging companies to cooperate in pooling test results and in sponsoring joint toxicological testing (Ref. 35) studies. A company may apply for an exemption to a testing rule if another company is already conducting the test. However, TSCA requires that the exempted company pay fair and equitable reimbursement to the company or companies which have developed the test data (Ref. 36). EPA has published a proposed rule on reimbursement (Ref. 37). The reimbursement obligation exists for any person who obtains an exemption before the end of the reimbursement period. This period of reimbursement starts when the data are submitted to EPA and ends 5 years later *or* at the end of a period of time equivalent to the time needed to develop the data, whichever is longer. This long reimbursement period will mean that manufacturers or processors who are evaluating a new venture involving a chemical subject to a test rule may have to reimburse those companies that developed data. This cost should be factored into the cash flow of the new venture. If the person submitting test data and those granted exemption cannot agree on the amount, then EPA will order the person granted the exemption to provide what an arbitrator considers fair and equitable reimbursement.

The final EPA rules on reimbursement will take into account market share, competitive position, and many other factors. EPA has concluded that reimbursement would be based on actual data development costs, plus cost of capital during the reimbursement period. It may allow some consideration of additional reimbursement to companies who have developed data in their own toxicology laboratories, but not allow for R & D costs associated with a product the development of which comes under a test rule. The rules on reimbursement will apply to TSCA Section 4(c,3), 4(c,4) and 5(h). EPA has made it clear that they would probably accept any system of reimbursement that is acceptable to industry. The Chemical Manufacturers Association has proposed a system to EPA.

As can be seen from the above discussion, the TSCA Section 4 rules will have a direct cost impact on industry. Testing costs, marketing and business decisions, and new venture possibilities related to priority list chemicals will all be impacted. To help address the impact of testing costs alone, Table 10 includes estimates of various toxicological tests. These estimates were developed by the Celanese Corporation Toxicology Department in 1982 and should only be used as guidelines.

B. Section 5 — Manufacturing and Processing Notices

Section 5 of TSCA became operative 30 days after publication of the chemical inventory on June 1, 1979. Any chemical not on the EPA Initial Inventory List of commercial substances was subject to Section 5 rules. It is expected that this section will continue to have far-reaching effects on both the chemical industry and the downstream industries that depend on chemicals. This section provides for the EPA Administrator to require notifications from companies that are planning to manufacture (including importation), or distribute in commerce, compounds which are not on the inventory list. This information is used by the EPA to assist the Administrator in determining if an unreasonable risk will result from the manufacturing and distribution of the subject chemical. The notification requires information on volumes to be manufactured or processed, assessment of end use exposure, potential hazards to human health and the environment, methods of distribution and disposal, and customer use information. It is expected that EPA will supplant the current interim rule with a final rule in early 1983 that may reduce the initial reporting burden as a result of past EPA experiences with the Premanufacture Notification (PMN) program. The information must be submitted to EPA 90 days prior to manufacturing, processing, or distribution. The EPA can delay ruling on the submission for an additional 90 days before making a ruling on risk. TSCA does not give the EPA rulemaking authority under Section 5 to require toxicological testing. However, the debate continues on what testing guidelines EPA may issue for use in preparing premanufacturing notifications under this Section. The issues here are whether the guidelines will reference the Section 4 testing rules, will there be base set and hierarchial testing, and how much flexibility will be demonstrated by EPA in reviewing PMNs.

Although Section 5 rules have not been finalized, Section 5 is in effect and EPA is reviewing PMNs. EPA has stated that it is not obtaining sufficient toxicity data with the initial PMN submission and this raises serious concerns with EPA about its ability to discharge its risk assessment responsibilities under the statute (Ref. 38). In fact, EPA recently delayed its first notice of review and told the submitting firm to supply additional information before the 90-day review period would start (Ref. 39). An interesting sidelight to the EPA lack of testing authority under Section 5 is that the U.S. chemical industry may be forced to begin premarket testing if they export chemicals to the European Community (EC) member nations. EC directive 67/548/EEC (Ref. 40) requires base set data for all new chemicals, and member nations adopted this directive in 1981. Exemptions for certain categories of polymers, site-limited intermediates, and small volume chemicals was proposed (Ref. 72) in mid-1982. Potential reduction in costs

Table 10ᵃ
TOXICOLOGICAL TESTING COSTS

	Cost ($, thousands)
Acute studies	
Dermal irritation	0.60
Eye irritation	0.90
Dermal sensitization	3.00
Dermal systemic toxicity	
Single dose screen	0.90
LD_{50}	4.00
Oral systemic toxicity	
Single dose screen	0.45
LD_{50}	2.00
In-vitro mutagenicity studies	
Ames test	0.50
Mammalian cell point mutation	4.00
Chromosome aberrations	4.00
Sister chromatid exchange	2.80
2-week subacute studies	
Oral or dermal	15.00
Inhalation	25.00
Metabolism studies	
Toxicokinetics (absorption and excretion rates)	approx. 40.00
Metabolite identification	approx. 70.00
Reproductive studies	
Teratogenicity	40.00
Dominant lethal test	20.00
One generation reproduction — parents exposed for 10—12 weeks prior to mating, pregnant females and offspring exposed until offspring weaned	
Oral or dermal	65.00
Inhalation	100.00
12-Week subchronic studies	
Oral or dermal	55.00
Inhalation	90.00
Combined chronic toxicity/carcinogenicity	
Mouse skin painting (18 months)	10.00
Chronic/carcinogenicity (24—30 months, rat and mouse)	750—1.500
Environmental studies	
Acute aquatic toxicity	0.65—1.0
On-site aquatic toxicity	20.00
Bioconcentration and persistency	25.00
Early life stage toxicity	15.00
Chronic fish toxicity	50.00
Avian toxicity	2.00
Avian reproduction	20.00

ᵃ   Costs estimated by the Toxicology Department, Celanese Corporation, New York, July, 1982.

associated with the exemptions would be expected, but would be highly sensitive to the scope and flexibility of the exemption criteria to be established by EPA. Testing may also be required for a PMN if a "new chemical substance" is included in the priority categories under Section 4. Therefore, if a Section 4 test rule has been written for this category, test data may have to be submitted on the PMN before manufacturing can begin.

Most chemical manufacturers do perform some kind of toxicity testing before exposing their own workers or their customers to new products. The questions raised in Section 5 PMN procedure are how much is needed, why, when is it needed, will it be cost effective, and on what basis will EPA make its decisions.

The proposed Section 5 rule was issued on January 10, 1979 (Ref. 41). Extensive comments were submitted to EPA. Industry and EPA have been working together to develop workable notification procedures which will meet both EPA and industry objectives of effectively and efficiently protecting the community and the environment from unreasonable chemical risks. However, many issues remain to be solved before the final rules are issued.

As stated previously, Section 5 of TSCA is in effect with interim status and the Administrator is authorized to take further action under this section. Under Section 5(e), if the administrator finds that the information available on a particular substance is insufficient to make an evaluation of unreasonable risk, he has the authority to prohibit or limit the manufacture, processing, distribution, or disposal of such a substance, or to prohibit or limit any combination of such activities.

Also, under Section 5(f), upon evaluation of all the data, the Administrator can find that the distribution or disposal of the chemical substance described in the PMN will present an unreasonable risk to health and the environment. After such a finding the Administrator can limit the amount of the suspect stubstance which may be manufactured or limit a significant new use for an existing product. "Significant new use" may include a new use in the conventional meaning, or an increased volume of production or placing a new population at risk of exposure. If a significant new use is contemplated for a chemical which is already on the inventory list and EPA has already established by rule boundary conditions of use, then a PMN will be required for these new uses. The proposed Section 5 rules published in 1979 did not cover significant new uses and proposed rules covering this area have not been issued.

Under Section 5(h), the Administrator has the right to exempt any person from the requirements of Section 5(a) and (b) and allow manufacture for test market purposes, or for which data have already been submitted. It is under this section where the issue of reimbursement, discussed under Section 4 above, comes into play. Significant here, where new products are at issue, is the EPA contention that R & D expenses are not part of the reimbursement formula.

The proposed rules mentioned above contain a form to be used for premanufacture notification. The form contains requirements for extensive submission of mandatory and optional information.

A. D. Little, Boston, Mass., under contract to EPA, performed a preliminary analysis of some of the economic impact of using this form to comply with requirements (Ref. 42). They estimated that the incremental costs of preparing the notification form were between $2500 and $14,000 for the minimum mandatory submission; $3700 and $22,000 for the maximum mandatory submission; and $9000 and $41,000 for filling out the combined mandatory and optional segments of the form.

A. D. Little estimated that 1000 chemicals ± 30% were introduced for commercial sale each year. Of these, about 300 would have sales above 1000 lb/year. Also, 50% of the chemicals introduced had annual sales of less than $50,000 after 5 years and 80% had sales of less than $150,000 after the same time in the marketplace. Therefore, they estimated that 50% of the chemicals recently introduced for commercial sale would not have been introduced if notification costs were about $10,000 per chemical. If notification costs were the upper limit of $40,000, then 90% of these new chemicals might never have reached market. The debate here should not be on whether this would keep hazardous materials out of the marketplace or whether we are disallowing many beneficial chemicals to be marketed. There may be truth on both sides. Rather, what this analysis does indicate is that there will be a serious decrease of new product R & D, innovation, and introduction due just to the costs of preparing forms. The cost for testing the products to determine their hazards have not been included in the A. D. Little study and would be an extra cost burden on the new product innovation.

It is also felt that this burden would fall disproportionately on smaller chemical businesses which rely more strongly on developing new, limited volume chemicals. The testing costs, combined with the PMN form costs, could be an extremely burdensome expense for small ventures. A deputy assistant Administrator of EPA indicated that subchronic toxicity tests under EPA PMN rules may cost tens of thousands of dollars. He added that if 2 or 3 tests are required together, for example, inhalation and skin toxicity studies, then the costs could be in the hundreds of thousands of dollars (Ref. 43).

The chemical industry, especially through the Chemical Manufacturers Association, made numerous comments on the PMN forms dealing with the length and complexity of the forms, issues of confidentiality, mandatory customer content and toxicity testing guidelines. This cooperative effort resulted in EPA making significant changes to the forms (Ref. 44). The form was shortened and mandatory customer contents and end-use contacts were reduced, but substantiation of confidentiality was required with submission of the PMN form. EPA has recently proposed relaxing the substantiation claims of confidentiality. Pragmatically this will likely have minimal impact on costs.

A. D. Little again estimated the costs for filling out the PMN forms (Ref. 42). The cost ranges for the mandatory sections, including confidentiality, now are between $1200 and $15,000. These costs are shown in Table 11. This compares to the $2500 to $22,000 for filling out the mandatory sections of the PMN form proposed on January of 1979. The optional sections were not estimated and filling them out would increase notification costs. Applying A. D. Little's previous criteria on the impact of new product introduction due to costs of completion of PMN forms, over 50% of new chemical products would not be introduced. It is interesting to note that the impact of Section 5 on new product introduction, which is

Table 11
TOTAL COST ESTIMATES FOR COMPLETION OF
REVISED PMN FORM

Part I:	General Information	$275—$2,125
Part II:	Human exposure and environmental release	
	Section A: Industrial sites controlled	
	by submitter	275—2,100
	Section B: Industrial sites controlled	
	by others	0—1,200
	Section C: Consumer exposure	0—800
	Total	275—4,100
Part III:	List of Attachments	
	a. Physical/chemical properties	150—600
	b. Health and environmental effects	
	data	300—1,400
	Total	450—2,000
Part IV:	Federal Register Notice	75—300
Clerical costs — all sections		80—400
Subtotal		1,155—8,925
Confidentiality costs — all sections		0—6,400
Grand Total		$1,155—15,325

described in the 1978 and 1979 A. D. Little analysis, is the same order of magnitude as that estimated in the 1975 F. D. Snell report previously discussed.

The Chemical Manufacturers Association submitted a proposed PMN for EPA consideration, and A. D. Little estimated the costs for filling out this form. The results are shown in Table 12. A. D. Little estimated that filling out the mandatory sections will cost between $1000 and $5500.

When one looks at the cash flow curve shown in Figure 2, one can understand the significant economic impact that Section 5 can have on chemical R & D. To avoid the risk of regulatory action at the time of commencing manufacture or commercial introduction, a number of actions must, and should, be considered during the R & D phase. More comprehensive analysis of potential product hazards; the extent, type, and cost of toxicity tests; form preparation; and the overall regulatory unknown will be superimposed on the known risks of any R & D project. All this can result in significant reduction in new product R & D, new product innovation, and consolidation in the chemical industry.

The impact of Section 5, in the positive side of the cash flow curve, can occur under significant new use rules. In these cases, additional testing and notification will be required if the existing product is to be marketed for a use which can be categorized as "significantly new" by the Administrator.

C. Sections 6 and 7 — Regulation of Hazardous Chemical Substances and Mixtures and Imminent Hazards

Section 6 of TSCA deals with the regulation of hazardous chemical substances and mixtures. The Administrator is required to regulate existing chemicals and mixtures when he has a reasonable basis to conclude that they will present an unreasonable risk of injury to health and the environment if they continue in commerce. Options available to the Administrator range from labeling and reporting, limited use provisions, disposal restrictions, and up to and including prohibition of manufacturing.

Polychlorinated biphenyls (PCBs) and chlorofluorocarbons are the only two materials now regulated under Section 6. The final rules on PCBs, which include the ban on their manufacture, were published in 1979 (Ref. 45). EPA estimated that the Section 6 regulation on PCBs would cost approximately $55 million annually for the first 5 years. This would decline to $42 million in the next year, and drop 7% thereafter as PCB-containing equipment is removed from service (Ref. 19, 46). EPA also estimated that fluorocarbon regulation will cost industry between $170 to $270 million annually in the initial years of the regulation. A National Academy of Sciences study on Polychlorinated biphenyls in the environment estimated the control costs on use and disposal of PCBs (Ref. 47). Their estimate of costs, required under regulation,

Table 12
TIME AND COST ESTIMATES FOR PREPARATION AND
SUBMISSION OF CMA FORMS[a]

	Clerical[b]	Technical	Managerial	Total
Mandatory				
Part I (hr)	2—10	6—44	3—13	11—67
Part II (hr)	6—30	15—108	4—14	25—152
Total time (hr)	8—40	21—152	7—27	36—219
Total cost ($)	80—400	525—3800	350—1350	995—5550
Optional				
Part III				
Time (hr)	0—40	0—204	0—38	0—282
Cost ($)	0—400	0—5100	0—1900	0—7400
Optional				
Part IV				
Time (hr)	0—20	0—128	0—14	0—162
Cost ($)	0—200	0—3200	0—700	0—4100

[a] Hourly labor rate estimates: $10—clerical, $25—technical, $50—managerial.
[b] Clerical time estimate developed separately for each part. Source: Arthur D. Little, Inc., Boston, estimates.

are shown in Table 13. This table indicates that the control costs for the Section 6 PCB rules would be over $700 million, which could be spent over an extended time period. However, the bulk of the costs will probably be incurred in the first decade. Further, the report estimates that additional costs for some of the proposed controls such as disposal of transformer fluids contaminated with over 50 ppm PCB and replacing and disposing of PCB-containing fluids in other equipment, could range between $617 and $770 million.

This significant economic impact results from the removal from the environment of a hazardous substance which meets the Administrator's criteria for unreasonable risk to health and the environment. Obviously, any action taken under Section 6 will result in significant immediate and future impacts to any cash flow curve. The possibility of any chemical substance or mixture being regulated under Section 6 will probably lead to increased hazard testing and evaluation at the early stages in the life of a product. This will be especially true if widespread exposure potential exists due to end use patterns. Furthermore, with the existence of a PCB rule, EPA has significantly increased its enforcement activities in the PCB area by using their authority under Section 16(a) of TSCA. This section provides for the assessment of civil penalties on persons who do not comply with a rule or order issued under Sections 4, 5, 6, and 7. Penalties can be up to $25,000 for each violation and for each day the violation continues. EPA regional offices recently filed a number of complaints with proposed penalties up to $131,000. Penalties of $80,000 for alleged violations at 3 different sites were issued to one firm (Ref. 48). This may be another reason for firms to have organizations in place to understand and expedite TSCA-related actions.

The use of Section 16(a) when a Section 6 rule is issued allows EPA to act quickly to effect remedial action in hazardous situations. This gives EPA an immediately effective regulatory tool whenever a Section 6 rule is final.

Section 7 provides the Administrator with the authority to commence a civil action to seize imminently hazardous chemical substances, mixtures, or articles containing such substances. He can also take action against persons who manufacture, process, distribute, or dispose of imminently hazardous substances. The definition of imminent hazard is left to the discretion and authority of the Administrator. He must judge that the substance may present an imminent and unreasonable risk of serious or widespread injury to the health or the environment. Also, the risk shall be considered imminent if injury is likely to result before a final Section 6 rule can prevent the risk which would arise from any combination of activities in the manufacturing, processing, and distribution in commerce of the suspect substance.

Obviously, severe economic dislocations can result from imposition of a Section 7 action. The Administrator has a great deal of discretionary authority under this section and he must act responsibly. Again,

Table 13[a]
PCB CONTROL COSTS FOR EPA
REGULATION ON USE AND DISPOSAL

	Option[b]	Costs, $MM
1.	Incinerate PCB transformer fluid	75.0
2.	Incinerate high and low voltage capacitors	493.0
3.	Flush and drain transformers after (1)	105.0
4.	Dispose of transformers in landfill	45.0
	Total	718.0

[a] Polychlorinated Biphenyls, 86-7, National Academy of Sciences, Washington, D.C., 1979.

[b] These are required options and will be operative as PCB-containing equipment is taken out of service. Some transformers, for example, may remain in service for up to 40 years.

it is important for affected persons to have sufficient data on the hazardous nature of their products in order to input to the process of determining unreasonable risk. Unfortunately, there are no clearcut definitions of unreasonable, significant or substantial risk as they are used in TSCA. Therefore, the mechanisms for defining risk must be based on sound scientific data, coupled with sound methods for risk/benefit and cost/benefit analysis. There is an on-going public, industrial, academic, and governmental debate on this issue and one hopes that workable methodologies will result.

D. Section 8 — Reporting and Retention of Information

Section 8 of TSCA provides for the numerous information reporting and record retention provisions of the Act. Most industries have established organizational procedures and systems to collect, review, retain, and send the required information to EPA. Two reporting rules, 8(b) and 8(e) were promulgated in final form in 1977 (Ref. 49) and 1978 (Ref. 50). The 8(b) rule was the first TSCA action that affected the total chemical industry. This rule required reporting of existing commercial chemicals to EPA, which were then published in the chemical substances inventory in June 1979. The inventory contained over 44,000 chemical substances and some additions were subsequently made to the inventory bringing the current 1982 total to near 60,000. EPA estimated that the 8(b) inventory reporting rule would cost industry $15 million (Ref. 19). To date, there is no comprehensive report from industry on the actual costs incurred to develop the chemical inventory. However, one large chemical company did develop firm cost data for submitting information under the 8(b) rule. Their cost, incurred up to March 1, 1978, was $638,000 (Ref. 51). Such economic impact studies, as described above, will better define the total industry costs resulting from the 8(b) rule. The completion of the inventory was a massive task for EPA as evidenced by the numerous delays before the June 1979 publication. The inventory, with its expected corrections and additions, is a necessary base document for TSCA since many of the other sections of the Act, e.g., Section 5, require an accurate inventory before they become operative.

The 8(e) guideline, notification of substantial risk, requires that if any person who manufactures, processes, or distributes a chemical substance or mixture receives information that it presents a substantial risk of injury to health or the environment, he must immediately inform the Administrator. This action must take place within 15 working days from the time the person obtains the information. This requirement has caused many industries to establish detailed communication and control procedures to help ensure that their employees can comply with the requirements of the 8(e) guidelines. EPA has the responsibility to review the 8(e) submission, make a finding on risk and, if necessary, trigger action via some of the sections of TSCA described above.

The section 8(a) portion of the Act requires maintenance of records and reporting of information on specific chemicals. The types of information which have been required are chemical names, generic end uses, volume manufactured and processed, by-products manufactured, number of persons exposed in the workplace, environmental release, and quantities disposed of. There are a number of sensitive issues raised by reporting this type of information. In this case, as in all the Section 8 procedures, confidentiality issues on customers, end-uses, chemical identity, volumes, etc., have been of significant concern to industry. The concern revolves around both the need for this information and the confidentiality provisions under the Act. Numerous comments were given to EPA before EPA published its first final 8(a) (Ref. 52) rule. The rule requires reporting only by manufacturers and importers on about 250 chemicals. Processors and small business are currently exempt from initial reporting requirements. EPA has estimated that the costs for complying with this information submission would be $6 million (Ref. 53). Industry figures are not available.

Section 8(c) requires the manufacturer, processor or distributor to maintain records of adverse reactions alleged to have been caused by a substance or mixture to their own or customer employees and the general public. These records are to be maintained for a period of 30 years. The EPA plans to issue an 8(c) rule sometime in 1982. Since this section has the potential for massive, long term, record keeping requirements, definition of what is a significant adverse reaction, chemicals covered, and what must be reported to EPA are necessary before the impact of the rule can be estimated.

Section 8(d) rules were proposed in 1979 (Ref. 54) These rules require manufacturers, processors, and distributors to submit health and safety data to the Administrator. The chemicals for which the data must be reported are those on the priority lists described under Section 4. The type of data to be reported includes acute and chronic toxicity studies, environmental fate tests, industrial hygiene exposure data, and may include medical records. Also, health and safety studies on chemicals used solely for R & D purposes must be submitted. Since portions of TSCA may be interpreted to mean that no health and safety information can be maintained confidential the disclosure of chemicals on which a company is performing R & D raises serious trade secret concerns.

EPA has estimated that the proposed reporting rule would cost industry $410,000. The chemical substances subject to the reporting rule number over 300. This is primarily due to the inclusion of categories on the priority lists. There are no estimates available from industry.

All of the Section 8 sections have a direct cost impact on all portions of the cash flow curve. The impacts for any one chemical will be small. However, one must also consider the organizational system which must be developed to comply with Section 8, the potential loss of confidentiality, and the ability to protect R & D activities. Also, reported information can trigger discretionary TSCA actions, such as extensive additional testing and reporting, and dislocation in end-use markets due to premature disclosure of partial information on risk. Both industry and EPA are sensitive to the broad issues raised by the information reporting requirements of Section 8. There is cooperation between them to develop efficient and effective rules which must meet the requirements of TSCA as well as necessary business needs.

E. Sections 12 and 13 — Exports and Entry into the Customs Territory of the U.S.

Both of these sections will have an impact on international trade. Section 12 gives the Administrator authority to require information on exported products (or impurities) for which a testing rule or action under Section 4, 5(b), 6, or 7 has been proposed or has been taken. The Administrator can then notify the government to which the product is going that it has been the subject of TSCA action. Even materials intended solely for export may be covered under this Section. The EPA has issued a Section 12 rule and the requirements apply immediately to chlorofluorocarbons, polychlorinated biphenyls, and asbestos.

EPA and the U.S. Customs Service are preparing a proposed Section 13 rule (Ref. 56). Under this Section, the Administrator, through the customs service, has the right to refuse entry to any material being imported if it does not comply with TSCA. Also, entry may be forfeited for violation of Section 5 or 6 rule and/or a civil action brought under Section 5 or 7. Importers will be required to certify that the imported materials meet all labeling, PMN, testing, reporting, control, or other special TSCA requirements.

Importers would be responsible for signing the certification and could be penalized with charges for storage, labor, cartage, or disposal of any articles which were refused entry. Also, procedural delays at the port of entry could result in further economic burdens to the importer, even if the delays are due to actions out of his control. Both importers and foreign governments are meeting with EPA to make their concerns known and to assist EPA in preparing the proposed rule. Since U.S. industry imports many specialty, low-volume chemicals, which are necessary for processing and formulating major product lines, delays and/or restriction of key imports could have a serious detrimental effect on the U.S. chemical industry. Searching for alternatives during the mature phase of the life cycle of a product will have an

Table 14[a,b]
PLANNED EPA ACTIONS ON TSCA

TSCA Section	Action	Date
4	Proposed rule on reimbursement for test data	June 1982
4	General exemption policy for test rules	September 1982
4	Health effects testing: GLP standards	Undetermined
4	Environmental testing: GLP standards	Undetermined
5	Premanufacture notification requirements and review procedure	March 1983
5	Site-limited and small volume exemption rule	June 1982
5	Polymer exemption rule	July 1982
6	PCB regulations [6(e)] — 3 regulations on electrical equipment, closed systems and manufacture and distribution	Fall 1982
6	Labeling rule for treated wood	February 1983
6	Proposed production restriction for chlorofluorocarbons	Undetermined
6	Rules restricting the commercial and industrial use of asbestos fibers	Undetermined
6	Asbestos-containing materials in school buildings	April 1982
8	Preliminary assessment information reporting 8(a) (250 + substances)	June 1982
8	Proposed preliminary assessment information reporting 8(a) (50 substances)	June 1982
8	Proposed small manufacturer exemption standards [8(a)]	June 1982
8	Health and safety study reporting 8(d)	Fall 1982
8	Significant adverse reactions 8(c)	Fall 1982
8	Asbestos use and substitutes reporting 8(a)	June 1982
12	Revision of 12(b) — export notification rule	Fall 1982

[a]   Source: Reference 57.
[b]   Note that EPA had scheduled 8 test rules to emerge in 1982. However the voluntary negotiated approach between EPA and industry has to date eliminated the need for formal rulemaking. Several such testing programs are currently underway.

immediate impact on a cash flow curve and could result in restrictions in specific end-uses which require specification testing on components and raw materials. EPA estimated only the direct cost impact due to certification by the importers. They estimated that 417,000 chemical shipments would require TSCA certification. General certification would cost $2 per shipment and special certification for controlled substances would be $14.25 for new shipments and $8.75 for repeats. The total direct cost impact was estimated to be $2.23 million per year.

Table 14 shows a timetable for a number of actions that EPA will prepare relative to the sections described above (Ref. 57).

IV. TSCA AND THE INNOVATION QUESTION

A. Reduction in R & D and Innovation

Over the past few years, there has been continuing debate over what has happened to innovation in the U.S. The debate centers around such questions as: What is innovation? How is it measured? Is it good? What affects the rate of innovation? and, What roles should government and industry play in stimulating R & D and innovation?

In 1978, President Carter announced the formation of a cabinet level interagency committee to review U.S. policies concerning domestic industrial innovation. He said that innovation provides a basis for the economic growth of the nation and is closely related to productivity and to the competitiveness of U.S. products in domestic and world markets. A number of publications discuss this innovation issue (Ref. 58—63). These references describe a number of events which support the conclusion that R & D activities and, thus, innovation is decreasing.

The American Chemical Society initiated a study in 1972 on incentives for R & D and to study trends in innovation and private investment in R & D (Ref. 58). The study indicated that for over 50 years the U.S. has been the world leader in innovation and technology. However, they see the lead narrowing because of such indications as larger trade deficits, increasing numbers of U.S. patents by foreign investors, a decrease in the number of new product introductions, and a shift in the emphasis to short-range research by U.S. firms. In 1983, U.S. industry will spend an estimated $42.6 billion of its own funds for R & D. The current 1982, 1981, and 1980 costs are $38.5, $34.4, and $30.4 billion, respectively.

In 1981, for example, industry R & D expenditures for chemicals and allied products amounted to $4.97 billion, accounting for 14.7% of all U.S. industry R & D spending. Approximately $2.3 billion are expenditures for industrial chemical R & D with the remaining $2.7 billion designated for drugs and other chemicals.

A recent National Science Foundation study (Ref. 58) indicates that in 1980, 67% of all industrial chemicals R & D funds ($2.3 billion total) came from firms with R & D budgets over $100 million, and 28% from firms with budgets between $10 to $100 million.

All U.S. basic R & D expenditures (including industry, government, and other private sources) have shown an increase in 1982 expenditures of 6% to $9.3 billion over 1981 ($8.8 billion). These figures, however, when adjusted to constant (1972) dollars, show a *decrease* of 2%. This is an indication that the increased emphasis on short range goals has eliminated much of the high risk, long range basic research.

Assuming that innovation is linked to R & D expenditures, R & D as a percent of GNP is down from 3% in the 1960s to 2% in the late 1970s (Ref. 63). There has also been a reduction in government R & D funding in the chemical sector. In 1967, R & D funding was $210 million, about 12% of the total outlay by industry. By 1976, Federal funding was $266 million, a drop to less than 9% of chemical industry R & D expenditures. Also, adjusted for inflation, the $266 million figure is equivalent to about one-half of the 1967 dollars (Ref. 59).

One other trend that is an indication of the slowdown in innovation is the rate of capital spending. In 1974 to 1976, capital spending, as a percent of sales, was 12% to 13%, but the percent for 1977 to 1979 was only 10%.

Of particular note were the results of a Chemical Specialties Manufacturers Association (CSMA) (Ref. 65) survey of their membership on the impact of TSCA regulatory activity (PMN) on innovation, particularly as it relates to smaller firms. This study dramatically points to a disproportionate burden on smaller businesses as regulatory costs increase.

Firms with annual sales in excess of $500 million reportedly would not pursue 2% of their PMN filings if the cost of preparation was $18,400. On the other hand, firms with less than $10 million in gross sales would not file PMNs for 58% of their new products (assuming the $18,400 preparation costs) which would be marketed without PMN requirements.

While government regulation has been singled out as the reason for these impacts on innovation, no one factor alone can be highlighted as the reason for a slowdown in overall R & D investment and in basic research. A number of economic factors, including government regulation, have made the cost of R & D much higher and lengthened the time necessary to introduce new products to the marketplace.

R & D has become higher risk and less profitable in recent years and a combination of factors could explain the decrease in new products, fewer U.S. patents by domestic investors and lower allocations of R & D dollars to basic research (Ref. 64). For example, one major factor which led to a reduction of basic R & D was the oil crisis in 1973, and the resulting rapid escalation of raw materials and energy costs. This led to a significant shifting of funds from basic R & D into process R & D. Also, research was directed to improving the profitability of established product lines. In this case, the chemical industry was very innovative in developing new processes which made more efficient use of energy and raw materials (Ref. 64).

B. Effect of Government Regulation

Many of the above references discuss the effect of government regulations on R & D and the process of innovation. In the past decade, two regulatory agencies have profoundly altered the operations of the chemical industry. The two are OSHA and EPA. During the lifetime of these agencies, 10 major pieces of legislation or amendments to legislation affecting R & D have come into being (Ref. 59). St. Clair (Ref. 63) quotes Dennison of the Brookings Institute, who stated that two areas of regulation alone, pollution control and health and safety, have cut the annual growth rate of productivity since World War II by almost 1/6 of the average rate of 2.1%. Weidenbaum (Ref. 62) states that the real effect of regulation on innovation is the reduced rate of new product and process introduction. He proposes that the longer it takes for an innovation to be approved by a Federal agency, the more costly the approval procedures will be, and this effect on cash flow can result in less new product activity.

The work of Ashford et al. (Ref. 61) states that technological innovation often creates the problems which lead to regulation. Also, technological innovation will be required to meet the standards mandated by Congress and implemented by the regulations. The major challenge is proper design of the regulation to meet the social goals expressed both by the regulation and by the need for industrial development and economic well-being. The regulation and innovation relationship is complex and dependent on many factors. These include both the nature of the regulation and the nature and ability of the regulated firm to handle the regulation. One of the conclusions is that we are going through a transition phase and it is premature to attempt to determine the ultimate long-term effects of regulation on technological change and innovation. In fact, there are two impacts of regulation; one is business innovation and the other is compliance innovation. There is suggestive evidence that regulation can have an enhancing or retarding effect on both areas.

A summary of the literature (Ref. 60) on regulation and innovation does not lead to a firm conclusion that innovation is restricted by regulation. However, there was no doubt that there is an effect. Many papers concluded that the drug and chemical industries were heavily impacted and the most direct effects were seen in the reduction of output of new products. What was also highlighted was that smaller firms, which are thought to be the technical innovators, would be more heavily impacted since they would have fewer resources to direct to compliance activities.

One concern raised in this review is that TSCA and the Federal Insecticide, Fungicide and Rodenticide Act, (FIFRA) were close to the content and design of the Kefauver-Harris amendments which affect the pharmaceutical industry (Ref. 66).

Another impact of regulation is the increased amount of what is referred to as "defensive R & D" (Ref. 59). These are the funds that are allocated to toxicological studies, pollution control, and other health-related regulations. A study of 15 top chemical companies showed that about 13% of R & D dollars were allocated to this area. The range was from 10 to 25% and there was not a concomitant increase in R & D funds for other areas.

Furthermore, the need for new toxicological facilities, many of them in anticipation of TSCA requirements, has also redirected funds from R & D capital and expense budgets. Existing laboratories at companies such as Dow, DuPont, and Stauffer have been increased by over 50%. Also, at least four new laboratories costing between $12 to $22 million have been built since 1979. The Chemical Manufacturers Association has toxicological studies underway on many commodity chemicals, ranging in cost from $200,000 to $3 million, each. Also, over 33 chemical firms sponsored the establishment of a new center for the advancement of the science of toxicology, the Chemical Industry Institute of Toxicology (CIIT).

While there may not be a strong consensus, pro or con, on government regulation impacts being technology forcing or a detriment to innovation, there is a perception that the effect is there and it is negative. The National Science Board Annual Report of 1976 indicated that 48 presidents and 78 vice presidents and directors of R & D responded that the reason for the lessening of innovation was government regulation and control. These same respondents highlighted the Toxic Substances Control Act as being a major concern. This was before TSCA was in effect and it remains to be proven if the perception turns into reality.

C. TSCA and Innovation

The full impact of TSCA will not be known until all rules and regulations are final and both industry and government build experience based on actual compliance actions. However, as discussed above, many companies have taken anticipatory action in increasing toxicological testing and toxicological testing capabilities. Most of the funding for this activity has come from R & D budgets. As described in Section III of this chapter, it is anticipated that new and existing product R & D will be affected by actions taken under Sections 4, 5, and 6, and by the concern about the confidentiality provisions in many of TSCA sections.

This latter concern on confidentiality was recognized by Ashford (Ref. 68) when he pointed out that the disclosure requirements under TSCA will demand a balance between the data needed to warn workers and consumers about toxic substances and the regulatory agency responsibility to adequately protect these same groups, as well as the trade secrets of the manufacturing firm. The concern that industry has in protecting their investment in time and expense associated with new product R & D activities is real, and if not addressed, will have an effect on the extent and type of R & D commitments which will be made to innovative, new product research.

Innovation in the chemical industry has resulted in numerous new drugs, pharmaceuticals, detergents, plastics, and lubricants. All have had a real societal benefit and a direct impact on our standard of living. Some technological innovations have also had detrimental impact on human health and the environment. Care must be taken when fully implementing TSCA to avoid "throwing out the baby with the bathwater".

Promulgation and implementation of Section 4 and, especially, Section 5 rules are anticipated to have significant impacts on innovation. The demands for toxicological testing under these sections will lead to technological advances for screening tests for carcinogens, for other toxicological effects, and for environmental effects (Ref. 67). The recognition of the unreasonable hazards from some chemicals and the impact of regulation can be technology forcing for new products and processes. However, these may be longer range phenomena and in the shorter term, the economic impact due to Section 5 of TSCA may have a negative impact on new product R & D. The added testing, administrative costs, potential for market delays, and other uncertainties will all combine to increase risks inherent in all R & D projects and reduce the number of new products. Smaller companies and small ventures would be the most impacted because of resource limitation. As described earlier under Section 5, the A. D. Little studies indicated that just the cost of filling out PMN forms would limit the introduction of new products. Most new chemical products are introduced in low volume, usually much less than 100,000 lb/year, while their end use markets are being tested. In fact, most mature chemical products never reach large volumes. In the TSCA chemical inventory report, over 70% of the substances listed have volumes of less than 100,000 lb/year. If extensive testing is required before EPA will approve a premanufacturing notice under Section 5, the impact will be considerable. A full battery of toxicity testing can cost approximately $2 million and some of the base set of tests being considered by EPA could mean a cost of $250,000 for each substance. This could mean that a particular product, based on the average rate of return for the chemical industry, would have to generate $3 million in revenue to pay out; or, at $1/lb, 3 million pounds of sales are required (Ref. 63) Recognizing that most products are low volume and never generate more than $50,000 in annual sales, the potential disincentive to new product R & D is obvious (Ref. 63).

Also, if the analogy described above (Ref. 66) between TSCA, FIFRA, and the drug law amendments is real, then we may see in the chemical industry what happened in the drug and pesticide area. The 15-year period before and after 1962, when new regulations on drugs were introduced, saw the number of new drugs introduced decrease from 641 to 247. Between 1951 and 1970, 20 new pesticides were introduced. However, only 2 or 3 new pesticides were introduced between 1971 and 1977 (Ref. 68). Obviously, other economic and sound toxicological reasons, as well as regulation, impacted the introduction of new drugs and agricultural chemicals. But regulation did play a role in limiting the number of R & D projects because of resource limitations. TSCA can be expected to limit similarly the number of new chemicals introduced after full promulgation of the key sections of the Act. The next few years will provide clearer indication of how TSCA will affect innovation in the chemical industry. Both industry and government are sensitive to this issue and are expected to closely follow this area.

The general concern on the status of innovation in the U.S. was addressed by President Carter (Ref. 69). In announcing a program which was designed to improve innovation in industry, he discussed a number of issues which evolved as a result of various studies on the impact of regulations on innovation. He indicated that all executive branch agencies concerned with environment, health, and safety would be required to prepare a 5-year forecast of their priorities and concerns. This would assist industry to better plan their activities and use their technical personnel to try and address the key concerns of the federal government, of industry, and the community-at-large. Other issues discussed were the need for better economic impact cost/benefit studies, better use of Executive Order 12044, and removal of regulations that inhibit the government from purchasing innovative products. In general, it appeared that the President did recognize that overly restrictive regulations were a disincentive to innovation in U.S. industry and was taking some steps to address this issue and to alleviate unnecessary regulatory burdens.

The current Administration has gone even further in recognizing and attempting to alleviate the excess burden of regulation. Executive Order 12291, Vice President Bush's Regulatory Reform Task Group and support of the Regulatory Reform Act are all positive approaches to the problem of innovation.

Congress appears to want to go further in addressing the innovation problem. There are a number of bills being considered in both the House and the Senate which propose to stimulate innovation. Also under consideration is creation of an independent agency, the National Technology Foundation, to promote technology for national welfare and coordinate federal sector activities in the area (Ref. 70).

In summary, it has been recognized that regulatory actions will impact innovation (this can be both positive and negative) and industry, the regulatory agencies, and other governmental sectors are attempting to define and address those actions which cause negative effects.

V. SUMMARY AND CONCLUSIONS

The Toxic Substance Control Act, though only partially implemented, has already had an economic impact on the U.S. There are the public costs for the EPA office of toxic substances of almost $320 million since 1977. Costs incurred by industry have been for the inventory reporting, Section 6 PCB and chlorofluorocarbon rules, 8(e) reporting, PMNs, establishing organizations to deal with TSCA, and TSCA-related toxicity testing costs. Also to be included are the many new and expanded toxicological test laboratories being built by industry and government (Ref. 71). The establishment of the Chemical Industry Institute of Toxicology, an organization dedicated to further the science of toxicology and the training of toxicologists, was a significant economic undertaking for industry. These costs can also be estimated in the hundreds of millions. There are also benefits which may be considered to have already resulted from TSCA actions to-date, such as restriction of PCBs and reporting of potential chemical risks.

The period of the early 1980s will be a significant test of TSCA in terms of its workability, effect on innovation, its economic impact on both the chemical industry and the U.S. economy as a whole, and its benefits. Numerous studies are in place to better define economic impacts and benefits, and we must wait for the results. TSCA is here to stay. It is a real challenge for both government and industry to work together to make it work efficiently and cost effectively in order to achieve the common objective of protecting human health and the environment from unreasonable risks due to chemicals.

REFERENCES

1. **Peterson, F.**, Private communication, Economic issues in TSCA, 1979.
2. **Dominguez, G.**, The Toxic Substances Control Act: a new challenge for business, *Environmental Regulation Analyst*, 1979, 1213.
3. **Rich, J. B.**, EPA memorandum on assistance to small businesses, Environmental Protection Agency, Washington, D.C., 1978.
4. **Clark, T. B.**, How one company lives with government regulation, *Natl. J.*, 1979, 772.
5. **Weidenbaum, M. L.**, On estimating regulatory costs, *AEI J. Gov. Society*, 1978, 115.
6. **Green, M. and Witzman, W.**, Business War on the Law: An Analysis of Benefits of Federal Health/Safety Enforcement, Ralph Nader, Washington, D.C., 1979.
7. **Anderson, A. and Co.**, Cost of Government Regulation Study for the Business Round Table, Chicago, 1979.
8. **Crandall, R.**, Is government regulation crippling business?, *Saturday Rev.*, 1979, 19.
9. Opinion Research Corporation, Government Regulation of Business: the New Outlook, the New Realities, special report from a national executive briefing, Washington, D.C., 1978.
10. **Schweitzer, G. E.**, Chemical Regulation in the 1980s, Cornell University Press, Ithaca, N. Y., 1979.
11. **Boram, M. S.**, Regulation of Health, Safety and Environmental Quality and the Use of Cost/Benefit Analysis, final report to administrative conference of the United States, Washington, D.C., 1979.
12. **Hoerger, F. D., Thomba, L. M., and Blair, E. H.**, Risk benefits and cost effectiveness methodologies in setting priorities and making decisions under the Toxic Substances Control Act, *Toxic Substances J.*, 1, 39, 1979.
13. **Morrall, J. F.**, Current Methodologies for Comparing the Costs of Controls with the Benefits; Examples from the Council of Wage and Price Stability, Analysis of OSHA Standard, presented at the AICHE 87th Natl. Meet., Boston, March, 1979.
14. **Miller, J. C.**, Occupational Exposure to Acrylonitrile; a Benefit/Cost Analysis, paper presented before the American Statistical Association, Washington, D. C., 1979.
15. **Howard, N. and Antilla, S.**, Dun's Review, 1979, 49.
16. **Ashford, N. A.**, The Role of Risk Assessment and Cost/Benefit Analysis in Decisions Concerning Safety and the Environment, paper presented at the FDA Symp. on Risk/Benefit Decisions and the Public Health, Colorado Springs, 1978.
17. EPA, Private communication, 1979.
18. Dow Chemical Company, The Impact of Government Regulation, Midland, Mich., 1979.
19. Council on Environmental Quality, Annual report on environmental quality, Washington, D.C., 1978, pp. 11, 180, and 196.
20. Council on Environmental Quality, Annual report on environmental quality, Washington, D.C., 1978, 427.
21. Foster D. Snell, Study on the Potential Economic Impacts of the Proposed Toxic Substances Control Act as Illustrated by Senate bill S. 776, Washington, D.C., 1975.
22. *Fed. Regist.* 44, 27334-75, 1979.
23. *Fed. Regist.*, 44, 44054-93, 1979.
24. *Fed. Regist.*, 42, 55026, 1977.

25. *Fed. Regist.*, 43, 50134, 1978.
26. *Fed. Regist.*, 43, 50631, 1978.
27. *Fed. Regist.*, 44, 31866, 1979.
28. *Fed. Regist.*, 44, 70664, 1979.
29. *Fed. Regist.*, 45, 35897, 1980.
30. *Fed. Regist.*, 45, 78432, 1980.
31. *Fed. Regist.*, 47, 22585, 1982.
32. *Fed. Regist.*, 46, 28138, 1981.
33. *Fed. Regist.*, 47, 5456, 1982.
34. *Chemical Regulation Reporter*, Bureau of National Affairs, Inc., Washington, D.C., 1980, 1585.
35. *Toxic Materials News*, 7, 1, 1980.
36. *Toxic Materials News*, 6, 310, 1979.
37. *Fed. Regist.*, 47, 24348, 1982.
38. *Chemical Regulation Reporter*, Bureau of National Affairs, Inc., Washington, D.C., 1980, 1586.
39. Chemical Regulation Reporter, Bureau of National Affairs, Inc., 1980, 1557.
40. *Toxic Materials News*, 7, 9, 1980.
41. *Fed. Regist.*, 44, 2242, 1979.
42. **A. D. Little, Inc.,** Impact of TSCA proposed premanufacturing notification requirements, EPA 230/2-12/78-005, Boston, 1978.
43. Toxic Materials News, October 24, 1979.
44. *Fed. Regist.*, 44, 59764, 1979.
45. *Fed. Regist.*, 44, 31514, 1979; EPA's Final PCB Ban Rule, Booklet TS-799, Office of Toxic Substances, Environmental Protection Agency, Washington, D.C., 1979.
46. **Versar, Inc.,** Microeconomic impacts of the proposed PCB ban regulations, EPA 560/6-77-035, Washington, D.C., 1978.
47. Polychlorinated Biphenyls, report prepared by the Committee on the Assessment of PCB's in the Environment, National Academy of Sciences, Washington, D.C., 1979.
48. *Chemical Regulation Reporter*, Bureau of National Affairs, Inc., Washington, D.C., 1979, 1154.
49. *Fed. Regist.*, 42, 64572, 1977.
50. *Fed. Regist.*, 43, 11110, 1978.
51. **Corey, W.,** Private communication, 1979.
52. *Fed. Regist.*, 47, 26992, 1982.
53. *Chemical Regulation Reporter*, Bureau of National Affairs, Inc., 1980, 1585.
54. *Fed. Regist.*, 44, 77470, 1979.
55. *Fed. Regist.*, 44, 56856, 1979.
56. *Chemical Regulation Reporter*, Bureau of National Affairs, 1980, 1557.
57. *Fed. Regist.*, 47, 15706, 1982.
58. **Anon.,** Facts and figures for chemical R & D, *Chem. Eng. News,* July 26, 1982, 38.
59. **Anon.,** Innovative R & D: gone with the wind?, Chemical Engineering, September 25, 1978, 73.
60. Environmental Regulation and Its Impact on the Development of Industrial technology, Contract No. 2311193503, report for the Office of Environmental Affairs, Battelle Memorial Institute, Columbus, Ohio, 1978.
61. **Ashford, N. A., Heaton, G. R., Priest, W. C., and Lutz, H.,** The Implications of Health, Safety and Environmental Regulation for Technological Change, Center for Policy Alternatives, Massachusetts Institute of Technology, Boston, 1979.
62. **Weidenbaum, M. L.,** Government Regulation and the Slowdown in Innovation, A presentation to the chemical forum, Center for the Study of American Business, Washington, D.C., 1977.
63. **St. Clair, J. B.,** Regulation and Innovation: A Challenge to Industry and Government, presented before the Chemical Manufacturers Association regional meeting on TSCA, Houston, October, 1979.
64. Innovation and Private Investment in R & D, *Chem. Eng. News,* April 30, 1979, 36.
65. **Heiden, E. J., Pittaway, A. R.,** Effect of TSCA on Innovation in Chemical Specialties, *Chemical Times and Trends,* 1982, 18.
66. **Schweitzer, G.,** *Regulation and Innovation: the Case of Environmental Chemicals* (Program in science, technology and society), Cornell University Press, Ithaca, N.Y., 1978.
67. **Smith, C. W.,** Impacts of TSCA on Innovation in the Chemical Industry, presented to the American Chemical Society Symposium on the Effects of TSCA, Washington, D.C., March, 1979.
68. **Ashford, N. A.,** Protection: How?, *Chemtech,* 1979, 608.
69. Regulatory Action Network Washington Watch, December, 1979, 4.
70. **Strickland, G.,** Innovation, private communication, 1980.
71. CPI Weighs in with Costly Toxicology Labs., *Chemical Week,* 1980, 38.
72. *Fed. Regist.*, 47, 33896, 1982.

Chapter 4

THE IMPACT OF THE TSCA INTERAGENCY TESTING COMMITTEE

Carl W. Umland and Robert Polack

TABLE OF CONTENTS

I. THE TSCA INTERAGENCY TESTING COMMITTEE

A. Statutory Provisions — Purpose and Composition

Section 4(e) of the Toxic Substances Control Act established an eight-member committee ''to make recommendations to the Administrator respecting the chemical substances and mixtures to which the Administrator should give priority consideration for the promulgation of a testing rule . . . .''

The members of this committee, now known as the TSCA Interagency Testing Committee (ITC), represent the following agencies:

- Environmental Protection Agency
- Occupational Safety and Health Administration
- Council on Environmental Quality
- National Institute for Occupational Safety and Health
- National Institute of Environmental Health Sciences
- National Cancer Institute
- National Science Foundation
- Department of Commerce

The members of the ITC, who are limited to no more than 4 years service thereon, are subject to strict conflict of interest rules. No member may be employed or accept any other compensation from chemical manufacturers or processors for 12 months after leaving the ITC, nor may any member, while serving on the ITC, own any chemical company stocks or bonds or have any other such pecuniary interest.

The significance of these tough rules is clear. Congress expected that recommendations of the ITC would carry great weight, could well subject recommended chemicals to expensive testing and severe market place pressures, and might otherwise adversely affect these products and their manufacturers. Thus, Congress felt the ITC members had to be insulated from any pressures regarding selection. (To date, these concerns do not appear to have been borne out, nor has the other implication inherent in this requirement, namely that the ITC would base its selections on accurate facts and science.)

In determining the chemical substances or mixtures for recommendation, the ITC is directed to consider all relevant factors, including:

1. The quantities in which the substance or mixture is or will be manufactured
2. The quantities in which the substance or mixture enters or will enter the environment
3. The number of individuals who are or will be exposed to the substance or mixture in their places of employment and the duration of such exposure
4. The extent to which human beings are or will be exposed to the substance or mixture
5. The extent to which the substance or mixture is closely related to a chemical substance or mixture which is known to present an unreasonable risk to health or the environment
6. The existence of data concerning the effects of the substance or mixture on health or the environment
7. The extent to which testing of the substance or mixture may result in the development of data upon which the effects of the substance or mixture on health or the environment can be reasonably determined or predicted
8. The reasonably foreseeable availability of facilities and personnel for performing testing on the substance or mixture

The ITC recommendations are in the form of a list published and revised semiannually of the individual substance or groups of substances (or mixtures) in ranked priority. (To date, no actual ranking has been done on the lists). Top priority is to be given to known or suspected carcinogens, mutagens, and teratogens. The total number of substances or mixtures on the list is limited, at any 1 time, to 50.[1]

B. Legislative History

The ITC concept was included in both the House and Senate versions of toxic substances legislation in

---

[1] This language has led to some policy disputes between EPA and industry. For example, is a ''group'' the same as a ''category'' (the term used in Section 2b(c) of the Act)? More important, if a group (or category) has several member substances, are all counted toward the limit of 50, or does only the category count as 1 (see II.A. below)?

the 94th Congress. There were a few differences, resolved in the Conference Committee; the result was 4(e) as it is now law.

The major compromise included the limitation of the list to 50 entries at any 1 time (much earlier versions of TSCA had this number as high as 300) and the requirement that if no action were taken by EPA it publish its reasoning (the House bill had no such requirement).

The ITC was set up to assist EPA in determining testing priorities; EPA provides administrative and support services to it.

C. EPA Responsibility

The Act requires that, within 1 year of a recommendation by the ITC, EPA must "initiate a rulemaking" under Section 4(a) or publish in the *Federal Register* its reasons for not doing so. (The Conference Committee Report states that in complying with this latter requirement, EPA should not divert from the regulatory activities of the Agency an inordinate amount of resources to justify the failure to require testing. In light of developments detailed below, this statement is particularly ironic.)

While the ITC recommendations are to be given great weight, the decision to order testing rests with EPA. Thus, regardless of the ITC recommendation, EPA must, prior to requiring testing, fulfill its obligations under 4(a). In other words, the ITC recommendation does not relieve EPA of the necessity to make 4(a) findings as a condition precedent for testing, nor does the ITC recommendation (necessarily) supplant or provide the research and information needed to make those findings.

II. ITC ACTIVITIES

A. Overview

When the first volume of this Guidebook on the Toxic Substances Control Act was published, there was little to be said about the ITC (known as the Interagency Priority Committee at that time). It was only possible to speculate on the formation of the Committee, what Congress intended it to do, what it would do, how the public could respond, and interactions of EPA.

With the passage of time, the picture is now clearer and a description of how the ITC approached its job, critical assumptions made, ranking systems used, lists generated, plus public and EPA responses is now possible. It is also possible to note some of the problems with the system and to predict future implications with greater certainty.

As of April, 1982, the ITC had submitted 10 reports to EPA recommending a net total of 25 chemical substances of 9 categories for testing. The recommendations are listed in Table 1 for easy reference. They, of course, represent far more than 50 substances contemplated by Congress. (See Footnote 1.)

The ITC performed a formidable task in arriving at these recommendations. It can reasonably be assumed by industry that these materials will receive active consideration for Section 4 testing and Section 8 information gathering rules by EPA. Appropriate preparation and response to these eventualities necessitates a clear understanding of exactly what ITC did and what data bases were used. No attempt is made here to repeat the detailed references and descriptions contained in the "Background Document on Preparation of the Preliminary List of Approximately 300 Chemicals" published in late July, 1977 by the Committee or the subsequent reports published for comment in the *Federal Register* by EPA (Ref. 1—7). However, a qualitative description of the culling process together with key assumptions made should be helpful to gain a general understanding along with a knowledge of where to look for more detailed information when needed. It also serves to highlight some of the inherent weaknesses in the selection of individual substances which may demand correction in subsequent proceedings, thus suggesting the kinds of information concerned industry members might appropriately submit to avoid unnecessary testing.

B. Initial List and Master File

The ITC first had to decide how to discharge its mandated responsibilities. It was faced with a large universe of chemical substances, the fact that no consolidated chemical information data base existed, lack of use and exposure data in many cases, and the frequent incompatability of such data systems as did exist.

The first significant decision of the Committee was to summarize and select candidates from existing priority lists of potentially hazardous substances developed by other agencies and organizations. This netted an "initial listing" of over 3600 substances and categories. (The Committee very early decided that the Administrator's authority under TSCA Section 26(c)(1) to promulgate Section 4(a) testing rules for cate-

TABLE 1
TSCA SECTION 4(E) ITC RECOMMENDATIONS
AS OF APRIL 1982

	Entry	Date of designation
1.	Acetonitrile	April 1979
2.	Acrylamide	April 1978
3.	Alkyl epoxides	October 1977
4.	Aniline and bromo-, chloro- and/or nitroanilines	April 1979
5.	Antimony (metal)	April 1979
6.	Antimony (sulfide)	April 1979
7.	Antimony trioxide	April 1979
8.	Aryl phosphates	April 1978
9.	Biphenyl	April 1982
10.	Chlorendic acid	October 1981
11.	Chlorinated benzenes, mono- and di-	October 1977
12.	Chlorinated benzenes, tri-, tetra-, and penta-	October 1978
13.	4-Chlorobenzotrifluoride	October 1981
14.	Cresols	October 1977
15.	Cyclohexanone	April 1979
16.	1,2-Dichloropropane	October 1978
17.	Ethyltoluene	April 1982
18.	Formamide	April 1982
19.	Glycidol and its derivatives	October 1978
20.	Halogenated alkyl epoxides	April 1978
21.	Hexachloro-1,3-butadiene	October 1977
22.	Hexachlorocyclopentadiene	April 1979
23.	Hydroquinone	November 1979
24.	Isophorone	April 1979
25.	Mesityl oxide	April 1979
26.	4,4'-Methylenedianiline	April 1979
27.	Methyl ethyl ketone	April 1979
28.	Methyl isobutyl ketone	April 1979
29.	Pyridine	April 1978
30.	Quinone	November 1979
31.	Toluene	October 1977
32.	1,2,4-Trimethylbenzene	April 1982
33.	Tris (2-chlorethyl)phosphite	October 1981
34.	Xylenes	October 1977

gories of chemical substances would extend to the ITC inclusion of categories as well as individual substances and mixtures in its recommendations to the Administrator.)

This list was reduced to a "master file" of about 1700 by purging substances: (1) used predominantly as pesticides, food additives, or drugs (and, therefore, not subject to TSCA regulation); or, (2) not in commercial production. (The assumption made here was that any substance without a Chemical Abstract Service identification number and which appeared only on the NIOSH Registry was unlikely to be in commercial production).

C. Preliminary List

A "preliminary list" of about 330 substances and categories was then developed on four factors mandated in Section 4(e)(1)(A) of TSCA. These were: (1) annual production quantity, (2) amount of environmental release, (3) number and duration of occupational exposures, and (4) extent of exposure to the general population. This approach required the use of both published data and the judgment of the Committee contractor. To accomplish this, the contractor developed a scoring system which made a couple of questionable and arbitrary assumptions which might have operated to include more substances than warranted at this stage.

First, they assumed that any substance for which an annual production value was not available in their reference sources should be scored at 300,000 lb/year; we now know from the TSCA 8(b) inventory that this is an unlikely figure since about 70% of all commercial substances were produced at annual volumes of under 100,000 lb.

Second, they assumed that the NIOSH National Occupational Health Survey (NOHS) covered 7000 substances, but that any substance which did not appear in the NOHS should automatically receive an occupational exposure score anyway. This is not a valid assumption even if the NOHS report were a sound data base.

Where use data did not exist, no scoring was attempted; this alone narrowed the "master file" to about 700 substances. Further eliminations and add-backs were then made by the Committee on the basis of its professional judgment if it found substances: (1) already stringently regulated or the hazard well characterized, (2) essentially inert or well characterized as having low toxicity, (3) covered by testing requirements under FDCA or FIFRA, (4) natural products to be deferred awaiting better characterization for testing purposes, (5) worthy of inclusion on the basis of the Committee knowledge of the substance and its uses, as a professional judgment. The latter is obviously an area for professional review on any substance ultimately subject to a testing rule.

All of the above and the source data are referenced in the Committee "Background Document" (Ref. 1). The latter was published by the Committee in late July 1977 for public comment by August 22, 1977. ITC was operating against a very tight schedule to meet its TSCA-mandated deadline of October 1 for its first report to the Administrator.

Overall, the effort was well geared to the task. However, it is necessary to recognize the weaknesses already noted plus the questionable basis of some of the data sources that were used such as some of the NIOSH registries and surveys; also, one would need to understand the basis for listing of each substance on any other list before accepting its validity. This observation is made even though repeated listing of the same substance by different agencies or organizations would tend to suggest a higher validity for the suggested hazard.

D. Preliminary Dossiers and Recommendations

The "preliminary list" was reduced to about 80 substances for more detailed review by extending scoring to three additional factors suggested in Section 4(e)(1)(A):

1. Relationship of a substance to another known to present an unreasonable risk
2. Existence of health or environmental effects data
3. Probability that testing would produce such data that were reasonably predictive

The Committee contractor scored each substance for seven biological activity factors: carcinogenicity, mutagenicity, teratogenicity, acute toxicity, reproductive effect or organ-specific toxicity, bioaccumulation, and ecological effects. This was done by individuals with some knowledge of the substances or related substances together with a review of a summary of information in the open literature. The Committee then selected the approximately 80 substances or categories for the drafting of dossiers and more detailed review. (This total procedure is discussed in the ITC initial report to the EPA Administrator. See Ref. 2.)

These total procedures have led to the recommendations made to EPA for testing rule consideration. There are some weaknesses, to be sure, which derive from opportunities for subjective judgment and incomplete knowledge. However, it is in these areas that industry comment and further EPA study (e.g., via further literature review, Section 8 rules) can help to overcome the weaknesses either prior to issuance of testing rules or during the rulemaking process itself. Starting in 1981, this process was improved by soliciting voluntary industry comments on preliminary lists of chemicals under consideration by ITC prior to final selection.

E. Comments

One area in which the ITC completely failed to meet its responsibilities was that of assigning priorities for EPA consideration. Instead, the Committee persisted in merely providing alphabetical listings with the statement that each listing "should be given equal priority in EPA's development of test rules." In fairness, this was originally done with the understanding from EPA that there would be test rules developed for effects rather than for individual substances or categories. This cannot be done, as EPA has since admitted because of variations in properties among substances and it is made even more difficult when an attempt is made to cover a category adequately and correctly.

A further criticism of the ITC effort is the incompleteness of the literature reviews done prior to making recommendations to EPA. This has placed additional burdens on EPA to accomplish a critically important prerequisite to any testing rule.

A practical feel for the kinds of issues and errors discussed above in general terms can be gained by reading the EPA official responses to the first two sets of ITC recommendations (Ref. 8, 9). These responses include public comments on the ITC reports received by the Agency. Random sampling of these and the Agency reactions bear out the need for more critical assessment of the accuracy of the data used and more detailed evaluation of correct information than the ITC has provided.

All of this has undoubtedly contributed to the delays which the National Resources Defense Council (NRDC) has found so unsettling as discussed below.

III. THE NRDC LAWSUIT

On May 8, 1979, over a year after the ITC had published its first two lists of recommendations, the Natural Resources Defense Council, Inc. (NRDC), an environmental organization, filed suit against EPA charging that EPA had failed to perform a nondiscretionary act under TSCA; i.e., either to initiate a rulemaking within 12 months of these recommendations or to publish its reasons for not doing so. At the time of the filing of this lawsuit, the only EPA response to the first ITC list was, essentially, that it was considering the information available about the chemicals thereon, had not reached any conclusions, and therefore had not initiated any Section 4 rules but was not foreclosing the possibility. No response had been published regarding the second ITC list.

Several organizations and companies intervened in this litigation: the Chemical Manufacturers Association, Synthetic Organic Chemical Manufacturers Association, American Petroleum Institute, Reilly Tar & Chemical Corp., and Kopper Co., Inc. (The two companies listed are manufacturers of chemicals on the first two lists.)

Despite the arguments of EPA and the intervenors that EPA had not violated TSCA, Judge Lawrence Pierce of the U.S. District Court for the Southern District of New York ruled on February 4, 1980 that EPA had defaulted on its legal obligations.

EPA was instructed to submit to the court a timetable for testing decisions on the ITC recommended chemicals. EPA did so and later submitted a revised response.

However, there is one clear result of the NRDC lawsuit. Henceforth, EPA will have to make a 4(a) decision with respect to testing within 1 year of the initial appearance of any chemical on the ITC list. This will speed up the Agency decision-making process and may result in more complex proposed rules under Section 4(a) which may have to include factors dealing with both the criteria therein and the actual proposed testing program.

IV. IMPLICATIONS OF ITC RECOMMENDATIONS

Although only limited test rule proposal action has been taken with respect to most of the chemical substances or categories of chemical substances recommended by the ITC for priority testing, it is nevertheless clear that the mere recommendation carries with it significant implications. This has led to development of voluntary test programs by industry in consultation with the agency as discussed below. As EPA responds more expeditiously to the ITC than was the case prior to 1981, it is within the realm of probability to predict certain eventualities.

The most notable and obvious result will be for EPA to make, within 1 year of designation, a decision as to whether the chemical meets the Section 4(a) criteria. These criteria, which must exist before testing can be ordered, are

I. a. The manufacture, distribution, processing, use, disposal, or combination thereof of a chemical substance or mixture may present an unreasonable risk of injury to health or the environment; or
    b. A chemical substance or mixture is or will be produced in substantial quantities and
        i. Enters or may enter the environment in substantial quantities, or
        ii. There is or may be significant or substantial human exposure; and
II. There are insufficient data upon which to reasonably determine or predict effects of manufacture, processing, etc.; and
III. Testing is necessary to develop these data.

Clearly, after recommendation is made by ITC, a producer (or processor, etc.) of a chemical will wish to show it does not meet these criteria. To do so, information will need to be collected and presented with respect to production volume, use and exposure patterns, disposal practices, and existing test data.

The last point is crucial. Many producers, upon an ITC listing, may prefer to undertake the testing suggested by ITC (if such data do not exist) in order to control or at least develop the test schemes themselves, rather than having EPA dictate it.

Another result of an ITC listing will be a Section 8 reporting requirement under TSCA. Already the ITC chemicals are among those subject to the first proposed Section 8(a) reporting rule; undoubtedly this trend will continue, as it provides EPA with a mandatory mechanism for adducing information about a chemical. Thus, an ITC recommendation will subject a chemical to some additional regulatory scrutiny and concomitant expense. However, the real effect will be that of Section 4(a) not 4(e).

V. INDUSTRIAL MANAGEMENT RESPONSE

From what has already been said, it is perfectly obvious that there are many opposing forces operative in the testing area which unquestionably will evolve from the ITC listings. It will be in the enlightened self-interest of industry to be sure that whatever is undertaken is built upon sound information. Sound information will also sometimes preclude unnecessary testing. Therefore, it becomes incumbent upon producers and users of nominated substances to assess opportunities for constructive input to the test development process. Industry has been specifically precluded from direct input to the ITC deliberations. However, there are several other chances to impact the process short of precluding listing by the ITC.

EPA practice has been to seek public comment on ITC recommendations on the occasion of each report. This is one opportunity. A second opportunity presents itself through informal discussions on substantive issues with EPA during the test-rule development period. A third opportunity arises during the formal rule-making proceedings. And, of course, there is the ultimate appeal to the judicial system if the potential for injury has not been satisfactorily mitigated in the EPA final rule.

These representations may be made either by individual companies or consortiums, or preferably by a combination of the two. The Chemical Manufacturers Association has given a great deal of thought to this matter which is well documented in its proposal to EPA for the handling of cost-sharing and reimbursement of testing costs (Ref. 10). It suggests a practical scheme for gaining the participation or exemption of affected parties. A significant prerequisite is undoubtedly going to be individual company anticipation and preparation since the likely time pressures will probably preclude protracted organizing efforts.

There are a number of examples of organized effort which can be looked to for guidance. The Chemical Manufacturers Association Special Program Panels are one; other groups have organized within the Synthetic Organic Chemical Manufacturers Association. There undoubtedly are others.

One of the more active initiatives in this regard has been the independent Pyridine Task Force. A case history is included in the Appendix.

VI. CONCLUSION

The impact of the ITC is only beginning to be felt. It will be real, however. The most effective management of the consequences must stem from good information. Both EPA and industry will need to collaborate early and closely to avoid unwise and unjustified commitment of scarce resources in an area where there is much to be done. It is not a case of doing nothing and it is impossible to do everything; therefore, sound management based on correct information is absolutely essential.

APPENDIX

Pyridine Task Force Response to ITC Listing: A Case History

In the second report of the ITC, dated April 1978, eight chemicals and categories were recommended for priority testing. Among these was pyridine. This is a brief history of the response to that recommendation and the full text of the comments filed with EPA.

The pyridine recommendation, like those of other ITC chemicals in 1978, was based in large part on work done by an EPA contractor.

When pyridine was recommended, it was apparent to the U.S. producers that the underlying data supporting the recommendation were seriously flawed. Therefore, to remedy these deficiencies and to ensure that EPA had the best information on pyridine available to it, a Pyridine Task Force was established.

The Task Force consisted of the three U. S. producers of pyridine, operating under the aegis of outside counsel. Principal consumers of pyridine were also included in some of the deliberations of the Task Force.

The overall objective of the Task Force was to develop and submit data on production, use, and exposure of pyridine. To accomplish this goal, the group devised a questionnaire which was sent to nearly all pyridine customers in the U.S. From this survey, a comprehensive profile of the pyridine industry was compiled. Indeed, the survey results covered over 80% of U.S. pyridine sales.

The full text of the Task Force comment is attached.

Since the submission of these comments to EPA, the Task Force has remained active. Among its activities have been the development and submission to EPA of exhaustive bibliographies on health and environmental effects of pyridine, reporting of previously unpublished data, correspondence with pyridine customers on the status of the chemical in the TSCA testing program, and monitoring of EPA activities in the testing area.

It is expected that the Task Force will remain in existence and play an active role as long as pyridine is subject to potential or actual scrutiny under TSCA.

APPENDIX I

CONFIDENTIAL

MANUFACTURERS QUESTIONNAIRE

Please complete this questionnaire fully. If there are any questions, contact Mr. Robert V. Zener, Pepper, Hamilton & Scheetz, 1776 F Street, N.W., Washington, D.C. 20006, 202/862-7560.

Please return to Mr. Zener by September 15, 1978.

Use figures for 1977.   Please submit one questionnaire for entire company. (However, if you have both manufacturing and processing operations, submit a manufacturer's questionnaire for the manufacturing operation, and a user's questionnaire for the processing operation.)

Name of Company   _____

Person Submitting
Report   _____

Title   _____

Address   _____

Telephone Number   _____

*   *   *   *

No. of Producing Sites   _____

At each site state:

   1. No. Persons Occupationally Exposed   _____

   2. Duration of Exposure:

No. of Employees whose Exposure Averages no more than

     40 hrs./wk.   _____

      8 hrs./wk.   _____

   3. Environmental Release

State methods used to dispose of pyridine (C_5H_5N)

   a) Publicly owned Treatment works   ___Yes ___No ___kg/yr.

b) NPDES   ___Yes ___No   ___kg/yr.

c) Other Water (well, groundwater,
   storm water, ocean)   ___Yes ___No   ___kg/yr.

d) Air   ___Yes ___No   ___kg/yr.

e) Solid Waste   ___Yes ___No   ___kg/yr.

Total Amt. you Produced in 1977   _____kg

Your Estimate of Total U.S. Production
(All Producers)   _____kg

Total Amt. you Exported in 1977   _____kg

Total Amt. you Sold in U.S.   _____kg

In 1977, Number of locations to which
you shipped Pyridine:

1 Drum or less   _____

More than 1 Drum   _____

CONFIDENTIAL

PYRIDINE USER QUESTIONNAIRE

Please complete this questionnaire fully. If there are questions, contact Mr. Robert V. Zener, Pepper, Hamilton & Scheetz, 1776 F Street, N.W., Washington, D.C. 20006, 202/862-7560.

Please return to Mr. Zener by September 15, 1978.

Use figures for 1977. Please submit one questionnaire for entire company.

Name of Company   _____

Person Submitting Report _____

Title   _____

Address   _____

Telephone Number   _____

*   *   *   *

Categories of Use of Pyridine	Total Amount of Use in '77 for each Category (Kg)	Total Amount of Use in '77 for Chemical Conversion into Another Product (Kg)
1. Export	1.	1.
2. Manufacture of Agricultural Chemicals and Their Intermediates	2.	2.
3. Manufacture of Food, Drug and Cosmetic Intermediates and Products	3.	3.
4. Manufacture of Rubber Chemicals and Their Intermediates	4.	4.
5. Manufacture of textile waterproofing agents and their intermediates	5.	5.
6. Industrial Solvent Use	6.	6.

Categories of Use of Pyridine	Total Amount of use in '77 for each Category (Kg)	Total Amount of Use in '77 for Chemical Conversion into Another Product (Kg)
7. Reagent Use (including resale)	7.	7.
8. Other (please specify)	8.	8.

Number of plant sites (i.e., geographically distinct locations) at which pyridine was processed or repackaged in 1977.   _____

At each site, state:

1. No. Persons Occupationally Exposed   _____

2. Duration of Exposure:

   No. of Employees whose Exposure Averages no more than

   40 hrs./wk.   _____

   8 hrs./wk.   _____

3. Environmental Release

   State methods used to dispose of pyridine (C_5H_5N)

   a) Publicly owned Treatment works   ___Yes ___No   ___kg/yr.

   b) NPDES   ___Yes ___No   ___kg/yr.

   c) Other Water (well, groundwater, storm water, ocean)   ___Yes ___No   ___kg/yr.

   d) Air   ___Yes ___No   ___kg/yr.

   e) Solid Waste   ___Yes ___No   ___kg/yr.

Please attach any comments regarding the accuracy, relevancy or validity of the studies and other information referred to in this dossier.

PYRIDINE TASK FORCE

c/o Pepper, Hamilton & Scheetz
1776 F Street, N.W.
Washington, D.C. 20006
(202) 862-7500

U.S. Environmental Protection
 Agency
Office of Toxic Substances (TS-793)
401 "M" Street, S.W.
Washington, D.C. 20460
Attn: Joan Urquhart

> Re: OTS — 04004
> Comments on Second Report
> by the TSCA Inter-Agency
> Testing Committee-Pyridine

Gentlemen:

We are pleased to submit this comment in response to EPA's call as published in 43 Federal Register 16684 (April 19, 1978) and 43 Federal Register 24907 (June 8, 1978). This comment deals with pyridine ($C_5 H_5 N$), one of the chemicals recommended for priority testing under the Toxic Substances Control Act (TSCA)[1] by the TSCA Inter-Agency Testing Committee (TITC), with specific reference to the dossier on pyridine prepared by Clement Associates (Contract No. NSF-C-ENV 77-15417; Section VII).

At the outset, we note that the dossier omits from the list of manufacturers Reilly Tar & Chemical Corporation of Indianapolis, Indiana, which is the largest producer of pyridine.

We desire to cooperate with and assist EPA in its response to the recommendation to require testing of pyridine. We believe that it is clearly in the public interest that EPA's decisions be based on fact, not on ignorance or erroneous information.

In reviewing the dossier preparatory to submitting this comment, we felt that it was essential that an accurate profile of pyridine production, use, disposal and exposure be developed and made available to EPA. Therefore, the three U.S. producers of pyridine (Reilly Tar & Chemical Corporation, Nepera Chemical Company, Inc., and Koppers Company, Inc.) with the assistance of counsel (Robert V. Zener of Pepper, Hamilton & Scheetz) formed a task force to obtain this information.

The task force devised two questionnaires, one for manufacturers, and one for users of pyridine.[2] These questionnaires were designed to obtain information on production, categories of use and amount of pyridine used in each, and occupational and environmental exposure. The questionnaires were distributed by counsel to manufacturers and users who returned them to counsel for compilation. Survey responses were received reflecting 84% of pyridine use in the U.S. Counsel has returned the original questionnaires to survey participants and has retained no copies of either the questionnaire responses or the customer lists provided by the manufacturers.

SUMMARY

Based on the available evidence, it appears unlikely that pyridine presents an unreasonable risk of injury to health or the environment. Also, there appear to be significant inaccuracies in the pyridine dossier particularly with regard to the amount of pyridine produced; the quantity released into the environment; the number of individuals occupationally exposed to pyridine; and pyridine release in coke oven emissions.

---

[1]   15 U.S.C. 2601, et. seq., (90 Stat. 2003, P.L. 94-469).
[2]   Copies attached as Appendix I.

The table below compares information contained in the dossier to that generated by the Pyridine Task Force survey.

Item	Dossier	Task Force Data
Total Production	27 million kg	<11.6 million kg
U.S. Sales	13.5 million kg	<6.0 million kg
Exports	13.5 million kg	<5.6 million kg
Uses	% (of U.S. Sales)	% (of U.S. Sales Covered by Questionnaire Responses)[3]
Agricultural Chemicals & their Intermediates	—	54%
Food, Drug & Cosmetics Intermediates & Products	26%	16%
Rubber Chemicals & their Intermediates	14%	7%
Waterproofing Agents & their Intermediates	14%	0%
Industrial Solvent Use	[	11%
	[32%	
Reagents	[	1%
Other	14%	10%
Persons Occupationally Exposed	249,000	813

Because these factors are principal components of the priority ranking scheme used by the TSCA Interagency Testing Committee, we recommend that the dossier and priority ranking of pyridine be re-evaluated by the TITC and that EPA refrain from acting on the TITC pyridine recommendation contained in their second report until such action is taken. We believe that such a re-evaluation and correction of the inaccuracies referred to above will significantly reduce the priority ranking ascribed to pyridine and result in the TITC's withdrawal of its recommendation for further consideration of pyridine at this time.

BACKGROUND INFORMATION

The manufacture of pyridine started at the turn of the century. It was originally isolated from the tar derived from coke-oven operations. During World War II, certain uses, namely biocides, waterproofing agents, sulfa-drugs, and rubber chemistry, resulted in the increased production of pyridine.

After the war new applications for pyridine were developed. Synthetic production from petrochemicals commenced to ensure adequate feedstocks for these end uses. Capacity was expanded to meet the demands, but since 1969, all significant expansions in capacity have been abroad. In view of the critical markets served by pyridine, it is of concern to the American industry that new restrictions on the domestic industry will encourage further growth from abroad.

The Koppers Company, Pittsburgh, Pennsylvania, produces pyridine from coal tar. Reilly Tar & Chemical Corp., Indianapolis, Indiana; and Nepera Chemical, Co., Harriman, New York, subsidiary of Schering AG, Berlin, produce pyridine synthetically from petrochemical feedstocks.

Pyridine is produced to a high quality specification. The majority of pyridine sold is greatly in excess of 99% purity. The high purity ensures that reactions using pyridine can be controlled to completion. These reactions produce a highly selective end product with a minimum of by-products and impurities.

---

[3]   Questionnaire responses accounted for 84% of U.S. sales.

For example, in chemical reactors pyridine:

a.   Chlorinates readily;
b.   Does not nitrate easily, and there are no commercial nitrations;
c.   Aminates to 80—90% yields;
d.   Does not sulfonate readily;
e.   Hydrogenates completely;
f.   Quaternizes completely;
g.   Does not readily oxidize, except in the formation of N-oxides;
h.   Is useful to scavenge acids (all large-scale applications involve the recovery and re-cycling of pyridine);
i.   Can be converted completely by methylation with alcohols; and
j.   Complexes readily with phenols.

It should be noted from the above that the commercial uses of pyridine are highly efficient chemical reactions. Moreover, as will be set forth in detail below, a high percentage of pyridine usage involves a complete chemical conversion to end products containing essentially no pyridine. This fact ensures minimal human and environmental exposure to pyridine. The high price of pyridine, in the $3.00/kg range, gives manufacturers and users a strong economic incentive to minimize production or effluent losses.

Generally, chemicals that react with genetic material are electrophiles. It has been demonstrated that carcinogenic forms of chemicals are usually, if not always, electropholic (i.e., electron-deficient) reactants. The electrophilic nature of the carcinogens was inferred from the structures of metabolic products, including the protein- and nucleic acid-bound derivatives, formed in vivo from a number of aromatic amines, amides, and nitro compounds as well as from various potential alkylating agents including the nitrosamines, nitrosamides, pyrrolizadine, alkaloids, polycyclic aromatic hydrocarbons and other compounds.[4]

Pyridine is not considered an electrophile. In most reactions, it may be considered nucleophilic (electrophobic). It can act as an electrophile in only limited specified reactions (such as with sodamide under severe reaction conditions). Therefore, it would be unlikely for pyridine to react with DNA (i.e., be mutagenic). This conclusion is substantiated in a battery of mutagenic assays referred to later in this comment.

DOSSIER TEST DATA

The dossier contains several references to studies on the potential toxic effects of pyridine. The references constitute a fairly comprehensive, but by no means exhaustive literature search. For example, there are several recent studies not included in the dossier (referred to below at page 20) suggesting the absence of mutagenic potential of pyridine.

We generally agree with the basic conclusions of the TITC, that pyridine is a moderately toxic chemical with limited potential to bioaccumulate or to produce severe chronic effects, and that the studies referred to in the dossier (particularly for those effects for which testing is recommended) are limited in scope and were (in many cases) not performed in accordance with currently accepted protocols.

We recognize EPA may consider these studies to be insufficient by contemporary standards. Therefore, although they should not be disregarded, they likewise should not be used to create a presumption of hazard.

Even assuming test data are insufficient, Section 4(a) of TSCA sets forth specific findings EPA must make before a testing rule can issue. EPA must find that either,

1.   Chemical activity (manufacture, use, processing, distribution, disposal, or a combination thereof) may present an unreasonable risk of injury to health or the environment, or
2.   That pyridine is (or will be) produced in substantial quantities and either,
     a. enters or may enter the environment in substantial quantities, or
     b. there is or may be significant or substantial human exposure.

---

[4]   Searle, Charles E., ed., Chemical Carcinogens, ACS Monograph 173 American Chemical Society (Washington, D.C.) 1976 — Chapter 16, ''The Metabolism of Chemical Carcinogens to Reactive Electrophiles and Their Possible Mechanisms of Action in Carcinogenesis'' by Elizabeth C. Miller and James A. Miller.

From the results of the task force survey and other information and experience available to us, we believe there is no evidence to support either finding.

PRODUCTION AND USES

The dossier estimates over 60 million pounds (27 million kg) of pyridine were produced in the U.S. in 1976. Our survey — all figures for which are based on 1977 — reveals that U.S. production was less than 11.6 million kg.[5] Imports are negligible (less than 1%). As can be readily seen, this figure is substantially below the production estimate in the dossier, on which the TITC recommendation was based.

Of total U.S. production, 48% was exported. Therefore, U.S. sales were less than 6 million kg. It is significant that a large amount of pyridine leaves the U.S. environment, and thus the jurisdiction of EPA.

Pyridine users (including the manufacturers) were asked to identify categories of use, the amount used in each, and the amount used which was chemically converted into another product. The responses received accounted for 84% of the quantity of pyridine sold in the U.S. The results follow (reported as percentage of U.S. sales covered by the questionnaire responses):

Category of Use	Total Amount Used in 1977 (%)	Total Amount (%) of Use Chemically Converted into Another Product[6]
Manufacture of Agricultural Chemicals & their Intermediates	54%	53%
Manufacture of Food, Drug & Cosmetics Intermediates & Products	16%	11%
Manufacture of Rubber Chemicals and their Intermediates	7%	7%
Manufacture of Textile Waterproofing Agents & their Intermediates	0%	0%
Industrial Solvent Use	11%	1%
Reagent use (including resale)	1%	0.5%
Other	10%	7%

The importance of this information is three-fold: first, it represents the most accurate use profile of pyridine known to exist; second, it indicates that many types of products made from pyridine are themselves subject to controls (e.g., pesticides, FDA products); and third, it shows that an extremely high percentage (80%) of the pyridine used in the U.S. is chemically converted to other products, disappears as pyridine, and hence environmental exposure is minimal.

The manufacturers also report that 22% of the locations to which they shipped in 1977 received 1 drum (200 kg) or less; while 78% received more than one drum. This suggests that pyridine was shipped in bulk to a small number of industrial consumers, each consuming a large amount.

---

[5]   With only two principal manufacturers it is difficult to state precisely total production without compromising vital commercial information and raising potentially troublesome anti-trust questions. The reported production figure was derived by this method: Both Reilly and Nepera separately and confidentially reported to counsel their actual 1977 production and their best estimate of total U.S. 1977 production. Counsel thus had three numbers: Actual total, Reilly estimate of total production, and Nepera estimate. Counsel took the largest of these three (the figure chosen being known only to counsel), and inflated it by 1 to 25% (said percentage being randomly selected and known only to counsel). Koppers' production was then added. It is this figure that is reported here.

[6]   The figures reported in this column are percentages of total U.S. sales covered by the questionnaire responses, rather than percentages of use within the particular category. Thus, for example, the 53% figure for agricultural chemicals and their intermediates represents 98% of reported use within that category.

OCCUPATIONAL EXPOSURE

Initially, it must be noted that pyridine is subject to OSHA controls (5 ppm on an 8-hour time weighted average; 15 mg/cu. meter of air).[7] In addition, the odor threshold of pyridine is a mere fraction of the OSHA level (0.021 ppm or approximately 68 μg/m³), and as the odor is disagreeable, this provides a built-in control.

Based on questionnaire responses from manufacturers and users, the total number of persons occupationally exposed to pyridine is shown below:

Number of Persons Occupationally Exposed	813
Number Whose Exposure Averages	
No more than 40 hrs/wk	256
No more than 8 hrs/wk	557

These figures do not take into account persons who may be occupationally exposed to coke oven emissions. A recent study done for OSHA estimates that about 30,000 persons are so exposed;[8] however, studies for EPA[9] do not list pyridine as a constituent of coke oven emissions, and as Appendix II shows, exposure to pyridine as such would be less than 1 microgram/cubic meter, or essentially nil.

We also recognize that there may be occupational exposure to pyridine in analytical laboratories where pyridine is primarily used as a Karl Fischer reagent. However, such exposure is neither substantial nor significant. Modern apparatus and design ensure minimal release, and the use of pre-packaged reagents keeps the level and duration of exposure extremely low.

Even adding in the figure of 30,000 for all possible laboratory uses, the occupational exposure does not come close to the figure of 249,000 cited in the dossier.

Manufacturers of synthetic pyridine routinely monitor worker exposure to pyridine and pyridine in the workplace environment. Results of personnel monitoring reported to the Task Force by the manufacturers and some major users show typical levels of the 8-hour TWA to be from 0.008-1.0 ppm, far below even the OSHA levels.

ENVIRONMENTAL EXPOSURE

As noted above, almost all (80%) pyridine sold in the U.S. is chemically converted to other products, and its cost mandates careful use and minimum waste or environmental discharge

Survey results from manufacturing and consuming sites reveal the total amount of pyridine disposed to be approximately 300,000 kg. We believe that this figure may be somewhat overstated, as some respondents may have failed to distinguish between input to a treatment system and ultimate environmental release. Details are summarized below:

Means of Disposal	% of Total	
Water	52%	
NPDES		2%
POTW*		50%
Air	10%	
Land Disposal/Solid Waste	1%	
Other	36%	

* Represents influent to POTW; POTWs contacted report no reason to believe a detectable amount of pyridine exists in their effluent.

[7] 29 C.F.R. 1910.1000 (Z-1).

[8] Inflation Impact and Analysis of Proposed Standard for Coke Oven Emissions, (NTIS PB-258-615) by D.B. Associates. (February 27, 1976).

[9] A Study of Industrial Data on Candidate Chemicals for Testing (EPA Report 560/5-77-006) by SRI International; An Assessment of the Health Effects of Coke Oven Emissions (EPA/ ORD, External Review Draft, April, 1978) (Pyridine absent in partial list of constituents, but reported in Czechoslovakia); Sampling and Analysis of Coke-Oven Door Emissions, by R. E. Barrett et al. (Battelle-Columbus Laboratories, Columbus, Ohio, October, 1977) NTIS No. PB-2-278-475. In addition, pyridine is not listed in table of Chemical Composition of Coal Tar Fumes in "Fumes Emissions from Coal-Tar Pitch," NTIS No. AD-AO22-844.

These data are at odds with the unsubstantiated dossier statement that pyridine is "often found in municipal waste water" (p. VII-1). The dossier, p. VII-12, Ref. 23, also states pyridine has been found in four water supplies. This, too, is inaccurate. The study referenced reported only two findings in the U.S. (Chattanooga Creek and Newport, Tennessee) but does not refer to water supplies, as such. The other two sites are not in the U.S., one is in the USSR, the other unspecified by the World Health Organization.

DEGRADATION

The dossier states that no information was found regarding bioaccumulation; however, based on the physical properties of the chemical, bioaccumulation would not be expected. Neely et al.[10] and Chiou et al.[11] have shown the relationships that exist between octanol/water partition coefficients and bioaccumulation, and water solubility and bioaccumulation. Chemicals having high water solubility and low octanol/water partition coefficients do not bioaccumulate in fish or other organisms. Pyridine certainly falls into this category since it is extremely soluble in water and has an octanol/water partition coefficient of 11.[12]

Other areas to consider regarding environmental effects are dissipation and persistence. Several studies have shown that soil microorganisms have been isolated which can readily degrade pyridine and utilize it as a food source.[13] Therefore, pyridine would not be expected to be biologically stable or persist in the environment. Also, pyridine has a relatively high vapor pressure and therefore can volatilize into the air where it can photodegrade.[14] This provides another route of environmental dissipation.

The dossier states that pyridine "has been reported to be present in the working area around coal furnaces and in agricultural crops and fish." The possible presence of pyridine around coke ovens has been discussed above. The reference cited[15] does not support this statement regarding the crops and fish. The reference gives a method for determining if pyridine is present in crops or fish; it says nothing about actually finding any pyridine present. Other studies[16] which report on pyridine uptake by plants only deal with cases where pyridine was intentionally applied to a crop. Therefore the statement in the dossier is misleading; it implies pyridine is found in agricultural crops and fish via environmental contamination when no supporting data are given.

It is unlikely that products derived from pyridine will biodegrade back to pyridine.[17]

GENERAL POPULATION EXPOSURE

As stated previously, workplace atmosphere concentrations of pyridine are very low; at or beyond plant property limits, the level of pyridine is lower yet.

The dossier (p. VII-1) estimates a small oral exposure to pyridine, primarily resulting from its presence in food. In this regard, we note pyridine does occur naturally in some foods,[18] and, more important, is listed by FDA as a substance generally recognized as safe (GRAS list — synthetic flavorings).[19]

[10] Neely, B. W., D. R. Branson and G. E. Blau. 1974. Partition Coefficient to Measure Bioconcentration Potential of Organic Chemicals in Fish. Environ. Science and Tech. 8(13):1113-1115.

[11] Chiou, C. T., V. H. Freed, D. W. Schmedding, and R. L. Kohnert. 1977. Partition Coefficient and Bioaccumulation of Selected Organic Chemicals. Environ. Science and Technol. 11(5):475-478.

[12] Leo, A., C. Hansch, and D. Elkins. 1971. Partition Coefficients and Their Uses. Chem. Rev. 71:515-616.

[13] Keith, W. G., and R. B. Cain, 1975. Microbial Metabolism of the Pyridine Ring. Metabolic Paths of Pyridine Biodegradation by Soil Bacteria. Biochem J. 146(1), 157-72; Shukla, O. P. 1973. Microbial Decomposition of Pyridine. Indian J. Exp. Biol 11(5), 463-5; Watson, G. K., and R. B. Cain, 1972. Metabolism of Pyridine by Soil Bacteria. Biochem J., 127(2), p. 44; Shukla, O. P., and S. M. Kaul. 1974. Constitutive Pyridine Degrading System in Corynebacterium Species. Indian J. Biochem. Biol. 11(3), 201-7.

[14] Abelson, D., E. Parthe, K. W. Lee, and A. Boyle. 1965. Spectral and Biological Change Induced in Nicotinic Acid and Related Compounds by Ultraviolet Light. Biochem J. 96, 840-52: See also dossier VII-5.

[15] Polishchuk, L. R., and L. A. Stempkovshaya. 1084. Determination of Pyridine in Farm Crops and Fish. Gig. Sanit. 39) 69-70. (Dossier reference 21, P. VII-12).

[16] Polischuk, L. R. 1975. Dynamics of the Level of Certain Organic Toxic Substances in Plants Irrigated With Waste Water from By-product Coke Manufacturing Plants. Gig. Sanit. 4, 117-119. (Dossier reference 22, p. VII-12); Polischuk, L. R. and S. V. Polischuk. 1977. Dynamics of Dissociation in Plants of Certain Organic Toxic Substances Introduced with Sewage Irrigation. Gig. Sanit. 10, 10-3.

[17] Goring, C. A. I. and Hamaker, J. W., ed., Organic Chemicals in the Soil Environment, Vol. 1, Marcel Dekker, Inc. (New York), 1972; Wedig, J. H., et al., Identification of Metabolites from Salts of Pyridine-2-Thiol-1-Oxide following Intravenous and Dermal Administration to Swine, Tox. and App. Pharmacology 43, 373-9 (1978); Adams, M. D., et. al., Urinary Excretion and Metabolism of Salts of 2-Pyridinethiol-1-Oxide Following Intravenous Administration to Female Yorkshire Pigs, Tox. and App. Pharmacology. 36, 523-31 (1976).

[18] See, Furia, T. E. and Bellanca, N., ed., Fenaroli's Handbook of Flavor Ingredients, Chemical Rubber Co. (Cleveland), 1971, p. 601 (reported pyridine found in coffee, tobacco, leaves and roots of Atropa belladonna).

[19] 21 C.F.R. 172.515.

We have no knowledge of any significant adverse environmental or human health incidents involving or caused by pyridine. Pyridine has been manufactured, transported and used for many decades without any such problems.[20]

Furthermore, safeguards of manufacturers and users (e.g., workplace monitoring, medical surveillance of workers, engineering controls) ensure strict limits on exposure and safe operating techniques. All plants making or using pyridine are covered by one or more environmental permits and by the OSHA standard.

BY-PRODUCTS

We do not believe there are any by-products from manufacture that pose an unreasonable risk to health or the environment. Reilly and Nepera have reviewed the chemicals proposed by OSHA and NIOSH for categorization under OSHA's proposed generic carcinogen standard and can certify that no by-products of their manufacture of pyridine appear on these lists, with the possible exception of trace amounts of benzene and toluene.

IMPURITIES

As noted above, most pyridine has a commercial purity of far greater than 99%. Therefore, impurities do not warrant serious concern.

TESTS RECOMMENDED BY THE TITC

Given the production, use, and exposure patterns reported above, we do not believe EPA can make the necessary findings to require testing. For example, human exposure is neither significant (i.e., level and duration) nor substantial (i.e., number of people). Pyridine is classified as a flammable liquid by the Department of Transportation,[21] hence its distribution in commerce (i.e., transportation) is also controlled.

Considering the limited test facilities and personnel available which EPA must take into account,[22] the potential cost of testing, and its effect on the producers,[23] there are other chemicals of higher priority than pyridine. Further, many final products containing pyridine are subject to other federal controls.

Although we do not believe pyridine should be subject to a testing rule, it is nevertheless desirable to comment on the appropriateness of the tests recommended by the TITC.

These tests cannot be viewed in the abstract, standing alone. The significance of production, use, and exposure patterns must be taken into account, as well as other factors such as chemical and physical properties and attendant structural-activity relationships. Specific comments follow.

Carcinogenicity — Human exposure is low level and number of persons is small. NCI is conducting bio-assay of pyridine; another such test may be duplicative and unnecessary. There is no previous indication of carcinogenic potential. Carcinogenicity testing should not be required based on results of short-term tests.

Mutagenicity — A recent Chinese hamster cell chromosomal aberration test produced negative results.[24]
Pyridine tested for frequency of sister chromatid exchanges in Chinese hamster cells caused a small frequency increase but no dose-related effect.[25]

Mutagenicity testing of pyridine using TRP reversion in E Coli showed 0.76 colony/CM^2 against 0.42-0.80 colonies/CM^2 in the control.[26]

[20] It is noteworthy that Section 4(a) of TSCA states testing is to be required if activity or exposure findings are made and that there are "insufficient data and experience "(emphasis added). Experience with pyridine is extensive and has shown the product to be safe when properly handled.
[21] 49 C.F.R. 172.101, 49 C.F.R. 173.118-119.
[22] See TSCA Section 4(b)(1)(c).
[23] See TSCA Sections 2(c) and 24(a)(1).
[24] Ishidate, M. Jr. and Odashima, S. "Chromosome Tests with 134 compounds on Chinese Hamster Cells in Vitro — A Screening for Chemical Carcinogens" Mutat. Res. 48, 337—354 (1977).
[25] Abe and Sasaki, J. of National Cancer Institute 58 (6), 1635-42, 1977.
[26] Pai et al. in "Biological Oxidation of Nitrogen" El Sevier, 1978, pgs. 375-382.

Both Ames test and cell transformation test of pyridine were negative.[27]
In view of these test results, no additional mutagenicity tests are needed.

Teratogenicity — Use and exposure levels are not sufficient to warrant this test on a priority basis.

Other chronic effects   If any long term test is run, these effects can and should be analyzed at the same time. Exposure levels and duration and experience indicate low potential for chronic effects.

With respect to health effects testing, the chemical and physical properties of pyridine and the results of the recent battery of mutagenicity assays suggest the absence of mutagenic potential. Therefore we do not believe further health effects tests are required.

Environmental effects — Due to pyridine's complete miscibility with water and low octanol/water partition coefficient, bioaccumulation is unlikely. Pyridine is not likely to be biologically stable or environmentally persistent. As the dossier itself indicates (p. VII-5), because of pyridine's high vapor pressure, it can be expected to volatilize to the atmosphere and photodegrade. In view of these properties, and the low level of environmental release, further environmental effects testing is unnecessary.

Epidemiology — Based on several factors, we believe an epidemiological study at production and consuming facilities would not produce meaningful results and that such a study is unnecessary.

Synthetic producers routinely conduct industrial hygiene surveys at manufacturing sites to measure pyridine vapor concentrations and determine employee exposure. Data gathered within the last 2 years indicate that breathing-zone concentrations of pyridine are consistently below the TLV.

Because of the flammability and disagreeable odor characteristics of pyridine and its potential to cause irritation of skin and respiratory tract, pyridine contact (both vapor and liquid) has always been minimized.

Because of the diverse nature of production and processing operations and employee mobility within these operations, very few employees have spent long periods in pyridine production. Major producers and users have approximately 256 employees who routinely work in pyridine production areas. Only a few employees have worked over 30 years in pyridine facilities. The small number of employees in pyridine production and processor facilities, their low exposures, the mobility of the work forces, and the difficulty in isolating pyridine as the sole or major agent in any workplace, combine to yield a cohort unsuitable for an epidemiology study. Routine medical surveillance programs have revealed no unusual medical problems among employees working in pyridine production facilities.

In summary, an epidemiological study would not produce meaningful results and is unnecessary because: 1) medical surveillance programs indicate no unusual medical problems among pyridine production employees, 2) these employees do not constitute a suitable cohort, and 3) industrial hygiene surveys show that pyridine exposures are consistently well below the TLV.

---

[27] Unpublished data by Central Toxicology Laboratory, Imperial Chemical Industries, (UK).

CONCLUSION

The data presented show that neither chemical activity nor exposure connected with pyridine suggest the threat of an unreasonable risk of injury to health or the environment. With respect to the TITC criteria for recommending testing, it should be emphasized that: a) the quantity of pyridine produced annually is substantially less than stated in the dossier, b) the amount of pyridine released into the environment is minimal, c) the number of individuals occupationally exposed is a small fraction of the dossier estimate, and d) the extent to which the general population is or will be exposed is negligible. Further, the necessary TSCA Section 4(a) findings cannot be supported by the evidence, and therefore testing should not be ordered. The specific tests recommended are duplicative, or inappropriate based on patterns of use and exposure or properties of the chemical.

Respectfully submitted for
the Pyridine Taskforce

REILLY TAR & CHEMICAL CORPORATION

By: _____

Thomas E. Reilly, Jr.
Vice-President
151 North Delaware Street
Indianapolis, Indiana 46204
(317) 638-7531

NEPERA CHEMICAL CO., INC.

By: _____

R. M. N. Molesworth
President
Route 17
Harriman, New York 10926
(914) 782-8171

KOPPERS COMPANY, INC.

By: _____

Alonzo Wm. Lawrence
Vice-President, Environmental
  Resources and Occupational
  Health
Rm. 1426, Koppers Building
Pittsburgh, Pennsylvania 15219
(412) 227-2741

OF COUNSEL:

PEPPER, HAMILTON & SCHEETZ

By: _____

Robert V. Zener
1776 F Street, N.W.
Suite 200
Washington, D.C. 20006
(202) 862-7560

APPENDIX II

POTENTIAL EXPOSURE TO PYRIDINE FROM
COKE OVEN EMISSIONS

With reference to the reported exposure of coke oven workers to pyridine, it is possible to estimate the level of exposure by relating it to the particulate polycyclic organic matter (PPOM) found on and near by-product coke ovens. Exposure to PPOM measured as benzene soluble particulate matter is the conventional basis for judging exposure to coke oven emissions.

In the book "Chemistry of Coal Utilization" by H. H. Lowry (John Wiley & Sons — 1945) pyridine bases in the gas at the ovens are stated to be 138 lbs. per 9,890,000 cu. ft. of gas, (page 1126) and pyridine is stated to be about 69% of the crude pyridine bases (page 1131). Also, the typical yields of coke oven gas and tar are 60,000 cu. ft. and 10 gal., respectively, per ton of coal. ["Coal, Coke and Coal Chemicals" by P. J. Wilson and J. H. Wills (McGraw-Hill-1950, page 187)]. Using the foregoing data and assuming that the benzene soluble particulate matter in coke oven emissions is one-third of the tar, the other two-thirds being either vapor or insoluble in benzene, then the level of exposure of coke oven workers to pyridine can be estimated as follows:

$$\frac{138 \times 10,000 \times 0.69}{9,890,000} = 0.1 \text{ lb pyridine}/10,000 \text{ ft}^3 \text{ of gas released from oven}$$

10×1.20 sp. gr. $\times 8.3 \times 0.33 = 33$ lbs. PPOM/10,000 cu. ft. of gas released from oven therefore $33/0.1 = 330$ PPOM/pyridine.

If a coke oven worker is exposed to 1 mg/M^3 of PPOM, (measured at benzene soluble particulate matter) then he would be exposed simultaneously to 3 μg/M^3 of pyridine. If a coke oven worker is exposed to PPOM at a level at or below the OSHA standard of 0.2 mg/M^3, then his exposure to pyridine would be at or below 0.6 μg/M^3.

In light of the foregoing information, it is evident that coke oven workers are exposed to very low levels of pyridine, the majority being below 1 μg/M^3. At these levels, it is difficult to believe that pyridine exposures are a significant factor in the health of coke oven workers and certainly no correlation would be expected between pyridine exposures and the epidemiology of coke oven workers because of the many other factors involved in their total work environment.

REFERENCES

1. Toxic Substances Control Act Interagency Testing Committee, Preliminary List of Chemical Substances for Further Evaluation, Washington, D.C., July, 1977.
2. Initial Report of Interagency Testing Committee to EPA Administrator, *Fed. Regist.*, 42, 55025, 1977.
3. Second Report of Interagency Testing Committee to EPA Administrator, *Fed. Regist.*, 43, 16684, 1978.
4. Third Report of Interagency Testing Committee to EPA Administrator, *Fed. Regist.*, 43, 50630, 1978.
5. Fourth Report of Interagency Testing Committee to EPA Administrator, *Fed. Regist.*, 44, 31866, 1979.
6. Fifth Report of Interagency Testing Committee to EPA Administrator, *Fed. Regist.*, 44, 70663, 1979.
7. Sixth Report of Interagency Testing Committee to EPA Administrator, *Fed. Regist.*, 45, 35897, 1980.
8. EPA Response to Interagency Testing Committee Initial Recommendations, *Fed. Regist.*, 43, 50234, 1978.
9. EPA Response to Second ITC Testing Recommendations, *Fed. Regist.*, 44, 28095, 1979.
10. Comments of the Chemical Manufacturers Association on EPA's ANPR Concerning the Reimbursement of Test Costs under Section 4 and 5 of TSCA, February 7, 1980, EPA Docket No. OTS-48001, Chemical Manufacturers Association, Washington, D.C.

Chapter 5

PREMANUFACTURE NOTIFICATION

Robert M. Sussman and Peter Barton Hutt*

TABLE OF CONTENTS

*   Mr. Sussman and Mr. Hutt are attorneys with the firm of Covington and Burling in Washington, D.C. They were responsible for preparing the comments of the Chemical Manufacturers Association (CMA) on EPA proposed and reproposed regulations under Section 5. Information was updated to September, 1982.

The Toxic Substances Control Act (TSCA) became effective on January 1, 1977. This complex legislation, which is administered by the Environmental Protection Agency (EPA) provides for comprehensive Federal regulation of "chemical substances" and "mixtures" insofar as they may affect human health or the environment. According to Section 2(b) of the Act, the purposes of the TSCA are to develop adequate data about the health and environmental effects of chemical substances and mixtures, and to regulate their manufacture, processing, distribution, marketing, use, and disposal where necessary to protect human health and the environment from unreasonable risk of injury. These goals are to be achieved, Section 2(c) declares, without unduly impeding or creating unnecessary economic barriers to technological innovation.

Section 5 is one of the central provisions of the TSCA. Its purpose is to authorize EPA to screen "new" chemicals in advance of manufacture so the Agency can identify their potential adverse effects before they can harm human health or the environment. Since Section 5 took effect, it has been unlawful for any "new chemical substance" to be manufactured or imported for commercial purposes unless, at least 90 days before the start of production, a premanufacture notice (PMN) describing the substance has been submitted to EPA. Section 5 also authorizes EPA to require, by rule, the submission of similar notices at least 90 days before the manufacture or processing of existing chemical substances for "significant new uses." Under specified circumstances, EPA may extend the 90-day review period while it continues its review of the substance, requires testing, or initiates regulatory action.

Section 5 thus gives EPA broad power to delay or prevent the manufacture and use of new and certain existing chemicals while it scrutinizes their health and environmental effects. Because of this power, EPA actions under Section 5 have a far-reaching impact on the future development of new chemicals and new applications for existing chemicals. Since Section 5 directly affects innovation in the chemical industry and the many other industries that depend on chemical products, it is critical for EPA to adopt procedures for premanufacture notification which are reasonable and realistic.

This chapter discusses the principal legal, regulatory, and practical issues that have emerged to date during EPA implementation of Section 5. The chapter is divided into seven different Parts, which address the following subjects:

- *Part I* provides an overview of the statutory provisions governing premanufacture notification and summarizes the progress of the EPA premanufacture notification program since it took effect on July 1, 1979.

- *Part II* discusses when the submission of a premanufacture notice is required under TSCA. Major issues discussed in this part include the differences between chemical substances and mixtures under TSCA; the use of the TSCA inventory to identify "new" chemical substances; the application of Section 5 to chemicals (like foods, drugs, cosmetics, and pesticides) regulated under other statutes; the status of chemical intermediates and research and development chemicals under Section 5; the proper treatment under Section 5 of by-products, impurities, and other noncommercial chemicals; and test market exemptions from Section 5.

- *Part III* analyzes the basic purposes and policies underlying Section 5. Based on the structure of TSCA and the legislative history of the Act, this Part demonstrates that, while Congress attached considerable importance to scrutinizing the health and environmental effects of chemicals prior to manufacture, it expected that Section 5 would function as a notification provision which alerted EPA to the impending manufacture of a new chemical, not as a broad safety substantiation provision under which manufacturers would be required to submit conclusive proof that every new chemical was safe.

- *Part IV* describes the role of premanufacture testing in the Section 5 review process. As the Part demonstrates, Congress did not authorize EPA to require mandatory premanufacture testing of all new chemicals under Section 5. Nevertheless, EPA has developed voluntary testing "guidance" for PMN submitters and, over time, the Agency may use its powers under Section 4 to require testing for an increasing number of new chemicals.

- *Part V* explores various issues relating to the information which must be submitted to EPA in a premanufacture notice. The Part discusses the statutory limitations on the content of PMNs and contrasts the draft PMN forms prepared by industry and EPA. In addition, the Part discusses the PMN submitter's obligation to obtain information that it does not already possess for inclusion in the PMN, and the extent to which PMN submitters must contact prospective customers for their products before a PMN is submitted.

- *Part VI* discusses the circumstances in which, and the procedures by which, EPA will treat PMN information as confidential. Addressed in this Part are such important matters as the PMN submitter's ability to prevent disclosure of the specific identity of a new chemical, the need to substantiate confidentiality claims at the time a PMN is submitted, and the supporting evidence which such substantiation must contain.
- *Part VII* analyzes a number of major unresolved legal and policy issues under Section 5. These include: the power of EPA to issue binding substantive rules implementing Section 5, the power of EPA to require supplemental reporting on new chemicals before and after the PMN review period, the Agency rejection of incomplete PMNs as "invalid", the application of Section 5 to export chemicals, EPA use of Section 5(e) to postpone manufacture of suspect new chemicals, and the scope of EPA authority to regulate existing chemicals under "significant new use rules."

I. OVERVIEW OF SECTION 5

The first step in understanding the premanufacture notification process created under TSCA is to identify the principal provisions of Section 5 and examine how they operate. Accordingly, the following discussion provides a brief overview of Section 5 requirements and the actions that EPA has taken to date to implement those requirements.

A. Contents of Manufacture and Processing Notices

Under Section 5(a), no "new chemical substance" may be manufactured for commercial purposes unless, at least 90 days before the scheduled start of production, a so-called "premanufacture notice" has been submitted to EPA. An identical notice must be submitted at least 90 days before an "existing" chemical substance can be manufactured or processed for a use that EPA has determined, by rule, is a "significant new use."

Pursuant to Sections 8(a)(2) and 5(d), these notices must contain a range of identifying information about the substance. The submitter must provide EPA with the trade name of the substance, chemical identity, molecular structure, actual and proposed categories of use, anticipated production volume for these uses, anticipated by-products, and expected methods of disposal. The submitter must also indicate the estimated number of employees that will be exposed to the substance and the duration of that exposure. Finally, the notice must be accompanied by all test data in the submitter's possession relating to the effects of the substance on human health or the environment, and a description of all such test data known to or reasonably ascertainable by the submitter. According to the House Report on TSCA, the purpose of the notice is to "give the Administrator an opportunity to review and evaluate information respecting the substance to determine if manufacture or processing should be restricted or delayed."[1]

B. Definition of "New" Substances and "Significant New Uses"

The mechanism for determining if a chemical is a "new chemical substance", and hence subject to notification requirements, is the preparation by EPA of an inventory of all chemical substances manufactured or processed for commercial purposes during the period preceding the effective date of rules governing the compilation of the inventory. Section 3(9) provides that any chemical substance omitted from the inventory will be considered a "new chemical substance". Under Section 5(a)(1)(A), 30 days after the inventory has been published in final form, manufacture of a chemical substance omitted from the inventory will be prohibited unless the requisite notice has been submitted to EPA 90 days before manufacture begins.

Section 5 prescribes a different procedure for identifying existing chemical substances subject to Section 5 review because they will have "significant new uses". Under Section 5(a)(1)(B), exercising its powers of informal rulemaking, EPA is required to determine that a particular use (or category of uses) is a "significant new use". Until such a determination has been made, notification requirements will not apply to a new use of an existing chemical substance. The Act provides no specific timetable for issuing regulations for "significant new uses". In identifying such uses, EPA is required by Section 5(a)(2) to consider various factors, including the projected volume in which the substance will be manufactured and processed for the new use and the extent to which the use will alter and increase human and environmental exposure to the substance.

[1] House Report No. 94-1341, 94th Cong., 2d Sess. 23 (1976).

C. *Federal Register* Publication of Premanufacture Notices

Not later than 5 days after a PMN or SNUN has been received, EPA is required under Section 5(d)(2) to publish a *Federal Register* notice that describes the identity of the substance in question, its intended uses, and any test data about the substance that may have accompanied the notice. In addition, at the outset of each month, EPA is required to list in the *Federal Register* each substance for which a notice has been received, including substances for which the 90-day review period has not yet expired and substances for which that review period has expired since the last *Federal Register* notice. According to the TSCA Conference Report, the purpose of these notices is to insure that "the public receives timely notification of any new chemical substance or any significant new use of an existing chemical substance" — presumably in order to advise EPA whether manufacture should be permitted to proceed.[2]

D. The Mandatory Development of Premanufacture Test Data

Under Section 5, there are two situations where a PMN or SNUN may be required to contain test data that the manufacturer has not voluntarily chosen to generate.

First, pursuant to Section 5(b)(1)(A), such test data will be necessary when a rule under Section 4 of TSCA requiring testing is in effect and is applicable to the substance. Generally, this situation will only occur when the EPA Section 4 test rule applies to an open-ended "category" of chemicals which includes the new substance. Under certain circumstances specified in Section 4(c) of TSCA, the manufacturer of the new substance may obtain an exemption from test requirements. In this event, however, the manufacturer may be required to reimburse the firm or firms that have performed testing in compliance with the EPA test rule. In addition, where another manufacturer is performing testing, the 90-day review period for the submitter's notice will not begin to run until testing has been completed and the test data are in the possession of the Agency.

Under Section 5(b)(2), the submission of test data will also be necessary when EPA has included the substance on a list of chemical substances or categories of substances that it has found present or may present an unreasonable risk of injury to health or the environment. Section 5(b)(2)(B) provides that, when a substance has been included on this "risk list" but an applicable test rule under Section 4 is not yet in effect, the manufacturer or processor must submit data which it believes wil establish that the substance will *not* present an unreasonable risk of injury to health or the environment.

The Section 5(b)(2) "risk list" is to be compiled in an informal rulemaking proceeding conducted pursuant to 5 U.S.C. § 553, in the course of which interested parties will have an opportunity to make an oral presentation. The statute prescribes a number of factors for selecting chemicals for inclusion on the list, including the health or environmental effects of the chemical and the magnitude of its human and environmental exposure. Under Section 5(g), when EPA receives a PMN or SNUN for a chemical on the "risk list" and fails to take regulatory measures against the chemical under Sections 6 or 7 of the Act, it must publish a *Federal Register* notice explaining the reason for its nonaction.

E. Postponement of Manufacture or Processing Pending Further Review

Section 5 provides EPA with 3 mechanisms to postpone manufacture or processing beyond the initial 90-day review period.

First, under Section 5(c), that initial review period may, for "good cause", be extended for an additional 90 days. EPA must publish this extension and the reasons therefor in the *Federal Register*. The Conference Report indicates that the EPA is intended "to have a large degree of flexibility" in exercising this power.[3] The decision to extend the review period is to be considered "final agency action" subject to judicial review.

Second, under Section 5(e), EPA may issue a "proposed order", which prohibits or limits the manufacture, processing, distribution, use, or disposal of the substance. Such an order is permissible when EPA has determined that (1) existing information is insufficient to permit "a reasoned evaluation" of the health and environmental effects of the substance *and* (2) there may be substantial human or environmental exposure to the substance *or* the substance may present an unreasonable risk of injury to health or the environment.

A Section 5(e) order must be issued in proposed form at least 45 days before the end of the PMN review

---

[2]   House Report No. 94-1679, 94th Cong., 2d Sess. 67 (1976).

[3]   House Report No. 94-1679, *supra,* at 67.

period. Either before or at the same time the proposed order is issued, EPA must inform the affected manufacturer or processor of the substance of the order in writing. Upon issuance of the proposed order, the affected manufacturer or processor will have 30 days in which to file objections specifying with particularity the grounds on which it considers the order unjustified. If no objections are filed, the proposed Section 5(e) order will automatically take effect at the conclusion of the 90-day review period. If objections are filed but are not accepted by the Agency, EPA may apply to a District Court for an injunction effectuating the order. The Agency may also apply for an injunction if it has not issued a proposed order but elected to proceed directly in the courts.

In determining whether relief should be granted, the courts are directed to apply the same test that EPA has applied in the first instance — whether existing information on the effects of the substance is insufficient and either the substance may present an unreasonable risk or will have substantial human or environmental exposure. If an injunction is granted, it will normally remain in effect until the submitter has provided EPA with test data sufficient to evaluate the health or environmental effects of the substance. If, however, EPA has commenced a rulemaking proceeding involving the substance under Section 6(a), the injunction will remain in effect until the effective date of any rule that EPA promulgates or the termination of the EPA rulemaking proceeding.

To prevent the production of the substance after the 90-day notification period has expired but before final judicial action can be taken, the courts are specifically authorized to issue temporary restraining orders and preliminary injunctions. According to the TSCA Conference Committee, such preliminary relief is intended to be "freely exercised . . . as is necessary to preserve the status quo in order to insure that the policy of this section can be fulfilled."[4]

Third, under Section 5(f), EPA can take action to prohibit or restrict manufacture, distribution, processing, use, or disposal of a new substance. The basis for such action is a determination by EPA that (1) "a reasonable basis" exists for concluding that the substance presents or will present "an unreasonable risk of injury to health or the environment" *and* (2) action pursuant to the normal regulatory procedures of Section 6 will be too late to protect against that risk. In contrast to Section 5(e), which is directed at situations where EPA lacks sufficient data to evaluate a new chemical, Section 5(f) authorizes the prevention of manufacture where the Agency has decided that the new chemical is dangerous.

EPA can prohibit or limit manufacture or processing of a chemical under Section 5(f) in one of three ways. First, it can propose a rule concerning the chemical under Section 6(a) of TSCA; such a rule will automatically take effect immediately and will only be appealable at the completion of the rulemaking proceeding. Second, EPA can issue a proposed order concerning the substance; in this case, the administrative and judicial procedures prescribed by Section 5(e) will apply. Finally, the Agency can proceed directly in the courts to obtain injunctive relief. If the Agency elects this course, the courts will make a *de novo* determination whether the chemical presents an "unreasonable risk."

If EPA desires merely to restrict but not to prohibit the manufacture or processing of the substance, it can utilize a proposed rule that becomes immediately effective. However, if EPA wishes to prohibit production, it must either directly apply to a district court for injunctive relief or employ the administrative order procedure prescribed by Section 5(e). Significantly, the Conference Report stresses that, even though technically not an outright ban, many limitations on the use of a substance may have that effect in practice and should be treated as a ban for purposes of implementing Section 5(f). H. Rep. No. 94-1679, *supra,* at 70.

F. Exemptions from Premanufacture Notification

Section 5(h) provides for a number of total and partial exemptions from premanufacture notification requirements.

Under paragraph (1), upon a showing that no unreasonable risk of injury will be presented, EPA may grant an exemption for manufacturing or processing a chemical substance for "test marketing purposes" and impose "appropriate restrictions."

Under paragraph (2), again upon application to the Agency, particular manufacturers and processors of products included on the Section 5(b)(2) "risk list" may be exempted from test data submission requirements. The standard for granting such an exemption is that data on the chemical substance have been or will be developed by others and additional data would be "duplicative." Companies that receive such an

---

[4]   House Report No. 94-1679, *supra,* at 69.

exemption may be required to assume reimbursement responsibilities on the same terms that govern the testing provisions of Section 4.

Paragraph (3) grants an exemption from premanufacture notification procedures for substances produced "in small quantities" solely for "scientific experimentation" or for "chemical research" related to new product development. All persons engaged in such experimentation research or analysis must, however, be notified by the manufacturer or processor of any risk that either EPA or the manufacturer or processor has reason to believe the substance may present.

Under paragraph (4), exercising its informal rulemaking powers, EPA may exempt any chemical substances from all or a part of the requirements for premanufacture notification when it determines the substance will not present an unreasonable risk of injury.

Finally, under paragraph (5), upon application to EPA, exemptions may be granted for so-called "intermediates" — e.g., chemical substances which exist temporarily during the manufacture or processing of another substance or mixture — to which there is or will be no human or environmental exposure.

G. Enforcement of Section 5

Under Section 15(1) of TSCA, it is unlawful to manufacture or import a new chemical substance for a commercial purpose without first submitting a premanufacture notice under Section 5. In addition, Section 15(2) makes it unlawful to process, distribute, or use a chemical substance that the person in question knew, or had reason to know, was manufactured in violation of the premanufacture notification requirements of Section 5. Thus, unless a new chemical substance has successfully completed premanufacture review, purchase, distribution, processing, or use of the substance, as well as manufacture, are prohibited. Under Section 16 of TSCA, any of these activities can be the basis for criminal or civil penalties. Under Section 17, EPA may seek injunctive relief and product seizure to prevent continued violations.

H. EPA Actions to Implement Section 5

On June 1, 1979, EPA published its Initial Inventory of chemical substances under Section 8(b) of TSCA. This inventory consists of all chemicals reported to EPA by manufacturers under its inventory reporting regulations. On July 1, 1979, 30 days after the publication of the Initial Inventory, EPA began implementing the premanufacture notification provisions of Section 5 with respect to all manufacturers and importers. Since that date, it has been unlawful to manufacture or import a chemical substance for commercial purposes unless a PMN was submitted to EPA at least 90 days before the start of manufacture or importation.

After publishing the Initial Inventory, EPA began preparation of a Revised Inventory consisting of chemicals that met the statutory criteria for inclusion on the inventory but had not been reported by manufacturers. For this purpose, on June 1, 1979, EPA initiated a 210-day reporting period during which processors, users, and certain importers could inform EPA of additional chemical substances eligible for inclusion on the inventory.[5] On July 30, 1980, EPA released a Cumulative Supplement to the Inventory which listed chemicals identified during the additional reporting period. 45 *Fed. Reg.* 50544. EPA has announced that thirty days later, it became unlawful to process or use a chemical substance which was omitted from the Revised Inventory and had not been the subject of premanufacture notification in accordance with Section 5.[6]

While it has been implementing Section 5, EPA has sought to develop regulations governing the submission and review of premanufacture notices. Proposed rules under Section 5 were published on January 10, 1979 (44 *Fed. Reg.* 2242) and reproposed on October 16, 1979 (44 *Fed. Reg.* 59764). Final rules have not yet been promulgated. To provide guidance to PMN submitters until that time, EPA published a Statement of Interim Policy under Section 5 on May 15, 1979 (44 *Fed. Reg.* 28564). A statement of Revised Interim Policy was published on November 7, 1980. 45 *Fed. Reg.* 74378.

Between July 1, 1979, when Section 5 took effect, and July 31, 1982, EPA received 1552 premanufacture notices. It also received 136 applications for test marketing exemptions under Section 5(h)(1). On nine occasions EPA issued Section 5(e) orders postponing manufacture; these orders were not contested by the PMN submitters. On one occasion, a submitter withdrew its PMN voluntarily without any formal EPA action, and 11 PMNs have been suspended voluntarily.

---

[5]   EPA announced the beginning of the reporting period for the Revised Inventory in a *Federal Register* notice published on May 15, 1979. *See* 44 *Fed. Reg.* 28588.

[6]   EPA published another cumulative supplement to the Inventory on July 19, 1982. 47 *Fed. Reg.* 25767.

EPA has also taken steps to implement the PMN exemption provisions of Section 5(h)(4). On June 4, 1982, the Agency promulgated a rule exempting certain manufacturers of instant photographic and peel-apart film articles from PMN requirements. 47 *Fed. Reg.* 24308. On August 4, 1982, EPA proposed an exemption rule for certain site-limited intermediate, low-volume chemicals, and polymers. 47 *Fed. Reg.* 33896. Final action has not yet been taken on this proposal.

II. CHEMICALS COVERED BY SECTION 5

For any firm contemplating manufacture or importation of a chemical, the first question will be whether that chemical is a "new chemical substance" subject to the premanufacture notification requirements of Section 5. If the answer to this question is "yes", manufacture or importation cannot begin until a premanufacture notice has been submitted to and reviewed by EPA — a process entailing substantial delay, expense, and uncertainty. On the other hand, if Section 5 does not apply, manufacture or importation will be able to begin forthwith, without compliance with any EPA requirements.

Whether a chemical is covered by Section 5 may often be a complex question, which turns on a variety of definitions and requirements imposed by statute or rule. The following discussion identifies the different criteria which will determine whether Section 5 does or does not apply, and provides some useful guidelines for determining the status of particular chemicals under these criteria.

A. Differences between Chemical Substances and Mixtures

Of major importance in determining the applicability of Section 5 is the difference between "chemical substances" and "mixtures". Section 5 only applies to "chemical substances". Thus, the manufacturer of a "new" mixture will not be subject to Section 5 unless, in creating that mixture, it also manufactures a new chemical substance.

In pertinent part, the term "chemical substance" is defined in Section 3(2) of TSCA as "any organic or inorganic substance of a particular molecular identity" — excluding "mixtures" but including (1) "any combination of such substances occurring in whole or in part as a result of a chemical reaction or occurring in nature" and (2) "any element or uncombined radical". EPA has properly emphasized the breadth of this definition. In the Preamble to its original inventory proposal of March 9, 1977, the agency observed that

the term "chemical substance" encompasses everything from the basic elements to the most complex organic chemicals. It also includes naturally occurring substances such as coal and wood and any chemical substances derived or extracted from them.

Section 3(8) of TSCA defines the term "mixture" as follows:

The term "mixture" means any combination of two or more chemical substances if the combination does not occur in nature and is not, in whole or in part, the result of a chemical reaction; except that such term does include any combination which occurs, in whole or in part, as a result of a chemical reaction if none of the chemical substances comprising the combination is a new chemical substance and if the combination could have been manufactured for commercial purposes without a chemical reaction at the time the chemical substances comprising the combination were combined.

In the Appendix to its inventory regulations, EPA has indicated that alloys, inorganic glasses, ceramics, frits, and cements, including Portland cement, are mixtures and, as such are not subject to Section 5. 42 *Fed. Reg.* 64585.

In general, EPA has advised, a combination of chemical substances will be considered a "mixture" if they have been combined by mixing, but will have the status of a "chemical substance" if the combination occurs in nature or is the result of a chemical reaction:

A combination of two or more chemical substances is itself a "chemical substance" for purposes of these regulations unless it falls within the specific definition of the term "mixture". In general, a combination of two or more chemical substances is a "mixture" if they have been combined by actually mixing them together.

If, however, the combination occurs in nature, it is a "chemical substance" and is not a "mixture". If, further, the combination is prepared by a chemical reaction, it is a "chemical substance" and not a "mixture", unless the combination could actually have been manufactured for commercial purposes at this time without a chemical reaction e.g., by mixing its separate components with each other. *Id.*

One nuance in this test for identifying a mixture is that certain combinations of chemical substances that result from a chemical reaction would still qualify as a mixture if the combination could have been manufactured merely by mixing. The EPA Preamble to its March 9, 1977, proposal explains this result as follows:

> In accordance with this definition, certain combinations of chemical substances would be encompassed within the term "mixture" even though there may be some chemical reaction among the component chemical substances. As clarified in clause (1) of the definition, the term "mixture" would include certain action-produced combinations of chemical substances which could have been manufactured for commercial purposes with a chemical reaction. To illustrate, a processor might combine two chemical substances, O and D, which do not react to produce a mixture consistent with the basic definition of the term. On the other hand, it may be more efficient to mix the precursor substances of C together with D in one step. Thus, a processor may combine chemical substances A, B and D. The final product, a mixture of C and D, would be considered equivalent to the mixture made by directly combining C and D. Whether the process of combining the precursor chemical substances involves a chemical reaction is not strictly relevant to the determination of whether the product is a chemical substance or a mixture. 42 *Fed. Reg.* 13131.

B. The Use of the Section 8(b) Inventory to Identify "New Chemical Substances"

1. Function of the Inventory

Once it is established that a chemical is a "substance" for purposes of TSCA, it will still only be subject to Section 5 if it is a "new" substance within the meaning of the Act. The mechanism for identifying "new chemical substances" subject to premanufacture review under Section 5 is the EPA "Inventory" of existing chemical substances. Section 8(b) of the Act, under which this Inventory is required, provides that it must contain all chemical substances manufactured or processed no more than 3 years before the effective date of EPA regulations for inventory reporting. The only substances included in this category but ineligible for the inventory are those substances that have been produced in "small quantities" solely for purposes of "scientific experimentation" or "product development".

Under Section 3(9) of the Act, any chemical substance omitted from the inventory will, by definition, be considered a "new chemical substance" and, as such, will be subject to premanufacture notification under Section 5. The Act contemplates that as new chemical substances complete the premanufacture review process they will be added to the inventory when they enter commercial production. Thereafter, it will be permissible for companies to commence manufacture or processing of these chemicals without submitting an additional premanufacture notice.

2. History of the Inventory

Under Section 8(a)(1), EPA is directed to prepare the inventory by promulgating reporting rules which apply to manufacturers and processors of chemical substances eligible for inclusion on the inventory. The Act provides that these rules must be promulgated in final form no later than 180 days after the effective date of TSCA. Nevertheless, because of a number of factors, including last-minute reversals of policy, EPA rules were not finalized until long after this deadline.

EPA first proposed inventory reporting regulations on March 9, 1977 (42 *Fed. Reg.* 13130). On April 12, 1977, EPA published a supplement to the proposed regulations (42 *Fed. Reg.* 19298) which provided detailed instructions for the use of a "Candidate List of Chemical Substances" that EPA planned to publish. Soon afterward, on April 28, 1977, EPA published a notice announcing the availability of such a Candidate List (42 *Fed. Reg.* 21639). The purpose of the Candidate List, which contains 30,000 listings and is 3 volumes in length, was to identify those chemical substances already known to EPA from existing data sources, thereby providing for the simplified reporting of such substances for the inventory. On July 8, 1977, EPA announced that the Candidate List could be obtained on computer-readable tape (42 *Fed. Reg.* 35183).

On August 2, 1977, EPA reproposed the inventory reporting regulations (42 *Fed. Reg.* 39182). This action was taken because of the Agency decision to require industry to report not merely the identities of existing chemical substances but information relating to their production volumes and sites of manufacture as well. On October 3, 1977, EPA published a supplement to the reproposal which elaborated on the reporting responsibilities of importers and enclosed draft reporting forms (42 *Fed. Reg.* 53805). Final inventory reporting regulations were promulgated on December 23, 1977 (42 *Fed. Reg.* 64572).

Under those regulations, the reporting process was divided into two broad phases: (1) the preparation of an Initial Inventory and (2) the refinement of that inventory and publication of a Revised Inventory. Reporting for the Initial Inventory was permissible at any time between January 1, 1978, and May 1, 1978. On June 1, 1979, EPA published the Initial Inventory, which contains approximately 40,000 chemicals. On the same date, it commenced a 210-day period during which additional chemical substances could be reported for the Revised Inventory. The Revised Inventory is contained in Cumulative Supplements published in 1980 and 1982 and lists over 58,000 chemicals.

The general approach of the EPA reporting regulations was to *require* companies engaged in significant chemical production or importation — based on either the size of their chemical operations or their total production of particular chemicals — to report for the Initial Inventory all chemical substances manufactured or imported in bulk during 1977. Chemical producers and importers engaged in less significant activities were *permitted* to report chemical substances manufactured during 1977 for the Initial Inventory. Finally, in their discretion, *all* chemical producers and importers were *permitted* to report for the Initial Inventory chemical substances manufactured or processed during 1975 and 1976. All reporting for the Revised Inventory was voluntary and was limited to processors, users, and importers of chemical substances as part of articles. Manufacturers and importers of chemical substances in bulk or as part of mixtures were only allowed to report for the Initial Inventory.

3. Using the TSCA Inventory to Determine Whether a PMN Is Required

Whenever a firm decides to manufacture, process, or use a chemical substance, it will need to consider whether that substance is "new" and, thus, must be the subject for a PMN under Section 5. To resolve this issue, it will normally be necessary to consult the EPA Inventory. If the chemical is included in the Inventory, no PMN will be required. Conversely, if the chemical is omitted from the Inventory, manufacture, processing, or use of the chemical without submission of a PMN may be unlawful.

The Inventory itself identifies chemical substances both by specific name and Chemical Abstracts Service (CAS) Registry number. In addition, the Inventory is accompanied by various indexes, contained in three volumes, which provide different access routes to the published Inventory list. Two of these volumes contain a Substance Name Index for chemicals included on the Inventory. The fourth volume contains a Molecular Formula Index for Inventory substances and an index of so-called "UVCP" chemicals, which could not be listed by molecular formula because they are of unknown or variable composition.

In addition to the four volumes of the Inventory itself, EPA has published a two-volume Trademark and Product Name List. This list, which was compiled from voluntary reports submitted by manufacturers, identifies the trademarks or other commercial designations of formulations whose constituent chemical substances were reported for the Inventory. Since trademark reporting was voluntary, not all formulation manufacturers reported their products to EPA. Thus, while formulations on this list fully comply with inventory reporting requirements, it cannot be assumed that formulations omitted from the list contain substances that do not appear on the Inventory itself.

4. Confidentiality of Inventory Chemicals

Under Section 710.7(e)(1) of EPA inventory regulations, firms were entitled to claim confidential treatment for the specific identity of a reported substance where disclosure of that identity "would reveal the trade secret fact that the particular chemical substance is manufactured or processed for commercial purposes." Companies making such claims were required to meet several conditions. First, they had to report the specific chemical identity of the substance in question and propose a generic chemical name for that substance which "is only as generic as necessary to protect the confidential identity of the particular chemical substance." In addition, they had to provide detailed, written substantiation of their claim of confidentiality and agree that EPA may disclose that the substance appears on the Inventory to "a person with a *bona fide* intent to manufacture the substance." Finally, persons claiming confidentiality had to prepare, and agree to furnish to EPA upon request, either an X-ray diffraction pattern or a mass spectrum for the chemical substance, a sample of the substance in its purest form, an elemental analysis, any additional or alternative spectra, and any other data necessary to resolve uncertainties about the chemical identity of the substance.

Under Section 710.7(f) of the inventory regulations, the EPA General Counsel reviewed all claims of confidentiality for the identities of chemical substances pursuant to the Agency procedures for confidential business information. If the General Counsel determined that inclusion of the chemical identity of a particular substance on the Inventory would disclose a trade secret, EPA omitted the substance from the Inventory. All confidential substances were then listed by generic chemical name in Appendix B to the Inventory. In preparing Appendix B, the inventory regulations authorized EPA to reject the generic names proposed by the company claiming confidentiality where these names were insufficiently descriptive. In this situation, EPA could either ask the claimant to develop other generic names or devise a generic name of its own.

If a chemical is not listed in the Inventory, companies interested in manufacturing that chemical should examine the list of generic names in Appendix B. At times, however, these generic names may not be sufficiently precise to enable the company to ascertain whether the chemical is a "new chemical substance" subject to Section 5. In such circumstances, the inventory reporting regulations allow EPA to disclose more precise information concerning the status of the chemical under Section 5. Under Section 710.7(g), EPA will disclose whether a particular chemical substance falls within one of the generic chemical names in Appendix B to the inventory if it receives an inquiry from a person with "a *bona fide* intent to manufacture the substance." To establish such an intent, the person must submit a signed statement that he intends to manufacture the substance in question for a commercial purpose, together with a description of his research and development activities with respect to the substance and the purpose for which it will be manufactured. In addition, the person must submit such chemical analyses and samples of the substance that EPA requests to ascertain its precise chemical identity.

Assuming EPA is satisfied that the inquiring firm has "a *bona fide* intent to manufacture" the substance in question, it will then obtain chemical analyses and samples from the manufacturer of the confidential substance and, based on all the information before it, determine that either (1) the substance proposed for manufacture is in fact encompassed by the generic name contained in Appendix B or (2) the substance does not appear in the Inventory and premanufacture notification will be necessary. This procedure, EPA has asserted, "will discourage fishing expeditions by persons without a *bona fide* intent to manufacture and thereby protect trade secrets from disclosure to competitors." 42 Fed. Reg. 64591.

EPA inventory regulations apply only to substances reported to EPA for inclusion in the Inventory. They do not apply to substances for which PMNs are submitted and which are thereafter added to the Inventory upon receipt of a notice that manufacture or import has commenced. Section 710.7(g) of proposed EPA regulations for premanufacture notification prescribes a procedure for disclosing the identity of confidential PMN chemicals to persons with a *bona fide* intent to manufacture such chemicals. Until EPA regulations are in effect, however, PMN submitters are not required to agree that EPA may inform persons with a *bona fide* intent to manufacture whether their particular substances are on the Inventory. To avoid confusion and hardship, the interim policy of the Agency for premanufacture notification nevertheless urges all PMN submitters to agree to such disclosure on a voluntary basis. See 44 *Fed. Reg.* 28570.

5. Late Submissions for the Inventory

EPA inventory regulations do not contain any formal mechanism by which otherwise eligible substances that were not reported to EPA may be added to the Inventory after the reporting period has closed. Nevertheless, in a *Federal Register* notice of May 15, 1979, EPA allowed manufacturers and importers to report substances inadvertently overlooked during initial inventory reporting until July 1, 1979. 44 *Fed. Reg.* 28558, 28563. Moreover, EPA continued to accept late inventory submissions after that date and added the chemicals in question to the Inventory. Nevertheless, in a *Federal Register* notice of April 18, 1980, EPA announced that, effective May 19, 1980, the Agency "no longer will accept late inventory reports from manufacturers, and from importers in bulk." According to EPA, as of that date, "any chemical substance manufactured in the United States, and any substance imported in bulk, that has not been reported for the Inventory is subject to Section 5 of the TSCA." EPA indicated that its Office of Enforcement "will strictly enforce this deadline." 45 *Fed. Reg.* 26452, 26453.

C. Chemicals Regulated under Other Statutes

Even if a chemical is omitted from the Inventory, it may not be subject to premanufacture notification if it will be manufactured or imported for a use exempt from regulation under TSCA. The definition of "chemical substance" in Section 3(2) of TSCA excludes various chemicals and articles presently subject to regulation under other Federal statutes. Unfortunately, however, the jurisdictional boundaries between TSCA and these statutes are not always clear or easy to apply. Discussed below are the principal issues with which PMN submitters must contend in this area and the guidance that EPA has provided on these issues to date.

1. Foods, Drugs, and Cosmetics

Section 3(2)(B)(vi) of TSCA exempts from the definition of "chemical substance" —

Any food, food additive, drug, cosmetic, or device (as such terms are defined in Section 201 of the Federal Food, Drug, and Cosmetic Act) when manufactured, processed, or distributed in commerce for use as a food, food additive, drug, cosmetic or device.

Both EPA inventory regulations and proposed PMN regulations explicitly incorporate these statutory provisions. These regulations also provide that the terms "cosmetic, device, drug, food, and food additive" will have the same definitions under TSCA that they possess under the Food, Drug, and Cosmetic Act and regulations that the Food and Drug Administration (FDA) has promulgated pursuant to that Act.

Under this approach, the exemption for FDA-regulated products is available only to the extent that, and only so long as, the chemicals in question are manufactured, processed, or distributed solely for use in those products. If such chemicals have multiple functions, their non-FDA uses are subject to regulation under TSCA and they would be subject to PMN requirements if not included in the TSCA Inventory. As the Appendix to EPA inventory regulations explains:

> As discussed in response to comment 37, if a substance has multiple uses only some of which are regulated under the FFDCA, the manufacturing, processing, distribution, and use of the substance for the remaining uses comes within the jurisdiction of TSCA. 42 *Fed. Reg.* 64586.

Moreover, when a substance is manufactured, processed, or distributed for "undifferentiated" uses, EPA has indicated that the substance would be presumed to be subject to TSCA and, on this basis, will be subject to Section 5. 42 *Fed. Reg.* 64585.[7]

During the comment period on the proposed inventory regulations, concern was expressed by industry about the status of precursor, intermediate, and catalyst chemicals used in the manufacture of FD & C Act products but not necessarily contained in finished articles sold to consumers. In the Appendix to its final inventory regulations, EPA confirmed that such chemicals are within the coverage of the FD & C Act and, hence, exempt from TSCA:

> Comment 41: Intermediates and catalysts intended solely for use in the production of a food, food additive, drug, cosmetic, or device are excluded from regulation under TSCA.
> Response: The Administrator agrees with this comment. The definitions of the FFDCA provide that chemical substances which are intended for use as a component of a food, food additive, drug, cosmetic, or device are encompassed within the meaning of such terms, respectively. The FDA considers intermediates and catalysts to be such components. Therefore, they are subject to regulation under the FFDCA. Any such substance is excluded from regulation under TSCA insofar as it is actually manufactured processed or distributed in commerce solely for use in the production of a food, food additive, drug, cosmetic or device. 42 *Fed. Reg.* 64586.[8]

EPA regulations do not address the question of whether substances presently regulated under the FD & C Act are subject to TSCA to the extent that their effects on health and the environment cannot be controlled by the FDA. Conflicting arguments on this issue can be made. On the one hand, the statutory language clearly links the availability of the exemption to the use for which the substance is manufactured, processed, or distributed. Thus, a strong claim can be made that any substance used solely for a use enumerated in Section 3(2) should be immune from regulation for *all* purposes. On the other hand, based on the broad objectives of the Act, EPA could argue that the exemption is limited to those aspects of an exempt substance directly regulated under other laws, with EPA retaining power over any additional effects of the substance on health and the environment that those laws do not control.[9] Until TSCA is construed and applied, it is impossible to predict with assurance how this aspect of the FD & C Act exemptions will be resolved.

There are many gray areas in the FD & C Act definitions of "food", "drug", and "cosmetic". For example, whether packaging materials are "foods" or "food additives" may depend on whether there is

---

[7] While EPA use of the term "undifferentiated is somewhat unclear, it may be referring to the situation where a substance is used both for FD & C and non-FD & C products and its manufacturer makes no effort to distinguish between chemicals destined for these different uses, either by modifying the composition of the chemical, producing it with different equipment, or packaging and marketing it in different ways. So long as the substance is in fact used for some nonexempt purpose, it is inconsistent with the definition of "chemical substance" in Section 3 to presume that it is subject to TSCA and, hence, is reportable. The danger in the EPA approach is that it suggests an intent on the part of the EPA to regulate the manufacture of a chemical substance for all purposes, both exempt and nonexempt, when the manufacture of those chemicals is not "differentiated" according to end-use.

[8] The EPA approach is consistent with explicit legislative history indicating that Congress believed that drug precursors, intermediates, and catalysts were "components" of drug products and, hence, should be exempt from TSCA. See H. Rep. No. 94-1341, *supra.*

[9] It is possible, of course, that the FD & C Act, as presently construed and applied by FDA, now effectively covers all effects of any exempt substance on health and the environment. *See* Environmental Defense Fund, Inc. v. Matthews, 410 F. Supp. 336 (D.D.C. 1976).

migration from the packaging to the food itself — a complex technical question in the case of many individual products. Care should be exercised to assure that borderline chemical substances that may fall outside the definitional provisions of the FD & C Act are the subject of PMN submissions if they have not been reported for the Inventory.

2. Pesticides

Section 3(2)(B)(ii) exempts from the definition of chemical substance "any pesticide (as defined in the Federal Insecticide, Fungicide and Rodenticide Act) when manufactured, processed, or distributed in commerce for use as a pesticide."

EPA inventory regulations and proposed PMN regulations explicitly incorporate this exclusion for pesticides and provide that the term "pesticide" shall have the meaning contained in the Federal Insecticide, Fungicide, and Rodenticide Act (FIFRA) and the regulations issued by EPA thereunder.

To the extent a pesticide has nonpesticide uses, these uses (like the nonexempt uses of FD & C Act products) will be subject to TSCA. (*See* 42 *Fed. Reg.* 64585.) In addition, a chemical substance will not be classified as a "pesticide" until an application for an experimental use permit or an application for registration has been submitted. Before this stage has been reached, chemical substances used by pesticide manufacturers will be subject to TSCA, even though those substances are being used solely for research to develop pesticide products. 42 *Fed. Reg.* 64585.[10]

In contrast to comparable materials used to manufacture FD & C Act products, EPA has indicated that raw materials, intermediates, and inert ingredients produced or used in pesticide manufacture are either chemical substances or mixtures which can be regulated under TSCA. 42 *Fed. Reg.* 64585. The Appendix to the inventory regulations indicates that EPA has adopted this approach because a substance will not be a "pesticide" subject to FIFRA unless it is an active pesticide ingredient; mere use in pesticide manufacture is insufficient. *Id.*[11]

3. Miscellaneous Chemicals Regulated under Other Statutes

Section 3()(B) contains, and the Inventory and proposed PMN regulations incorporate, several other exemptions for chemicals regulated under other statutes. These chemicals include: any source material, special nuclear material, or by-product material as defined by the Atomic Energy Act of 1974 and regulations issued thereunder; any pistol, firearm, revolver, shells, or cartridges; and tobacco or any tobacco product, but not including any derivative products.[12] These chemicals, too, are exempt from premanufacture notification requirements.

D. Research and Development Chemicals

Under Section 5(h)(3) of TSCA, PMN requirements will not apply to any chemical substance which is (1) manufactured or processed only in small quantities; (2) solely for purposes of scientific experimentation or analysis, or chemical research in connection with the development of a product; or (3) if all persons engaged in such experimentation or research are notified of any risk to health which may be associated with the substance.

The Conference Report on TSCA notes that the statutory exemption for small quantities for purposes of scientific experimentation in Section 5(h)(3) was taken from the House bill. H.R. Rep. No. 94-1679, *supra*, at 71—72. The House Report, in turn, states that:

> In limiting the exemption to chemicals manufactured or processed in "small" quantities, your Committee recognizes that the term "small" cannot be viewed in an absolute sense. The amount of the chemical substance which must be manufactured or

---

[10] Nevertheless, as EPA points out, virtually all such substances will be considered to be produced "in small quantities for research and development" and, hence, would be exempt from both inventory reporting and premanufacture notification under Section 5, 42 *Fed. Reg.* 64585.

[11] EPA has also pointed to specific legislative history which purports to evidence a congressional intent to subject chemicals used in pesticide manufacture to TSCA unless they are active ingredients of the final pesticide. *See* Committee on Interstate and Foreign Commerce, 94th Cong., 2d Sess., Legislative History of the Toxic Substances Control Act, 232 (1976).

[12] The original March 9, 1977, inventory proposal elaborated on the tobacco exemption by providing that the exemption — "...extends only insofar as a chemical substance is manufactured, processed, or distributed in commerce as tobacco or a tobacco product such as cigarettes or cigars. A chemical substance, such as nicotine, which is derived from tobacco and is processed or distributed in commerce as an industrial or other commercial product would not be exempt from the definition of chemical substance." 42 *Fed. Reg.* 13122.

processed for research and analysis may vary in relationship to the kind of use for which the chemical substance is being developed. Obviously, a manufacturer must be able to produce a chemical in sufficient quantities during the developmental period to adequately test and evaluate the chemical for its intended use. For instance, laboratory reagents may be tested in terms of grams, while textile fibers or paper processing materials may have to be manufactured in much greater quantities in order to be adequately evaluated.

The Committee also recognizes that a manufacturer may not be able to fully evaluate a potential product in-house. For example, a manufacturer may have to use an outside testing laboratory or make the product available to a potential industrial user to complete the analysis or experimentation. The fact that the other industrial user may pay for the costs of the substance does not necessarily signal the end of the development period. So long as the purchaser is continuing the research and evaluation of the substance by individuals technically qualified to analyze and evaluate the physical, chemical and performance characteristics of the substance, the exemption continues to apply. H.R. Rep. No. 94-1341, *supra*, at 29—30.

EPA inventory regulations embody the current views of the Agency on the scope of the R & D exemption from Section 5.[13] Section 710.4(c)(3) exempts from inventory reporting "any chemical substance which is manufactured, imported, or processed solely in small quantities for research and development." The term "small quantities for research and development", in turn, is defined by Section 710.2(6) as:

. . . quantities of a chemical substance manufactured, imported, or processed or proposed to be manufactured, imported, or processed that (1) are no greater than reasonably necessary for such purposes and (2) after the publication of the revised inventory, are used by, or directly under the supervision of, a technically qualified individual(s).

In devising this test, EPA considered, and then expressly rejected, establishing upper numerical limits for the quantities of a substance devoted to research and development that would be deemed "small", 42 *Fed. Reg.* 64586. As EPA explained, the Agency "found that different values might have to be assigned for different groups of substances depending upon their physical/chemical characteristics and intended uses." At the same time, however, the note to Section 710.3(y) establishes a presumption that any chemical substance manufactured, imported, or processed in quantities of less than 1000 lb annually is devoted to research and development. Such a substance will be considered eligible for the Inventory only if the reporting company can certify that the substance is in fact used for some purpose other than research and development.

EPA has confirmed that substances distributed commercially will still be within the R & D exemption if their sole end-uses are research and development. In its original inventory proposal of March 9, 1977, EPA pointed out that the exemption from reporting applied to those chemical companies which "supply or sell chemical substances to chemical companies, universities, or other institutions solely for research and development purposes." 42 *Fed. Reg.* 13133. EPA has also confirmed that, to preserve their exempt status, research and development chemicals need not be confined to laboratory use. Rather, the "research" exemption covers all phases of product development, including product evaluation which may occur outside the manufacturer's facility, 42 *Fed. Reg.* 64586-87.

Under Section 710.2(y) of the inventory regulations, the exemption for research and development will remain valid only if the substance in question is used by, or directly under the supervision of, a "technically qualified individual". Such an individual, in turn, is defined by Section 710.2(aa) as a person:

. . . who because of his education, training, or experience, or a combination of these factors, is capable of appreciating the health and environmental risks associated with the chemical substance which is used under his supervision, (2) who is responsible for enforcing appropriated methods of conducting scientific experimentation, analysis, or chemical research in order to minimize such risks, and (3) who is responsible for the safety assessments and clearances related to the procurement, storage, use, and disposal of the chemical substance as may be appropriate or required within the scope of conducting the research and development activity. The responsibilities in clause (3) of this paragraph may be delegated to another individual, or other individuals, as long as each meets the criteria in clause (1) of this paragraph.

As this definition recognizes, the responsibility for overseeing the safe use of a research chemical may be delegated to other persons, providing those persons themselves meet the criteria for a "technically qualified individual".[14]

---

[13] Under Section 8(b) of TSCA, EPA may not include on the inventory "any chemical substance which is manufactured or processed only in small quantities . . . solely for purposes of scientific experimentation or analysis or chemical research on, or analysis of, such substance or another substance, including such research or analysis for the development of a product."

[14] As an example of such division of responsibility, EPA has suggested that "one person may be responsible for analyzing the properties of the chemical substance used as a glue to back a rug, while a second person may have responsibility for determining how to dispose of the rug samples that contain the experimental glue. Similarly, there may be a duly-authorized individual responsible for procurement of research chemical substances who is different from the technically qualified individuals who conduct the experiments with those chemical substances." 42 *Fed. Reg.* 64587.

The provisions concerning the R & D exemption in the EPA proposed inventory regulations are similar to the parallel provisions in the Agency inventory regulations. Thus, it can be expected that the approach embodied in those regulations will be largely carried forward in the Agency program for implementing Section 5. Nevertheless, in a number of respects, the EPA PMN proposal broadly expands the procedures for invoking the R & D exemption and arguably exceeds EPA authority under the Act.

Proposed Section 720.14(a)(2) of the EPA PMN regulations has the effect of conditioning the availability of the R & D exemption on proper notification by the manufacturer or importer of "all persons engaged in the manufacture, processing, use (including use in research and development), transport, storage or disposal of the substance of any risk to health which may be associated with the substance . . . ." This provision would appear inconsistent with Section 5(h)(3), which only permits EPA to require notification of "persons engaged in such experimentation, research, or analysis for a manufacturer or processor." Thus, the statute apparently does not require notification of persons engaged in transporting, storing, or disposing of the substance. Nor does it require the manufacturer to assume the burden of notifying *all* persons engaged in manufacturing, processing, use, or research and development involving the substance. Such an expansion of the exemption holder's notice responsibilities, moreover, would seem unwise as a matter of policy. The obligations of the R & D manufacturer are centered on the health and safety of those who are using the chemical at its direction; extending that obligation to unrelated persons would complicate and expand the activities of the R & D manufacturer and detract from the goal of fostering innovation which underlies the R & D exemption.

The second basic defect in proposed Section 720.14(a)(2) is that it requires notification of risks to health which "may be associated" with the substance. The statute, in contrast, requires notification for risks that the manufacturer "has reason to believe" may be associated with the substance. This standard is less amorphous and uncertain than that of the EPA provision and, thus, defines the obligations of the R & D manufacturer with more precision.

EPA has also arguably exceeded its statutory authority by broadly requiring under proposed Section 720.14(c) that the manufacturer or importer evaluate all information, as well as all test data, in his "possession or control" before making the notification. Section 5(h)(3) of the statute does not contemplate a file search encompassing all companies associated with the manufacturer or importer with respect to chemical substances shipped for research and development purposes, and EPA should not require one.

There is an even more serious defect in proposed Section 720.14(d). This provision requires the manufacturer to provide to EPA "any information" that it has evaluated in deciding to make the notification required by Section 5(h)(3) of the Act. The statute nowhere prescribes such a duty as a condition to the R & D exemption, and the creation of that duty would have the effect of opening the files of the manufacturer to EPA in any situation where the manufacturer has chosen to provide notification of the possible risks associated with an R & D chemical. If EPA feels it needs the information underlying that notification, it should proceed in accordance with the procedures and criteria for inspections and subpoenas under Section 11 of the Act.

E. Intermediates

The legislative history of the TSCA recognizes that Section 5 applies to "intermediates" — i.e., chemicals intentionally present during a manufacturing or processing operation but consumed or altered before that operation reaches its conclusion. (*See* H.R. Rep. No. 94-1431, *supra* at 30.) Nevertheless, under Section 5(h)(5), EPA may exempt such intermediates from premanufacture notification requirements when (1) they "exist temporarily as a result of a chemical reaction in the manufacturing or processing of a mixture or another chemical substance" and (2) they have no "human or environmental exposure".

In construing the term "intermediate", the consistent EPA approach under TSCA has been to exempt compounds from regulatory requirements when they are "non-isolated" — i.e., not intentionally removed from the equipment in which they are manufactured. The inventory regulations illustrate this approach. In Section 710.2(p) of its regulations, EPA has defined the term "manufacture for commercial purposes" to mean manufacture "for use as an intermediate." Nevertheless, in Section 710.2(h), the Agency has then defined "intermediate" as:

> . . . any chemical substance (1) which is intentionally removed from the equipment in which it is manufactured, and (2) which either is consumed in whole or in part in chemical reaction(s) used for the intentional manufacture of other chemical substances or mixture(s), or is intentionally present for the purpose of altering the rate of such chemical reaction(s).

To be reportable under this definition, an intermediate must be intentionally removed from the equipment in which it is manufactured.[15] Moreover, this removal must involve separating the intermediate from the chemical substance that it is used to produce. As the note to Section 710.2(n) indicates, the term "equipment in which it was manufactured" includes:

> . . . the reaction vessel in which the chemical substance was manufactured and other equipment which is strictly ancillary to the reaction vessel, and any other equipment through which the chemical substance may flow during a continuous flow process, but does not include tanks or other vessels in which the chemical substance is stored after its manufacture.

Thus, insofar as a chemical is not intentionally removed from the process in which it plays a part, it is not an "intermediate" for inventory reporting purposes.

Section 720.2 of proposed EPA PMN regulations contains a general definition of "intermediate" and a separate definition of "non-isolated intermediate". According to proposed § 720.13, non-isolated intermediates — i.e., those which are not intentionally removed from the equipment in which they are manufactured — will be exempt from premanufacture notification under Section 5. This approach is essentially an extension of EPA regulations governing reporting for the TSCA Inventory.

Nevertheless, in the preamble to its Section 5 proposal, EPA has indicated that it is considering narrowing the exemption of intermediates from premanufacture notification so it includes only those chemicals to which there is in fact no exposure. 44 *Fed. Reg.* 2248. Industry groups have vigorously opposed any such narrowing. As EPA itself has recognized, nonisolated intermediates are extremely difficult to identify. While miniscule amounts of these substances may occasionally escape from plant equipment, the resulting human and environmental exposure will be extremely small. Totally eliminating the possibility of any such exposure, however, will normally be impossible. The EPA suggested approach — identifying all non-isolated intermediates and reporting them under Section 5 — will thus impose substantial burdens and result in intensive regulatory requirements for chemicals which present little or no risk to health or the environment. Accordingly, there are important considerations which militate in favor of retaining the approach to "intermediates" embodied in EPA inventory regulations and PMN proposal.

F. Byproducts, Impurities and Other Noncommercial Chemicals

EPA has exempted a number of other chemical substances from inventory reporting and premanufacture notification requirements on the ground that they are not manufactured for distribution in commerce as chemical substances *per se* and have no commercial purpose separate from the substance, mixture, or article of which they are a part. These categories of chemicals are as follows:

1. Impurities

Impurities are excluded from inventory reporting by Section 710.4(d) of the EPA inventory regulations, and from premanufacture notification by Section 720.13(e) of the EPA proposed PMN regulations. Under Section 710.2(m) of the inventory regulations, "impurity" is defined as "a chemical substance which is unintentionally present with another chemical substance." This definition is carried forward in Section 720.2 of the proposed PMN regulations.

2. Byproducts

Section 720.13(3) of the EPA PMN proposal also excludes from PMN requirement "any byproduct." Under Section 720.2 of that proposal, a "byproduct" is defined as a "chemical substance produced solely without commercial intent during the manufacture or processing of another chemical substance(s) or mixture(s)."

Under this definition, a substance that was a waste material in one manufacturing process but which a company reused or recycled for another process would not be considered a "byproduct". Rather, under the EPA PMN proposal, such materials would be considered "coproducts", a term defined in proposed

---

[15] However, it should be noted that EPA proposed inventory regulations defined "intermediate" to include all chemical substances intentionally produced in the course of manufacturing other substances and mixtures when their presence is reasonably ascertainable and they could be isolated and identified under conditions practically encountered in the environment. *See* proposed Section 710.2(6); 42 *Fed. Reg.* 39191. Under this definition, manufacturers would have been responsible for reporting chemical substances that were merely "isolatable", even though they were never removed during the manufacturing process and, hence, were never encountered in the environment.

Section 720.2 as a "chemical substance produced for a commercial purpose during the manufacture, processing, use or disposal of another chemical substance(s) or mixture(s)."

There may be situations where a coproduct of the manufacturing processes of a company, while commercially useless to the company itself, may have value for municipal or private organizations engaged in reclaiming waste materials. Under Section 710.4(d)(2) of EPA inventory regulations, a coproduct which had commercial value only to these organizations which (i) burn it as a fuel, (ii) dispose of it as waste, including in a landfill or for enriching soil, (iii) extract component chemical substances which have commercial value, could have been reported for the inventory at the option of individual processors. This approach is carried forward in the EPA proposed PMN regulations. Under proposed Section 720.13(d), coproducts which meet these criteria are exempt from premanufacture notification requirements. Nevertheless, a manufacturer of such coproducts may submit a PMN if it chooses to do so. In this event, the coproduct in question will be added to the inventory and in the future would have the status of an existing chemical substance.

It is important to bear in mind the distinction between "intermediates" on the one hand, and "byproducts" and "impurities" on the other, since the two latter categories of substances are not reportable for the Inventory while the first is. Like an intermediate, a byproduct may be formed and then consumed during the reaction sequence which creates another chemical substance. However, as EPA has observed, a byproduct is unintentionally produced during the course of a reaction sequence, whereas an intermediate is an expected part of that sequence which performs a deliberate and necessary function in bringing it to completion. 42 *Fed. Reg.* 64588. Similarly, an impurity is a chemical substance which is unintentionally present with another chemical substance and does not play a necessary and intended role in its manufacture. Intermediates, in contrast, while often present as trace impurities in a final product, also serve an intentional purpose in the reaction sequence by which that product is created. *Id.*

3. Chemical Substances Derived from Reactions during the Exposure, Storage, or End-Use of Other Chemical Substances, Articles, or Mixtures

Sections 710.4(d)(3)-(5) of the inventory regulations and proposed Section 720.13(e) of the EPA PMN regulations exclude from reporting all chemical substances that result from chemical reactions that occur incidental to the exposure to environmental factors, storage, or end-use of another chemical substance, mixture, or article. These three classes of substances are described as follows:

> (3) Any chemical substance which results from a chemical reaction that occurs incidental to exposure or another chemical substance, mixture, or article to environmental factors such as air, moisture, microbial organisms, or sunlight.
>
> (4) Any chemical substance which results from a chemical reaction that occurs incidental to storage of another chemical substance, mixture, or article.
>
> (5) Any chemical substance which results from a chemical reaction that occurs upon end use of other chemical substances, mixtures, or articles such as adhesives, paints, miscellaneous cleansers or other housekeeping products, fuels and fuel additives, water softening and treatment agents, photographic, films, batteries, matches, and safety flares, and which is not itself manufactured for distribution in commerce or for use as an intermediate.

4. Reaction Products That Are Created during the Manufacture of an Article

Section 710.4(d)(6) of the inventory regulations and Section 720.13(e) of the PMN regulations exempt certain chemical substances that result from reactions which occur during the manufacture of articles. This category of substances is defined as follows:

> Any chemical substance which results from a chemical reaction that occurs upon use of curable plastic or rubber molding compounds, inks, drying oils, metal finishing compounds, adhesives, or paints; or other chemical substances formed during manufacture of an article destined for the marketplace without further chemical change of the chemical substance except for those chemical changes that may occur as described elsewhere in this § 710.4(d).

While somewhat unclear, this language seems to encompass substances that come into being during the finishing process of an article but experience no further chemical change. The Appendix to the final inventory regulations emphasizes that paragraph (d)(6) excludes from reporting only reaction products that exist solely during the manufacture of an article, not raw materials or processing aids purchased for use in the manufacturing process. According to EPA: "[d]yes and fire retardants are chemical substances manufactured for distribution in commerce as chemical substances, and therefore do have a separate commercial purpose . . . . The exclusion in § 710.4(d)(6) is for chemical substances formed when the dye or fire retardant reacts with fibers of the garment or other article upon end-use of those substances by a processor." 42 *Fed. Reg.* 64588.

5. Incidental Substances Produced When Processing Aids Function as Intended

Section 710.4(d)(7) of the inventory regulations and proposed Section 720.13(e)(7) of the EPA PMN proposal also exclude from reporting those chemical substances which are produced when certain processing aids intended solely to impart a specific physical-chemical characteristic to a product function as intended. This class of substances is described as follows:

> Any chemical substance which results from a chemical reaction that occurs when (i) a stabilizer, colorant, odorant, antioxidant, filler, solvent, carrier, surfactant, plasticizer, corrosion inhibitory, antifoamer or de-foamer, dispersant, precipitation inhibitor, binder, emulsifier, de-emulsifier, dewatering agent, agglomerating agent, adhesion promoter, flow modifier, pH neutralizer, sequesterant, coagulant, flocculant, fire retardant, lubricant, chelating agent, or quality control reagent functions as intended or (ii) a chemical substance, solely intended to impart a specific physico-chemical characteristic, functions as intended.

6. Legal Effect of EPA Exclusions

Under Section 5(i), as used in connection with premanufacture notification, the terms "manufacture" and "process" are defined to mean "manufacturing or processing for commercial purposes." Thus, if a chemical is not manufactured "for commercial purposes", it is automatically excluded from PMN requirements under Section 5.

Both in its inventory regulations and its proposed PMN regulations, EPA has taken the position that the above categories of chemicals are manufactured "for commercial purposes", and thus have been exempted from Section 5 solely as a matter of administrative discretion. In proposed regulations under Section 8(a), EPA has explained the legal basis for this conclusion as it applies to byproducts and impurities. According to EPA, whether or not such byproducts or impurities have commercial value in themselves, they are "nonetheless produced for the purpose of obtaining a commercial advantage since they are part of the manufacture of a chemical product for a commercial purpose." 45 *Fed. Reg.* 13646, 13656 (Feb. 29, 1980).

Portions of the TSCA legislative history, however, suggest that this expansive definition of "commercial purpose" is unwarranted. For example, the Senate Report on TSCA contains the following discussion of the "commercial purpose" requirement in Section 5:

> Three are mixtures such as adhesives, paints and inks which can produce chemical substances upon end use ... These types of substances would not be covered under the premarket notification provisions because they are not manufactured for a commercial purpose *per se*. S. Rep. No. 698, 94th Cong., 2d Sess. 19 (1976).

Thus, Congress clearly intended that chemicals created upon the end use of other substances, mixtures, or articles would not be considered manufactured "for commercial purposes."

Nor did Congress intend to include other incidentally produced substances, such as byproducts and impurities, within the scope of that term. The Senate Report continues:

> Manufacture is defined under section 3(a)(7) to mean to "import, produce, or manufacture for commercial purposes." *...[M]inor reactions occurring incidental to the mixing process or upon storage of a mixture such as the cross-linking of polymers, would not constitute a basis for subjecting such mixtures to the premarket notification provisions intended for new chemical substances because the resulting substances are not manufactured for commercial purposes.* S. Rep. No. 698, *supra*, at 19 (emphasis added).

The House Report contains a similar analysis:

> Any commercial purpose, such as use as a chemical intermediate in a manufacturing process, is sufficient to bring the manufacture or processing of a substance within the ambit of section 5. *The Committee realizes that there are certain minor reactions occurring incidental to the mixing process or upon storage of a mixture, such as the cross-linking of polymers.* Such a minor reaction may result in what would technically be considered a "new" chemical substance. However, *since the "new" substance is not manufactured for commercial purposes per se, it would not be subject to the notification provisions of this section.* H. R. Rep. No. 1341, *supra*, at 30—31 (emphasis added).

While these passages apply to chemicals produced incidentally during the storage or mixing of other chemicals, the reasoning that they embody would appear applicable to byproducts and impurities as well. Byproducts and impurities are not produced intentionally during a commercial process. Instead, they are either created incidentally during the manufacture or processing of other chemicals or are present during such activities as an unintended residue of a prior reaction sequence. Thus, it would seem equally improper to conclude that byproducts and impurities are manufactured for a "commercial purpose".

EPA insistence on treating byproducts and impurities as manufactured "for a commercial purpose" is presently of little practical significance because the Agency has decided to exempt these categories of chemicals from Section 5 as a matter of administrative discretion. Should EPA decide to subject some or all of these chemicals to Section 5 at some point in the future, however, there would be a sound basis for arguing that the Agency has exceeded its statutory authority.

G. Test Market Exemptions

Under Section 5(h)(1), EPA may exempt a new chemical substance from premanufacture notification requirements in order to permit the substance to be manufactured "for test marketing purposes." To qualify for such an exemption, the applicant must make a "showing . . . satisfactory to the Administrator" that its proposed test marketing activities "will not present any unreasonable risk of injury to health or the environment." EPA is empowered to impose any restrictions on the applicant's activities which it "considers appropriate." Under Section 5(h)(6), EPA must publish a *Federal Register* notice immediately upon receiving an application for a test marketing exemption (TME). After giving interested persons an opportunity to comment, EPA must act on the application within 45 days and announce its action in an additional *Federal Register* notice.

Section 710.2(bb) of the EPA inventory regulations and Section 720.2 of the EPA proposed PMN regulations define "test marketing" as follows:

> "Test marketing" means the distribution in commerce of no more than a predetermined amount of a chemical substance, mixture, or article containing that chemical substance or mixture, by a manufacturer or processor to no more than a defined number of potential customers to explore market capability in a competitive situation during a pre-determined testing period prior to the broader distribution of that chemical substance, mixture or article in commerce.

The distinction between "test marketing", which is subject to Section 5 unless specifically exempted, and "research and investigation", for which a specific exemption is unnecessary, is often a subtle one. In connection with the Inventory, EPA has indicated that this distinction should turn on "the greater degree of control maintained by the manufacturer and the greater technical qualification of those handling and supervising the use of the substance during the research and development phase." 42 *Fed. Reg.* 64588. Based on this approach, distribution of a product to consumers and other members of the public will be considered "test marketing", as opposed to "research and investigation", when use and disposal of the product have been removed from the direct supervision of the manufacturer.

Section 720.15(b) of the EPA proposed PMN regulations governs the submission of applications for test marketing exemptions. This provision requires the submission of extensive information regarding the chemical substance for which an exemption is sought, including all existing data regarding its health and environment effects, the maximum quantity of the substance to be manufactured for test marketing purposes, the maximum number of persons that may receive the substance for those purposes, and the maximum number of articles containing the substance that will be distributed during test marketing activities.

Not surprisingly, industry representatives have criticized these requirements as unduly onerous. EPA itself has stated that qualifying for a TME "will require the same information, and impose similar reporting burdens upon manufacturers, as premanufacture notices." 44 *Fed. Reg.* 2247—2248. Even if EPA grants a TME, moreover, the manufacturer will be required to submit a normal PMN before commencing full-scale commercial manufacture of the chemical in question. For this reason, and because of the relatively stringent requirements of the EPA proposed regulations, one would expect relatively few TME applications. Accordingly, it is surprising that, as of July 31, 1982, EPA had received 136 TME applications.

On several occasions, EPA has exercised its statutory authority to impose conditions on the manufacture of the TME substance. For example, EPA has limited TMEs only to the applicant manufacturer; limited customers for the TME substance to those identified in the TME application; required the TME applicant to maintain production and shipment records for the TME chemical and to provide those records to EPA upon request; required the applicant to inform all customers of the restricted uses of the TME chemical; and limited the amount of the chemical which may be manufactured.

In addition, EPA has criticized TME applicants for failing "to provide sufficient information and data in exemption applications for EPA to make the required finding that the test marketing will not present any unreasonable risk of injury to health or the environment." 45 *Fed. Reg.* 37520, 37521 (June 3, 1980). Where inadequate information is provided, EPA has announced it "may decide to deny the applications, because to approve an application the Agency must make an *affirmative* finding that the test marketing activities will not present any unreasonable risk of injury to health or the environment." 45 *Fed. Reg.* at 37522 (emphasis in original). As EPA has stated, "if EPA has significant uncertainty concerning the risks presented by the test marketing activities due to a lack of data or information in the application, the Agency will not approve the application." *Id.*

III. PURPOSES AND POLICIES OF SECTION 5

Since the enactment of the TSCA, there has been considerable debate concerning the purposes and objectives of Section 5 and the scope of the EPA authority in implementing that Section. On the one hand, it is clear that Congress attached considerable importance to scrutinizing the health and environmental effects of chemicals prior to manufacture in order to anticipate and prevent serious harm before it could occur. On the other hand, the legislative history of the TSCA demonstrates that Section 5 was intended to be a *notification* provision which alerted EPA to the impending manufacture of a new chemical — not a *safety-substantiation* provision which required manufacturers to submit conclusive proof that every new chemical was safe. That there are tensions between these goals is clear from the legislative debate preceding passage of the TSCA, which was marked by protracted controversy concerning the scope of EPA powers under Section 5.

A. Preventive Purposes of Section 5

The House Report on TSCA reflects recognition by Congress that, in the past, too many chemicals were introduced into commerce without adequate scrutiny of their health or environmental effects:

> This vast volume of chemicals have, for the most part, been released into the environment with little or no knowledge of their long-term health or environmental effects. As a result, chemicals currently in commercial and household use are now being found to cause or contribute to health or environmental hazards unknown at the time commercial use of the chemical began. For example, vinyl chloride was the 23rd most produced chemical when it was discovered to cause cancer, and the chemical has now been implicated as causing birth defects as well. Asbestos, widely used in items ranging from talcum powder to brake linings to wallboard, is now known to cause cancer and other debilitating illnesses. However, such effects were not discovered until hundreds of workers had developed a rare form of lung cancer as a result of exposure to the substance. Polychlorinated biphenyls (PCBs) had been used for forty years and approximately 390,000 tons had been released into the environment before they were recognized as an enduring environmental poison. Unfortunately, such recognition came too late to prevent contamination of such major water systems as the Great Lakes and the Hudson River. *As the preceding examples indicate, it is often many years after exposure to a harmful chemical before the effects of its harm become visible. By that time it may be too late to reverse those effects. As indicated, hundreds of people may have been exposed to a carcinogen or an entire river system may have been polluted.* H.R. No. 94-1341, *supra*, at 3.)

The Senate Report emphasizes a similar point:

> In order to protect against these dangers, the proposed Toxic Substances Control Act would close a number of major regulatory gaps, for while certain statutes, including the Clean Air Act, the Federal Water Pollution Control Act, the Occupational Safety and Health Act, and the Consumer Product Safety Act, may be used to protect health and the environment from chemical substances, *none of these statutes provide the means for discovering adverse effects on health and environment before manufacture of new chemical substances.* Under these other statutes, the Government regulator's only response to chemical dangers is to impose restrictions after manufacture begins.
>
> The most effective and efficient time to prevent unreasonable risks to public or the environment is prior to first manufacture. It is at this point that the costs of regulation in terms of human suffering, jobs lost, wasted capital expenditures, and other costs are lowest. Frequently, it is far more painful to take regulatory action after all of these costs have been incurred. S. Rep. No. 94-698, *supra*, at 4 (emphasis added).

The Conference Report explains how these concerns would be addressed by premanufacture notification:

> Section 5 sets out the notification requirements with which manufacturers of new chemical substances and manufacturers and processors of existing substances for significant new uses must comply. The requirements are intended to provide the Administrator with an opportunity to review and evaluate information with respect to the substance to determine if manufacture, processing, distribution in commerce, use or disposal should be limited, delayed or prohibited because data is insufficient to evaluate the health and environmental effects or because the substance or the new use presents or will present an unreasonable risk of injury to health or the environment.

*The provisions of the Section reflect the conferees recognition that the most desirable time to determine the health and environ-
mental effects of a substance, and to take action to protect against any potential adverse effects, occurs before commercial production begins. Not only is human and environmental harm avoided or alleviated, but the cost of any regulatory action in terms of loss
of jobs and capital investment is minimized.* For these reasons, the conferees have given the Administrator broad authority to act
during the notification period. H.R. Rep. No. 94-1679, *supra*, at 65 (emphasis added).

B. Limits on Statutory Powers of EPA

At the same time that it emphasized these concerns, however, Congress made it clear that Section 5
was not intended to function as a system of intensive premarket clearance for new chemicals, but merely
required EPA to be informed of such new chemicals before manufacture so the Agency could take regulatory
action in those cases where it was appropriate.

In contrast to other regulatory statutes, Section 5 does not authorize EPA to "approve" new chemical
substances. Nor does it require manufacturers and importers of such substances affirmatively to demonstrate
their safety.[16] Rather, the only precondition to beginning manufacture under Section 5 is the submission
of a notice to the Agency at least 90 days prior to the start of production. Under Sections 5(d)(1) and
8(a)(2), this notice must merely provide EPA with certain limited information identifying the substance
and include existing data relating to its health and environmental effects. A specific requirement for the
development of new data concerning the substance is the infrequent exception rather than the norm.

Once a PMN has been received, EPA does not have an indefinite period to review the substance. The
Agency must act within 90 days or, if there is "good cause" to extend this review period, within a
maximum of 180 days. At the end of this period, manufacture of the substance may proceed unless the
Agency has taken affirmative steps to prevent it. If the Agency elects to take action respecting the chemical,
it must either seek an injunction or issue an order prohibiting manufacture under Sections 5(e) or 5(f) or,
pursuant to Section 5(f), propose a rule under Section 6(a) which applies immediately. Under either
provision, the burden is on EPA to justify its action. Moreover, both Sections 5(e) and 5(f) afford manufacturers and importers considerable opportunity to contest the EPA actions. Thus, the purpose of a PMN
is not to provide "safety substantiation" for a new chemical substance. Rather, the notice essentially alerts
EPA that the manufacture of the chemical is imminent and the burden then shifts to the EPA to show why
manufacture should not proceed.

C. Relationship between Section 5 and Other Provisions of TSCA

The interrelationship between Section 5 and other provisions of the Act further illuminates the role that
premanufacture notification plays in the total statutory scheme. Section 4 of the Act provides EPA with
authority to require testing of chemical substances which it believes may have adverse effects on health
or the environment. Section 6 of the Act authorizes EPA to prohibit or limit the manufacture of substances
which present an "unreasonable risk of injury", and Section 7 empowers the Agency to take immediate
action against those substances which pose an "imminent hazard". According to the Senate Report, these
Sections include "the major regulatory provisions" of the Act. S. Rep. No. 94-698, *supra*, at 10.

Moreover, the Agency has broad power under the Act to obtain additional information about a chemical

---

[16] In contrast with TSCA, Congress has on several occasions enacted statutes which explicitly require manufacturers to submit test
data substantiating the safety of their products before marketing can commence. For example, Section 505 of the FD & C Act,
21 U.S.C. § 355(a), provides that no new drug may be sold "unless an approval of an application. . . is effective with respect
to such drug." To obtain such approval a new Drug Application (NDA) must contain "full reports of investigations which have
been made to show whether or not such drug is safe for use and whether such drug is effective in use . . . ." 21 U.S.C. §
355(b)(i). FDA may refuse to grant approval to a proposed new drug when, among other things, the manufacturer's tests "do
not show that such drug is safe for use" (21 U.S.C. § 355[d][2]), or when those tests lack "substantial evidence that the drug
will have the effect it purports or is represented to have." 21 U.S.C. § 355(d)(5). Thus, the FD & C Act squarely requires the
manufacturer to develop affirmative proof that a new drug is safe and effective. Unless the manufacturer bears that burden, the
NDA must be denied.

The Federal Insecticide, Fungicide, and Rodenticide Act (FIFRA), employs a similar approach for the regulation of new
pesticides. Under Section 12, 7 U.S.C. § 136j, it is unlawful to sell any pesticide which is not registered pursuant to Section 3.
In turn, Section 3, 7 U.S.C. § 136a, requires all applicants for registration to file a statement which supports the safety of their
products and includes relevant test data. EPA is explicitly empowered to publish guidelines which specify the test data which
are necessary for registration and the agency may deny registration when, based on these data, it determines that the manufacturer
has failed to demonstrate that the pesticide "will perform its intended function without unreasonable adverse effects on the
environment." Thus, as in the new drug provisions of the FD & C Act, the burden of proof rests with the manufacturer. Unless
it has developed and submitted data demonstrating that the pesticide is safe, the Agency must withhold approval of its use.

substance after the commencement of manufacture. Section 8 of the Act permits EPA to promulgate rules which require manufacturers and processors to maintain specified records and report specified information. Under Section 8(c), the Agency may require the maintenance of records of significant adverse reactions. Section 8(d) permits EPA to issue rules which require the submission of lists of health and safety studies as well as copies of those studies themselves. Under Section 8(e), manufacturers and processors of a chemical substance are required to inform EPA of all information which reasonably supports the conclusion that the substance presents a "substantial risk of injury." Finally, Section 11 of the Act gives EPA power to conduct inspections and issue subpoenas.

As this enumeration of EPA powers under TSCA demonstrates, the Agency has several statutory mechanisms for acquiring further information about the uses and effects of a new chemical substance after its manufacture has begun. It also has several mechanisms for taking regulatory action if subsequent information indicates that a new chemical substance has harmful effects on health or the environment. Thus, Section 5 is not the exclusive method for controlling new chemical substances.

Moreover, TSCA is itself only one of several statutory schemes for regulating the effects of chemicals on health and the environment. EPA is empowered to eliminate contaminants from the national water supply under the Safe Drinking Water Act, 42 U.S.C. § 300f. EPA may limit or prevent air and water discharges of chemicals under the Clean Air Act, 42 U.S.C. §§ 1857, *et seq.*, and the Federal Water Pollution Control Act, 33 U.S.C. §§ 1251, *et seq.* Under the recently enacted Resource Conservation and Recovery Act, 42 U.S.C. § 6974, moreover, EPA may control the treatment, storage, or disposal of solid wastes.

Other regulatory agencies are authorized to control additional facets of the manufacture and use of chemicals. The Consumer Product Safety Act, 15 U.S.C. §§ 2049, *et seq.*, for example, empowers the Consumer Product Safety Commission to limit or eliminate unsafe chemicals which are contained in consumer products. The Occupational Safety and Health Administration, exercising its powers under the Occupational Safety and Health Act, 29 U.S.C. § 651, has comparable responsibilities in the area of work place safety. Under the Hazardous Materials Transportation Act, 49 U.S.C. § 1801, the Department of Transportation is authorized to control the movement of potentially unsafe chemicals from one location to another. This comprehensive system of statutes and agencies further underscores the limited nature of Section 5 — and TSCA generally — in controlling new chemical substances.

D. Legislative History of TSCA

The function which premanufacture notification was intended to serve in the total statutory scheme is also demonstrated by the legislative history of TSCA. Bills to regulate toxic substances were passed by both houses of Congress in 1972 and 1973 but were never enacted into law. A major reason for the failure of those Congresses to enact toxic substances legislation was persistent controversy over the scope of the TSCA premanufacture review provisions. As described below, TSCA evolution through the 92nd, 93rd, and 94th Congresses was marked by growing opposition in both houses to comprehensive premanufacture screening. As Congress continued to consider TSCA, increasing support emerged for a system of premanufacture notification which placed the burden on EPA to regulate selectively those new chemicals which present a basis for concern.

Section 103 of S. 1478, 92d Cong., 2d Sess. (1972), the toxic substances legislation passed by the Senate in 1972, would have required EPA to issue regulations to prescribe uniform test protocols for all chemical substances. At least 90 days before commencing manufacture, Section 103 of the bill would then have required all manufacturers of new chemical substances to provide EPA with test data developed pursuant to those regulations. The Senate Report on S. 1478 leaves no doubt that the purpose of this provision was to require across-the-board approval of all new chemicals before they were introduced into commerce. According to the Report, one objective of S. 1478 was "that new chemical substances should be tested by their manufacturer prior to their commercial production and that the test results should be reviewed by the Environmental Protection Agency prior to such production." S. Rep. No. 92-783, 92d Cong., 2d Sess. 15 (1972). The Report stated that "there is a compelling need. . . to provide for a regulatory review of test results prior to the entry of [a new] substance into the market." *Id.* at 16.[17]

---

[17] Prior toxic substances bills were based on the concept of "premarket notification", a concept which became transformed into "premanufacture notification" as the legislation evolved. Portions of the text which discuss these earlier bills will refer to "premarket notification".

Important members of the Senate spoke out against the concept of premarket clearance embodied in these provisions. Senator Baker was particularly emphatic:

> Mr. President, it is this fine balance between the appropriate control of the use of chemical substances, but without impeding their beneficial uses, to which the President referred which is of concern to me and which prompted me to write supplemental views in the report on this legislation and to offer two amendments to the bill on the floor. Specifically, I am concerned over the inherent danger made possible by the broad language of section 103 and by the premarket screening provisions of section 104, which might have the result of depriving the public of the benefit of many new chemical products and of stifling development, particularly on the part of small chemical companies, of new products which may pose no hazard to health or the environment. 118 Cong. Rec. 19157—19158 (May 30, 1972).

As Senator Baker stressed, premarket clearance would discourage the introduction of important new chemicals:

> Mr. President, I think the burden of the debate can best be summarized by saying that the question at hand is: As a matter of public policy should a substance be proven innocent before it can be marketed or should it be proven guilty before it can be marketed? The bill as drafted requires that a substance be proven innocent before it can be marketed. That might be simple, but we are dealing with at least 2 million chemical substances, and to prove innocence, even by the most modern, efficient proceedings, and even by the most premiere and efficient Washington bureaucracy, is overwhelming. So proving innocent before a chemical can be marketed is not practical and it is the reason for this amendment.

> *   *   *

> The variation, inventiveness, and resourcefulness of the human mind, of industry the future of our Nation depends on that inventiveness and that we will want to stimulate the development of new chemicals and not restrict that development. But if we are going to say that no chemical substance can be marketed until it is proven innocent, we will deprive ourselves of many chemicals. I doubt many chemicals that are in use today could be proven innocent. I wonder if the burden of proving innocent or free of environmental consequences, gasoline, for instance, could be carried by this bill. *Id.* at 19160.

Senator Spong expressed similar views:

> After hearings on the bill and on the amendments which I introduced, I became convinced that the weight of the evidence confirmed the fears of many in the chemical industry, that bureaucratic delay could hamper them if explicit approval was required, that there would be a tendency on the part of approving something if there were any doubt at all. *Id.* at 19158.

While these sentiments did not prevail in the Senate in the 92nd Congress, they had considerable impact on the thinking of the House. After its passage by the Senate, S. 1478 was reshaped in the House and, as revised, embodied a dramatically different approach to premanufacture review. Under Section 5 of S. 1478, 92nd Cong., 2nd Sess. (1972) as enacted by the House in 1972, EPA was required to publish a list of chemical substances likely to pose "substantial danger to health or environment." If, but only if, a new chemical substance appeared on this list, the manufacturer would be required to submit data substantiating its safety to EPA. New chemical substances which were absent from the list would not be subject to any premanufacture notification at all.

The reasons for the highly selective approach adopted by the House are articulated in the House Report. The Report emphasizes the drawbacks of a comprehensive scheme of premarket screening:

> The committee has carefully limited the scope of this Section to respond to fears, expressed at the hearings by Russell Train, Chairman of the Council on Environmental Quality, that requirements for premarket screening of the full spectrum of chemical substances may entail "stifling and cumbersome administrative procedures and dilute the ability of the Administrator to effectively and efficiently regulate the most significantly hazardous materials." By restricting the pre-market screening provisions to those substances likely to pose *substantial* danger to health or environment, the bill will enable the Administrator to focus his attention and resources on the most serious and dangerous problems without requiring him to dilute his effort by dealing with the whole range of chemical substances. H. Rep. No. 92-1477, 92d Cong., 2d Sess., 8—9 (1972) (emphasis in original).

The aversion of the House to premarket clearance was motivated in part by the views of spokesmen from the Council on Environmental Quality and EPA. Mr. Ruckleshaus, EPA Administrator, expressed the matter as follows in explaining why government registration of all chemicals was unwise:

> This proposal was considered extensively and rejected as being both unnecessary because a number of new compounds are minor variants of existing compounds known to be safe, and a cursory review of such new compounds often would be sufficient to indicate their safety. It was considered unworkable at least at the present time, because based on the experience with the pesticide

and drug registration systems it is likely that a new registration system dealing with a large number of very heterogeneous substances would deteriorate into a bureaucratic quagmire. Environmental Impact Statement for the Toxic Substances Control Act of 1971, March 13, 1971.

These views were repeated when toxic substances legislation was again proposed during the 93rd Congress. As passed by the Senate, Section 5 of S. 426 required premarket testing of new chemical substances only if those substances were already subject to test requirements under Section 4. In all other instances, the manufacturer was merely required to submit a notice containing limited identifying information concerning the chemical that it intended to produce.

The reasons for this change in the Senate approach are clearly discernible from the legislative record. First, the Administration introduced toxic substances legislation, S. 888, which lacked *any* premanufacture notification requirement. Equally important, while supporting the concept of premanufacture notice, EPA again strongly opposed premarket approval of new chemical substances prior to their introduction into commerce. The rationale for the EPA position was presented in a lengthy letter to Senator Magnuson, dated June 18, 1973, which argued that broad premarket screening requirements were undesirable "partly because of the burden they would impose in connection with the production and distribution of chemicals, and partly because of the unwarranted administrative burden they would impose on this Agency."[18] The letter stresses the importance of selectivity in regulating new chemical substances as follows:

> We prefer the more selective provisions of S. 888 which would allow the administrator to focus all of his resources and attention on those substances or classes which we suspect may pose health or environmental hazards. *Id.* at 46.

The impact of these concepts on Senate thinking is evident in the Senate Report on S. 426 as finally adopted. The Report explained the Senate shift to a far more modest system of premanufacture review as follows:

> Concern was expressed to the committee by the chemical industry that S. 426 as originally introduced would have a severe impact on the chemical industry. . . .
> As reported by the committee, the bill contains safeguards against this possibility. First, EPA would be directed to routinely require the submission of test results prior to commercial production only for those new chemical substances which there is reason to believe may pose unreasonable threats to human health or the environment. For these chemicals, EPA must make a prior determination that unreasonable risk is at least suspected. The data would be submitted in accordance with testing procedures developed by EPA.
> Manufacturers of all other new chemical substances which are not so specified would only be required to give notification to EPA ninety days in advance of commercial production. *Id.* at 4.

In keeping with these concerns, the bill passed by the House in 1973 contained even greater limitations on the scope of premanufacture review. Under Section 5 of H.R. 5356, 93d Cong., 1st Sess. (1973), as under the prior bill enacted by the House, premanufacture screening was required only for those chemical substances which were "substantially dangerous." The remainder were not subject to premanufacture notification at all.

Thus, when toxic substances legislation was considered again in the 94th Congress, there was a consensus in both houses that comprehensive premarket screening of all new chemical substances was undesirable. Accordingly, Section 5 of S. 776, 94th Cong., 1st Sess. (1976), much like the final Act, required the development of test data only for new substances covered by a rule under Section 4 and the submission of a brief identifying notice for all other new substances. Legislation proposed in the House embodied an even more narrow concept of premanufacture notification. For example, under H.R. 7229, 94th Cong., 1st Sess. (1975), introduced by Mr. Eckhardt, submission of a premanufacture notice was necessary only for those new substances included on the EPA list of chemicals which "may pose an unreasonable risk." Of these substances, premarket testing was required only when a Section 4 rule was in effect. New substances omitted from the EPA list could be manufactured without *any* prior notice to EPA.

H.R. 14032, the bill finally enacted by the House, expanded on H.R. 7229 and H.R. 7664 by requiring premanufacture notification for all new chemical substances. However, during hearings before the House Subcommittee on Consumer Protection and Finance, EPA representatives were sharply questioned about the need for premanufacture notification concerning all new chemical substances. In response, they assured Subcommittee members, particularly Congressmen Eckhardt and McCollister, that premanufacture noti-

---

[18] The letter is reprinted in S. Rep. No. 93-254, 93d Cong., 1st Sess. 45, 47 (1973).

fication would place insignificant burdens on most manufacturers. For example, Deputy Administrator Quarles testified that:

> We do not feel that we are dealing with a situation that is just totally out-of-hand and that there are unlimited numbers of chemicals that pose a terrible and imminent threat to public health.
>
> The need here is to be able to identify those chemicals that really pose serious problems and go after them with requirements of testing and if appropriate, requirements of control. Hearings on H.R. 7229, H.R. 7548, and H.R. 7664, before the Subcommittee on Consumer Protection and Finance of the House Committee on Interstate and Foreign Commerce, 94th Cong., 1st Sess. 212—213 (1975).

In response to Congressman Eckhardt's concern that across-the-board premanufacture notification might impose serious "administrative burdens", Mr. Quarles stated that:

> [W]e do not have the manpower to look closely at every new chemical that is going to come out and I am certain that in the practical administration of the program, we would need to determine some general criteria for selecting certain chemicals that are coming on the market for really extensive scrutiny. *Id.* at 218.

Later in the questioning, Congressman McCollister described the EPA concept of premanufacture notification as one where:

> [A]nything [would be] premarket screened but that premarket screening for a great number of chemicals is going to be very perfunctory and it is that other group that he is going to get down to that they are going to give more attention to. *Id.* at 221.

Mr. Quarles agreed that Congressman McCollister's description was "helpful" and emphasized that "[t]here is a distinction between screening and notification." *Ibid.*

Further questioning by Congressman Eckhardt elicited additional statements from EPA. Terming universal premanufacture screening "pretty drastic", Mr. Quarles indicated that he would only have "no objection at all to notification which does not tie up distribution and manufacture." *Id.* at 222. Attempting to allay this concern, Mr. Quarles indicated that, for the great majority of new chemicals, the manufacturer would notify EPA of basic identifying information and manufacture would routinely commence after 90 days:

> Mr. Quarles. Well, as I have understood these provisions right as they stand now they do not tie up that manufacturer. In other words, under section 5 there is a prohibition on manufacturing until the notice is given. That is just another way of saying that you must give the notice 90 days in advance of the time when the manufacturer intends to begin production.
>
> But the manufacturer having given that notice does not have to wait until EPA gives them a green light to go ahead. The manufacturer is free to go ahead. So if the bureaucracy bogs down and does not act on the subject or does not come to a decision, the manufacturer is not held back.
>
> Mr. Eckhardt. That is right.
>
> Mr. Quarles. *So really what we are trying to talk about here is notification.* We feel that there can be real advantages to having a Government agency receiving notifications of all the new chemicals that are going into the society. There are important advantages in being able to look across the board at all of these, having some awareness of the new chemicals, and then being able to look closely into certain ones that for one technical reason or another raised some warning flag.
>
> *But that notification is essentially all that we are asking for in regard to the vast array . . .*
>
> *Then we come to the process of scrutinizing a chemical and determining that testing is required. Those are separate steps and we would have to act individually in those cases and act on a basis that could withstand judicial review. Ibid.* (emphasis added).

After further discussion, Mr. Quarles reiterated this point:

> The point is you really, at this stage, are talking about notification and even though the section is entitled screening, you are really talking about notification, because what that provision really does is to simply require a manufacturer, before he goes into production, to give EPA notice that he intends to do so with the description of whatever figure he may have.
>
> *So this is a disclosure provision. It is not immediately regulatory.* To get to the regulatory provisions, then we would have to go on to the next section or wherever it is where there is authority in the Administrator to begin imposing some requirements. Either for more testing, or in case it is warranted, some restrictions on production. And that would require the foundation. *Id.* at 229 (emphasis added).

To insure that there would be no misunderstanding, Mr. Quarles reemphasized the minimal nature of premanufacture notification for most new chemicals.

> I think as we wrestle with what notification should be required, we have to keep asking ourselves what is the consequence of requiring the notification. *In our judgment, the consequence is — and should be — fairly minimal. It really just gives us a chance to get started with our analysis. Id.* at 230—31 (emphasis added).

In accordance with these EPA assurances, Congressman Eckhardt later emphasized to the full House that, while requiring notification for all new chemical substances, the version of Section 5 adopted by the Committee accommodated "the concern of industry that we not adopt such a busybody policy as to interminably investigate, thus inordinately delaying the final decision so that chemicals could not flow to the marketplace." 122 Cong. Rec. H. 8811—8812 (August 23, 1976). According to the Congressman, the Committee had explicitly rejected the concept that "no chemical could go into the market unless it had gone through the governmental screen." *Ibid.*

Thus, in shaping Section 5, Congress recognized the pitfalls of requiring full safety substantiation and protracted premarket review for every new chemical substance. While it considered Section 5 an important "advance warning signal" for chemicals with potential adverse effects on health for the environment, it wished to minimize the delays and burdens entailed by premanufacture review. Reflecting this concern, the statutory scheme that Congress finally devised is one which gives EPA broad power to delay manufacture where a chemical may pose harm to health or the environment, but requires the Agency to exercise that power on a highly selective basis by minimizing its demands on PMN submitters and utilizing other regulatory provisions of TSCA where appropriate.

IV. TESTING OF NEW CHEMICALS

Perhaps the most controversial issue which has emerged during the EPA implementation of Section 5 is the nature and extent of the testing which new chemicals should undergo. The following discussion examines various legal, policy, and scientific aspects of premanufacture testing. That discussion (1) considers whether EPA can prescribe mandatory testing requirements under Section 5, (2) explores the pros and cons of EPA's voluntary testing "guidance" under Section 5, (3) analyzes the potential use of Section 4 test rules which are applicable to categories that include new chemicals, (4) discusses the potential use of the Section 5(b)(4) "risk list" to require data development for new chemicals, and (5) examines the situations in which Section 5 obligates the PMN submitters to provide existing test data to EPA.

A. EPA Lack of Statutory Authority to Require Premanufacture Testing

Both the language and legislative history of TSCA demonstrate that, except in certain limited circumstances, the PMN submitter has discretion to determine how much testing a new chemical substance will receive. If EPA believes that the submitter's test data are insufficient, its remedy is to invoke Section 5(e), under which the Agency may postpone manufacture under certain circumstances pending further testing of the new chemical. EPA itself has recognized that it "does not have authority to require premanufacture testing of all new chemical substances under Section 5." 44 *Fed. Reg.* 16240, 14243 (March 16, 1979).

1. Contrasts between TSCA and Other Statutes That Require Testing

It is instructive to contrast TSCA with other statutes which explicitly require manufacturers to submit test data substantiating the safety of their products before marketing can commence. For example, Section 505 of the FD & C Act, 21 U.S.C. 1 355(a), provides that no new drug may be sold "unless an approval of an application . . . is effective with respect to such drug." To obtain such approval, a New Drug Application (NDA) must contain "full reports of investigations which have been made to show whether or not such drug is safe for use and whether such drug is effective in use . . . ." 21 U.S.C. 1 355 (b)(i). FDA may refuse to grant approval to a proposed new drug when, among other things, the manufacturer's tests "do not show that the drug is safe for use," 21 U.S.C. 1 355(d)(2), or when those tests lack "substantial evidence that the drug will have the effect it purports or is represented to have." 21 U.S.C. 1 355(d)(5). Thus, the pharmaceutical manufacturer is explicitly required to develop affirmative proof that a new drug is safe and effective.

FIFRA employs a similar approach for the regulation of new pesticides. Under Section 12, 7 U.S.C. § 136j, it is unlawful to sell any pesticide which is not registered pursuant to Section 3. In turn, Section 3, U.S.C. § 137a, requires all applicants for registration to file a statement which supports the safety of their products and includes relevant test data. EPA is explicitly empowered to publish guidelines which specify the test data which are necessary for registration. The Agency may deny registration when, based on these data, it determines that the manufacturer has failed to demonstrate that the pesticide "will perform its intended function without unreasonable adverse effects on the environment." Thus, as in the new drug provisions of the FD & C Act, the burden of proof rests with the manufacturer to develop and submit data demonstrating that the pesticide is safe.

Although it describes premanufacture notification in great detail, Section 5 of TSCA contains no provision which requires the testing of all new chemical substances. Moreover, the provisions of Section 5 which relate to the submission of test data suggest that the PMN submitter will normally be free to determine how much testing a new chemical will receive.

Section 5(d) of TSCA specifies two circumstances where a PMN must contain test data relating to health and environmental effects of a new substance:

(1)The notice required by subsection (a) shall include —

*   *   *

(B) in such form and manner as the Administrator may prescribe, *any test data in the possession or control of the person giving such notice* which are related to the effect of any manufacture, processing, distribution in commerce, use, or disposal of such substance or any article containing such substance, or of any combination of such activities, on health or the environment, and

(C) a description of any other data concerning the environmental and health effects of such substance, *insofar as known to the person making the notice or insofar as reasonably ascertainable.* (Emphasis added.)

These requirements encompass only test data that (1) either exist already or (2) the manufacturer has voluntarily chosen to generate. There is no reference to new test data developed in accordance with criteria established by EPA — an omission which demonstrates that, in the first instance, manufacturers have discretion to determine when test data should be developed for new chemicals and how extensive those data should be.

Two provisions of Section 5 specify when a PMN *must* contain new test data. First, Section 5(b)(1)(A) provides that such test data will be required when a test rule under Section 4 is applicable to the substance for which the PMN is submitted. In this event, the PMN submitter will be responsible for conducting the testing that the EPA test rule prescribes. Second, under Section 5(b)(2), test data must be included in the PMN when EPA has included the substance on a list of chemical substances which present or may present an unreasonable risk of injury. Manufacturers of new substances on this "risk list" must submit data which affirmatively show that no unreasonable risk of injury exists. Having identified two classes of substances for which testing would be specifically required, Congress could easily have added a broad provision authorizing EPA to adopt testing requirements for all other new substances if it so desired. The absence of this requirement indicates that Congress intended to limit mandatory testing to those new substances which have been expressly selected for close scrutiny in the limited circumstances specified in Sections 5(b)(1)(A) and 5(b)(2).

Yet another provision of TSCA which points in the same direction is Section 4(g). Under this provision, a chemical manufacturer which plans to submit a PMN under Section 5 may petition EPA to prescribe standards for the development of test data for a particular new chemical. If testing criteria were already in place which applied to all new chemical substances, Section 4(g) of the Act would become academic. The very existence of this provision thus demonstrates that Congress believed that testing decisions for new chemicals would ordinarily be committed to the discretion of the manufacturer and that EPA would develop a testing program for specific new chemicals prior to reviewing a PMN only when the manufacturer has expressly requested the Agency to do so.

2. The Use of Section 5(e) to Require Data Development

That is not to say that the sufficiency of the data in a PMN is unreviewable and EPA is powerless to delay manufacture of a new chemical which has not been adequately tested. Under Section 5(e), the Agency is explicitly authorized to prohibit or limit manufacture of a new chemical when it determines that there are "insufficient data to evaluate its health and environmental effects." In issuing an order under Section 5(e), EPA can specify the data which the PMN submitter must develop before the Agency will reconsider whether manufacture of the new chemical can proceed.

It is significant, however, that, under Section 5(e), insufficient test data alone will not justify an order preventing manufacture of a new chemical. Rather, in addition to having insufficient information to evaluate the chemical, EPA must possess evidence (1) that the chemical may present an "unreasonable risk" to health or the environment or (2) that the chemical will be produced in "substantial quantities" *and* will have significant human or environmental exposure. Except in these situations, EPA is not authorized to prevent or limit manufacture of a new chemical even if it has not been tested at all. Thus, Section 5(e) contemplates that testing of new chemicals will be mandated only when there is evidence suggesting that the chemical may be hazardous or that the human and environmental exposure which the chemical is likely to receive will be substantial.

Another significant feature of Section 5(e) is that it reserves evaluation of the "sufficiency" of the test data for a new chemical until after, rather than before, a PMN has been submitted. Thus, Section 5(e) contemplates that, instead of issuing test requirements which apply to all new chemicals, EPA would assess the testing needs of new chemicals on a case-by-case basis, relying on the manufacturer to determine in the first instance whether a chemical would be tested and how much testing it would receive. For this reason, Section 5(e) assigns initial responsibility to the manufacturer to make testing decisions for a new chemical based on such individualized factors as its uses, probable level of exposure, biological activity, expected production volume and economic importance. EPA is to review the manufacturer's testing decisions after the fact, requiring additional testing only where the particularized criteria of Section 5(e) can be met.

3. The Portions of TSCA Legislative History Relating to Premanufacture Testing

The legislative history of the TSCA contains significant additional evidence that EPA is not authorized to require premanufacture testing for all new chemicals under Section 5. For example, responding to fears that Section 5 would impose excessive burdens on industry, the Senate Report states that:

> While the EPA administrator must be given the authority to act during the premarket notification period to gather more data or to take appropriate restrictive action, the notification burden itself should not be onerous. *Unless testing has been otherwise required, notification only consists of reporting routine information which should be in the hands of the manufacturer in the first place.* Included is information as to the identity of the product, categories of use, estimates of the amount to be produced and, insofar as reasonably ascertainable, to be produced for each of the categories of use, a description of byproducts, lists of existing test data, and estimates of the number of persons who will be exposed in their places of employment. S. Rep. No. 94-698, *supra,* at 10—11 (emphasis added).

A similar understanding was expressed by EPA itself in its written response to a questionnaire from the House Subcommittee. When asked how EPA would define the term "reasonably ascertainable," Deputy Administrator Quarles responded as follows:

> The information required as part of premarket notification of a new chemical should be information which is readily available to the manufacturer. *We do not anticipate that the manufacturer will be required to perform additional research and testing to determine any of the information requested.* A copy of a preliminary draft of a combined annual and premarket notification report form is attached. *"Reasonably ascertainable" in-formation would include that information that the manufacturer has already developed or obtained.* Hearings on H.R. 7229, H. R. 7548, and H.R. 7664, before the Subcommittee on Consumer Protection and Finance of the House Committee on Interstate and Foreign Commerce, 94th Cong., 1st Sess. at 242 (1975) (emphasis added).

EPA thus expected that the PMN would only include test data which the manufacturer had voluntarily chosen to develop.

Other portions of the TSCA legislative history confirm that Congress rejected the concept of mandatory premanufacture testing. For example, in response to a question of Congressman Moss, Russell Train, then the Chairman of the Council on Environmental Quality, testified that, because of the wide variations in toxicity among new chemicals, uniform testing requirements would be unnecessary and inappropriate:

> Now, you indicate in your testimony that the pattern of premarket testing contained in the Senate bill would require, I think it uses the words, "stifling and cumbersome administrative procedures." I wonder if you can expand on that within the realm of the possibility or probability that the House and Senate will have to discuss these differences.
>
> Mr. Train. Yes, sir; I suspect this may be the main issue involved in this legislation, so I think it is useful to focus on it.
>
> At first glance, I think that, recognizing that we are dealing with a class of products the toxicity of which we are concerned about, this would lead to the conclusion that certainly they should all be tested and approved before being introduced into use.
>
> *On the other hand, as I point out, there are many thousands of these substances already in use, many hundreds developed and introduced each year, of which presumably only a few are really a problem, not the generality of them.* Hearings on H.R. 5276 and H.R. 10840 before the Subcommittee on Commerce and Finance of the House Committee on Interstate and Foreign Commerce, 92d Cong., 2d Sess. 76 (1972) (emphasis added).

EPA was also outspoken in challenging the need for universal premanufacture testing. In a lengthy letter to Senator Magnuson, dated June 18, 1973, the Agency pointedly argued that selective testing of new chemical substances was justified by the limited resources of industry and EPA, the gradations in toxicity among new chemicals, and the lack of governmental experience with toxicological testing:

We do not feel that it is necessary to automatically require some testing of all chemicals except for those chemicals or classes which the Administrator can positively identify as being of no [sic] environmental or public health threat. Past Federal regulation and experience in the toxic substances area, particularly with respect to testing, has been limited. The rather sweeping testing requirements of the Staff Working Draft would tax the resources of the Agency and of the chemical industry before the Agency has had the opportunity to assess the need for and the benefits of a broad-based testing program. We prefer the more selective provisions of S. 888 which would allow the Administrator to focus all of his resources and attention on those substances or classes which we suspect may pose health or environmental hazards. As reprinted in S. Rep. No. 93-254, 93d Cong., 1st Sess. 45—46 (1973) (emphasis added).

A subsequent letter from EPA to Senator Tunney reiterates the importance of selectivity in testing new chemical substances:

There will be chemicals, perhaps many, that will not warrant any statutory testing at all. We intend to administer the Toxic Substances Control Act selectively, with attention to the types of tests and information suitable to require of the manufacturer and for us to review for different chemicals and chemical groups. Principal focus will be on those that appear to have the greatest potential for harm. *Id.* at 51 (emphasis added).

The Senate Report on S. 426 as adopted explained the Senate shift to a far more selective system of premanufacture testing as follows:

Concern was expressed to the committee by the chemical industry that S. 426 as originally introduced would have a severe impact on the chemical industry. Primarily the industry was concerned that too many new chemicals would have to be tested and that EPA will be unreasonable in its demands for testing of both new and existing chemicals.

Originally, S. 426 provided that testing must be made of all new chemicals unless EPA excluded them from coverage because of no unreasonable risk to human health or the environment or because they could more efficiently be dealt with by testing of components. *The chemical industry feared that EPA would be reluctant to exclude chemicals and that the same testing requirements would apply to all new chemicals regardless of risk.*

As reported by the committee, the bill contains safeguards against this possibility. First, EPA would be directed to routinely require the submission of test results prior to commercial production only for those new chemical substances which there is reason to believe may pose unreasonable threats to human health or the environment. For these chemicals, EPA must make a prior determination that unreasonable risk is at least suspected. The data would be submitted in accordance with testing procedures developed by EPA.

Manufacturers of all other new chemical substances which are not so specified would only be required to give notification to EPA ninety days in advance of commercial production. S. Rep. No. 93-254, *supra*, at 4 (emphasis added).

During hearings before the House Subcommittee on Consumer Protection and Finance in the 94th Congress, EPA representatives were sharply questioned about the need for premanufacture notification concerning all new chemical substances. In response, they assured Subcommittee members, particularly Congressmen Eckhardt and McCollister, that premanufacture notification would place insignificant burdens on most manufacturers. For example, Deputy Administrator Quarles testified that:

We do not feel that we are dealing with a situation that is just totally out-of-hand and that there are unlimited numbers of chemicals that pose a terrible and imminent threat to public health.

The need here is to be able to identify those chemicals that really pose serious problems and go after them with requirements of testing and if appropriate, requirements of control. Hearings on H.R. 7229, H.R. 7548, *supra*, at 212—213.

Later in the questioning, Congressman McCollister described the EPA concept of premanufacture notification as one where:

[A]nything [would be] premarket screened, but that premarket screening for a great number of chemicals is going to be very perfunctory and it is that other group that he is going to get down to that they are going to give more attention to. *Id.* at 221.

Clearly, both Mr. Quarles and Congressman McCollister intended that standards for the development of test data would be imposed on a chemical-by-chemical basis and only after EPA decided that the substance presented a basis for concern.

Further questioning by Congressman Eckhardt elicited even more explicit statements from EPA. Terming universal premanufacture testing "pretty drastic," he indicated that he would only have "no objection at all to notification which does not tie up distribution and manufacture." *Id.* at 222. Attempting to allay this concern, Mr. Quarles indicated that, based on the test data and other information which the manufacturer chose to submit, EPA would identify those few chemicals for which further testing was necessary:

Mr. Quarles. *So really what we are trying to talk about here is notification.* We feel that there can be real advantages to having a Government agency receiving notifications of all the new chemicals that are going into the society. There are important advantages in being able to look across the board at all of these, having some awareness of the new chemicals, and then being able to look closely into certain ones that for one technical reason or another raised some warning flag.

But that notification is essentially all that we are asking for in regard to the vast array. . .

Then we come to the process of scrutinizing a chemical and determining that testing is required. Those are separate steps and we would have to act individually in those cases and act on a basis that could withstand judicial review. Ibid. (emphasis added).

After further discussion, Mr. Quarles reiterated that testing requirements would be based on a consideration of the specific safety concerns associated with particular new chemicals and would be imposed only after EPA had reviewed the PMNs for those chemicals:

The point is you really, at this stage, are talking about notification and even though the section is entitled screening, you are really talking about notification, because what that provision really does is to simply require a manufacturer, before he goes into production, to give EPA notice that he intends to do so with the description of whatever figure he may have.

So this is a disclosure provision. It is not immediately regulatory. To get to the regulatory provisions, then we would have to go on to the next section or wherever it is to begin testing, or in case it is warranted, some restrictions on production. And that would require the foundation. *Id.* at 229 (emphasis added).

Emphasizing the same point, Mr. Schweitzer of EPA assured Congressman McCollister that proposed Section 5 only contemplated the submission of test data that the manufacturer had generated *voluntarily:*

Mr. McCollister. You would not seek to influence that requirement for test data, to describe it, to define it?

Mr. Schweitzer. We are talking about test data which the manufacturer would have *voluntarily developed.* Hearing on H.R. 7229, H.R. 7548, and H.R. 7664, *supra,* at 227 (emphasis added).

Significantly, Mr. Quarles then added that Section 5 would not authorize the promulgation of additional test requirements:

Yes, I think that there is a very detailed provision in the bill that relates to the requirement that the Administrator might impose for developing test data. And there are procedures we would be required to go through to impose requirements that manufacturers developed test data.

Mr. McCollister. But that is not in connection with the premarket notification?

Mr. Quarles. That is correct, that is something separate. I think I am clear on this. *We would certainly not expect somebody to grab hold of the premarket notification requirements and use that to develop additional testing requirements. Ibid.* (emphasis added).

These views were reiterated by Mr. Quarles in his written response to the Subcommittee questionnaire. As he pointedly stated: "We do not anticipate that the manufacturer will be required to perform additional research and testing to determine any of the information requested" by Section 5. *Id.* at 242.

Similar concepts of Section 5 were expressed by members of Congress on several occasions during the floor debate on TSCA. When introducing S. 776, for example, Senator Tunney stated that:

[This bill] requires that certain [new] chemicals be tested if EPA makes a determination that further examination is necessary due to potential health or environmental threats.

The results of these tests then must be furnished to EPA 90 days in advance of manufacture.

In addition, for all other new chemicals for which EPA cannot make a prior determination that testing is necessary, manufacturers must give notice of their impending intention to market the product. *Armed with this type of information, EPA could, where necessary, then take action to impose restrictions or conclude that further tests must be conducted.* 121 Cong. Rec. 3780 (February 20, 1975) (emphasis added).

Congressman Eckhardt, one of the House sponsors of H.R. 14032, also believed that testing could be required under TSCA only by the promulgation of test rules tailored to particular substances:

Mr. Chairman, if the gentleman will yield, actually I think the author of the amendment may be under the misconception that merely by identifying a chemical the EPA may require testing. *Actually of course there is a rulemaking process which must determine that the chemical poses the risk and should be tested in order to determine whether or not the risk is in actuality.*

I am inclined to think that the terms of the act may be even stricter than the gentleman may conceive in his amendment because we do not permit the EPA merely by putting the chemical on the list to require its testing. *The EPA must identify the chemical and by rule direct testing in the case.* 122 Cong. Rec. H 8816 (daily ed., August 23, 1976) (emphasis added).

Senator Durkin made the same point more directly when he observed that, in its final form, TSCA "does not require premarket testing of all new chemicals." 122 Cong. Rec. S. 16808 (daily ed., September 28, 1976).

As the above analysis demonstrates, Congress recognized that the universe of new chemicals subject to Section 5 would be extremely diverse and would contain wide variations in potential toxicity, use, production volume, and potential profitability. For this reason, Congress decided that testing criteria which applied to all new chemical substances were unwarranted. The approach which Congress adopted was one which allowed each manufacturer to tailor its own testing program based on its familiarity with the hazard potential, uses, and profitability of the new chemical. If the manufacturer's testing decisions are insufficient and additional test data prove necessary to evaluate the safety of the chemical, Congress expected that test requirements would be imposed under Section 5(e) *after* a premanufacture notice had been submitted.

B. The Controversy over Premanufacture Testing Guidance

1. The Test Data Included in PMNs Received by EPA to Date

There has been significant controversy concerning the adequacy of the test data contained in the PMNs submitted to EPA to date. Environmentalists have argued that too little testing has been done. The industry, on the other hand, believes that the testing has been adequate to assess the chemicals involved and that additional testing would have been unnecessary and not cost-effective.

EPA tabulations[19] show that, of the 659 valid PMNs submitted in 1981, 374 (57%) contained some toxicity data. Of these PMNs, 290 contained the results of acute tests; 96 contained data from mutagenicity screening; 65 contained ecological effects test data; and 2 contained the results of subchronic testing. In addition, 504 PMNs contained physical and chemical properties data.

These figures are high from a number of standpoints. According to CMA's contractors, RRS, out of a sample of 345 PMNs, 123 were submitted for polymers and 106 were submitted for intermediates.[20] Both polymers and intermediates typically present low levels of risk — polymers because they have limited biological activity and frequently are comprised of known monomers; intermediates because they are used under highly controlled conditions. Thus, industry is in fact generating test data on chemicals that can reasonably be presumed to present very little risk.

Strengthening this conclusion is the large number of PMN chemicals manufactured in low production volumes. For example, out of a sample of 203 PMNs, RRS found that 63 were expected to be manufactured in quantities of 1000 lb or less per year. It is a reasonable assumption that most of these chemicals received physical and chemical properties testing and that at least some were subjected to toxicity testing in addition. Consequently, here as well, industry is conducting testing on chemicals that can reasonably be viewed as presenting little risk.

It is also significant that, in all but a small number of cases, EPA has been able to conclude that the PMN chemicals it has reviewed will not present a significant hazard to human health or the environment under their anticipated conditions of manufacture and use. As of July 1982, EPA had issued only nine proposed orders under Section 5(e) of TSCA, and PMN submitters had either withdrawn their PMNs or agreed to suspend the PMN review period in only ten other instances. For the great majority of PMNs, EPA had no basis for concluding that the PMN chemical might present an unreasonable risk or be manufactured under conditions of significant human exposure or substantial environmental release — the criteria that will justify postponing manufacture pending further testing under Section 5(e). Former EPA Administrator Costle recognized this in a letter to then-Senator Muskie dated March 18, 1980. According to that letter, "none of the new substances that have completed our review process . . . will present an unreasonable risk of injury to health or the environment from their projected production and use patterns."

2. Significant Issues Involving the EPA Premanufacture Testing Policy

On March 16, 1979, EPA published a *Federal Register* notice soliciting comment on a number of policy and technical issues concerning the development of premanufacture testing guidance under Section 5. 44 *Fed. Reg.* 16240. The Agency published a final Premanufacture Testing Policy on January 27, 1981. 46 *Fed. Reg.* 8986. This Policy establishes a base set of tests that all submitters of premanufacture notices

---

[19] Environmental Protection Agency, Status of Premanufacture Notices (PMNs) (July 30, 1982).

[20] Other PMNs in this sample may have also been for these categories of chemicals, but their uses could not be ascertained because they were claimed confidential.

are recommended to conduct. Included in this base set are tests to determine (1) physical, chemical, and persistence characteristics, (2) acute toxicity, (3) repeated dose or subchronic toxicity, (4) potential mutagenicity, (5) ecotoxicity, and (6) degradation and accumulation potential. This battery of tests has been provisionally endorsed by working groups of the Organization for Economic Cooperation and Development (OECD) as providing Minimum Premarket Data (MPD) for assessing the human health and environmental effects of new chemicals.

In its Policy, EPA acknowledged that "Section 5 does not require that any particular test be performed on all new chemical substances before submission of premanufacture notices." 46 *Fed. Reg.* at 8986. According to the Agency, "it cannot use testing guidance published under Section 5 to establish a *de facto* general testing requirement for new chemicals." 46 *Fed. Reg.* at 8987. In addition, the Agency stressed that its recommended base set is merely a "starting point for testing" and that the particular circumstances of individual chemicals "may justify deletion, substitution, or addition of data components, resulting in either more or less testing than reflected by the base set of data." *Id.*

To illustrate this approach, EPA pointed out that technical considerations or available data on physical/chemical properties may "make some tests inapplicable for certain chemicals." It also indicated that, in "circumstances of very low human exposure or environmental release, a lesser amount of testing may be warranted." 46 *Fed Reg.* at 8988. Conversely, EPA observed, additional testing might be necessitated by a variety of considerations such as structure/activity analysis, which may suggest the need to test for carcinogenicity, or "circumstances of high potential human exposure or environmental release," which may suggest the need for a 90-day subchronic test and tests for teratogenic and reproductive effects. 46 *Fed. Reg.* at 8988. EPA stressed that the "screening-level base set data also may indicate the need for follow-up testing." *Id.*

EPA policy requested all PMN submitters "to explain the scientific rationale for any deletions, substitutions or additions to the base set." *Id.* Nevertheless, even this explanation would appear to be optional. According to the Agency, it "will not automatically initiate 5(e) actions if a manufacturer declines to utilize the recommended base set of deviates from the base set." 46 *Fed. Reg.* at 8989.

Industry has been encouraged in that EPA has recognized its lack of authority to require premanufacture testing and acknowledged the need for flexibility in applying a base set of tests to all new chemicals. Nevertheless, industry representatives have questioned the need for any testing guidelines under Section 5 and expressed concern about the possible misuse of EPA's base set in day-to-day Agency operations. The principal concerns voiced by industry are as follows:

1. As discussed above, except in certain limited circumstances, Section 5 provides that manufacturers will themselves determine the type and quantity of test data that their PMNs will contain, subject to EPA's after-the-fact review of the PMN in accordance with the criteria of Section 5(e). Any testing guidelines which attempt to delineate in advance when and to what extent new chemicals must be tested could be inconsistent with these Congressional policies, even if those guidelines are nominally voluntary.

2. There will inevitably be wide variations among the new chemical substances which EPA will review under Section 5 of TSCA. Because of these variations, the decision whether to test and the type and amount of testing to conduct will differ from chemical to chemical, often dramatically. The factors which will shape testing decisions will include the assurance of safety available from existing data or similarities to other chemicals, structure/activity relationships, physical and chemical properties, exposure factors, and use factors.

3. For any new chemical, the degree of certainty which further testing can contribute to the evaluation of safety must always be weighed against its costs and the manufacturer's ability to bear those costs. The manufacturer may decide in light of the relatively low risk of a chemical that the added certainty provided by additional testing simply cannot be justified, given his prior expenditures on the chemical's development, its expected production volume, and the monetary return it is anticipated to provide. Depending on the chemical's short and long-term benefits to society and its technological importance, a decision to forego additional testing under these circumstances may be prudent and proper.

It is disturbing that the EPA Policy neither contains a meaningful analysis of the economic impact of the Agency's base set nor attempts to assess the likely effects on innovation of performing such tests for all new chemicals. Equally disturbing, the Policy contains no recognition of the economic realities that govern the development and commercialization of new chemicals, and nowhere does EPA state that PMN submitters can and indeed should, take economic considerations into account in making testing decisions for new chemicals.

4. EPA's Testing Policy fails to recognize the circumstances in which all or some of the base set tests will be unnecessary. For example, while EPA recognizes that "circumstances of very low human exposure or environmental release" may warrant a "lesser amount of testing" (46 *Fed. Reg.* 8983), the Agency fails to acknowledge that there may be many low-exposure chemicals, such as site-limited intermediates, for which no testing is warranted. In addition, EPA does not even mention other factors which may justify performing fewer tests or no testing at all. High molecular weight polymers, for example, have low levels of biological activity and, where they are comprised of monomers known to be safe, can often be presumed to present little risk. As another example, many new chemicals may be close structural analogs to existing chemicals for which extensive test data are available.

5. EPA's Testing Policy stresses that the Agency's base set of tests is identical to the MPD testing battery currently under consideration by the OECD. 46 *Fed. Reg.* at 8987. However, this emphasis on the similarity between the EPA and OECD testing guidelines is misleading.

The MPD concept under consideration by OECD evolved from the premarket data requirements that have been adopted by the European Economic Community (EEC) in its Sixth Amendment of September 18, 1979. While the Sixth Amendment prescribes minimum testing requirements for new chemicals, these requirements do not apply to (1) polymers comprised of known monomers, (2) chemicals that are not offered for sale either in bulk form or as components of other products, and (3) chemicals that are marketed in quantities of 2200 pounds or less per year. These exemptions from testing afford an important degree of flexibility that the EPA base set lacks, since they make testing unnecessary in those situations where testing costs would be unduly burdensome or testing would provide limited benefits because of a chemical's low risk potential and small exposure.

Moreover, the OECD Council has to date failed to adopt a proposed Decision endorsing the MPD, and the United States has questioned whether such a Decision would be consistent with TSCA. Under these circumstances, the precedential value of the OECD test guidelines is dubious.

6. In its Policy, EPA asks PMN submitters who depart from its testing recommendations "to explain the scientific rationale for any deletions, substitutions or additions to the base set." 46 *Fed. Reg.* at 8989. The message implicit in this statement is that EPA will look with disfavor on departures from its base set and that the burden will be on the PMN submitter to provide a scientific justification for such departures that EPA finds acceptable. As a result, PMN submitters could be reluctant to apply EPA's base set in a flexible and cost-effective manner.

*　　*　　*

Despite the existence of EPA testing guidelines under Section 5, PMN submitters will continue to possess discretion to determine the type and quantity of test data that will be generated for new chemicals. This decision will obviously require a particularized examination of the economic and safety considerations applicable to specific chemicals. Nevertheless, firms should bear in mind dissatisfaction by the EPA with PMNs that totally lack test data. Failure to submit any data at all could increase the risk of a Section 5(e) order. This increased risk will be avoidable where a battery of acute tests is economically feasible and the results of those tests are not likely to be adverse.

In general, no PMN should be submitted unless a firm has considered whether the PMN chemical is similar to other compounds which EPA or other Federal agencies have identified as suspect — e.g., by an Interagency Testing Committee (ITC) recommendation under Section 4(e), submission of a Section 8(e) "substantial risk" report, or initiation of a long-term bioassay by the National Cancer Institute (NCI). In such situations, the PMN submitter should either conduct testing of its own or be prepared to present scientific arguments which allay the concerns of the Agency about the PMN chemical. If it is economically infeasible to conduct adequate testing, careful consideration should be given to not submitting a PMN, since issuance of a Section 5(e) order will place a cloud on future development of the PMN chemical and similar compounds.

C. Utilization of Section 4(a) Rules to Require the Testing of New Chemicals
1. Inclusion of New Chemicals in Categories

Under Section 4(a) of TSCA, EPA is empowered to promulgate rules that require the testing of chemical substances or mixtures. The statute assigns the responsibility for conducting and financing such testing to the manufacturers and/or processors of the chemicals involved. Under Section 26(c)(1) of TSCA, EPA may frame its actions under Section 4 to apply to a "category" of chemical substances or mixtures rather than to a single chemical. Thus, in an appropriate case, EPA is authorized to promulgate Section 4 test rules which cover an entire class of chemicals. Acting under Section 4(e) of TSCA, the ITC has already recommended a number of categories of compounds for testing under Section 4.

When EPA test rules are applicable to entire categories of chemicals, testing requirements may fall on new chemical substances which are included in those categories but are not in commercial production at the time the EPA test rule takes effect. Before manufacture of these chemicals may begin, data developed in compliance with EPA requirements must be provided to the Agency. According to Section 5(b)(1)(A), unless the submitter has been granted an exemption from test requirements, PMNs for new chemicals subject to a Section 4 rule must include the test data which that rule prescribes. Moreover, Section 5(b)(1)(B) provides that, whether or not the PMN submitter is exempt from testing, the PMN review period will not begin until the required data are in fact submitted to EPA. Thus, manufacturers of new chemicals subject to test requirements will be unable to submit their PMNs until the testing mandated by EPA has been completed and submitted.

2. Exemptions from Test Requirements

Under Section 4(c)(2)(A), EPA must grant exemptions from test requirements if it determines that (1) the chemical for which an exemption is sought is "equivalent" to a chemical for which testing is either underway or completed, and (2) submission of data by the applicant would be "duplicative" of data that have already been, or will be, submitted to EPA. The House Report on TSCA makes it clear that, when the EPA test rule applies to an entire category, each individual category member need not be tested separately and different chemicals within the category after it can be considered "equivalent". According to the Report, once data are being obtained or have been submitted on one chemical within the category, "such data may provide a basis for manufacturers and processors of other chemical substances to apply . . . for an exemption from the test requirements and thereby unnecessary time and expense could be saved by the affected manufacturers and processors by not having to test each minor modification of substances within the category." H.R. Rep. No. 94-1341, *supra*, at 61.

Based on these principles, new chemicals which fall within a Section 4(a) category will often qualify for exemptions from test requirements if the PMN submitter can show that testing is being conducted by manufacturers of other category members and differences between the new chemical and the chemicals being tested are insufficiently important from a toxicological standpoint to justify additional testing. Once the PMN submitter is exempted from test requirements, it will be required under Section 4(c)(3) to provide reimbursement to those companies that are bearing the cost of testing. If reimbursement terms cannot be negotiated voluntarily, Section 4(c)(3) authorizes EPA itself "to provide fair and equitable reimbursement." Under Section 4(c)(3)(A), EPA reimbursement decisions must be based on rules which reflect a consideration of "all relevant factors", including "the effect on the competitive position" and "share of the market" of the businesses involved.

Of necessity, a PMN submitter will want to determine whether its new chemical is eligible for an exemption from test requirements as soon as feasible. If it is not, the submitter will be required to complete testing in accordance with the EPA rule before submitting a PMN — a costly and time-consuming process. Thus, prospective PMN submitters should apply for exemptions as far in advance of the PMN submission as possible.

3. Criteria for Establishing Categories

The text and legislative history of TSCA demonstrate that the EPA power to promulgate test rules for "categories" is a limited one and can be exercised only in highly circumscribed situations. Under Section 26(c)(2)(A), EPA can regulate entire chemical "categories" only when those categories are "suitable for classification as such for purposes of this Act." Thus, whenever EPA seeks to require testing of a group of chemicals, the Agency must show that its definition of that group comports with the basic policies of TSCA. This principle means that EPA cannot use its authority to regulate "categories" as a mechanism to require the testing of chemicals which would not otherwise meet the test rule criteria of Section 4(a). Rather, in any case where EPA proposes to issue a category-based test rule, it must show that the different chemicals within the category qualify for testing under the applicable statutory standards. In addition, to provide clear guidance to industry and to insure that the category is properly defined, proposed and final rules must identify each individual chemical which falls within the category.

This does not mean that, in its proposed rule, EPA must necessarily demonstrate that testing on every chemical in a category is needed under the criteria of TSCA. At a minimum, however, the EPA proposal must identify each such chemical and present a rational basis for concluding that the members of its category meet the criteria of Section 4. It also must give industry an opportunity to rebut that tentative conclusion with respect to any particular chemical within the category.

Industry representatives have argued that this approach requires EPA to allow manufacturers or processors of particular chemicals that fall within a proposed category to respond to the proposed Agency test rule with evidence that these individual chemicals do not qualify for testing under the criteria of Section 4(a). In its notice promulgating a final rule, EPA must then explicitly address this evidence either by producing contrary evidence of its own which demonstrates that each such chemical qualifies for testing or by excluding it from test requirements. The prior EPA test rule proposals under Section 4 indicate that it substantially accepts this approach.[21]

4. Procedures for Deleting New Chemicals from Categories

New chemicals, by definition, will not have been manufactured for commercial purposes at the time a Section 4 rule is proposed. For this reason, manufacturers and processors of new chemicals will have no opportunity to participate in the rulemaking process by presenting evidence that those chemicals are ineligible for testing under Section 4(a). Unless this opportunity is afforded at a later date, test data would be required to accompany the PMNs of new chemicals even though they may not qualify for mandatory testing under the requirements of TSCA.

To correct this situation, industry representatives have argued that each Section 4 test rule covering a chemical category must contain a mechanism by which manufacturers of new chemicals within that category can subsequently obtain exclusions from testing requirements. At a minimum, such a mechanism should enable manufacturers of new chemicals to provide EPA with evidence, before submitting their PMNs, that those chemicals do not meet the Section 4 criteria for testing. For example, if EPA had based its test rule on a finding that all chemicals within a category are manufactured in substantial quantities and have substantial human or environmental exposure within the meaning of Section 4(a)(1)(B)(i), the manufacturer of a new chemical within the category should be allowed to show that the expected production volume of the chemical is too small to satisfy this test. Unless EPA could produce contrary evidence, it would then be required to exclude the new chemical from test requirements, and the PMN for that chemical would not need to contain the data required by the Agency test rule.

EPA has not yet responded to this industry argument, largely because it has not to date proposed a Section 4 rule for an open-ended chemical "category" that could include new chemicals. It is likely that the issue will be faced soon, however, because EPA has stated that, utilizing its power to regulated chemical categories, its long-term goal is to issue Section 4 test rules which cover "a substantial proportion of new chemicals." 44 *Fed. Reg.* 1624, 16243 (March 16, 1979).

D. The EPA Section 5(b)(4) "Risk List"

Another mechanism which Section 5 affords EPA for requiring testing on new chemicals is the preparation of a "risk list", authorized by Section 5(b)(4). This list is to be comprised of chemicals which EPA finds present or may present an unreasonable risk of injury to health or the environment. In view of Section 26(c) of TSCA, such a list could include not just individual substances but chemical categories as well, and could thus cover new chemicals.

The preparation of the EPA "risk list" must occur in a notice-and-comment rulemaking proceeding, in which interested persons are afforded an opportunity for the oral presentation of data, views, or arguments. The EPA inclusion of chemicals on the list is to reflect its consideration of all relevant factors, including the effects of the chemical on health in the environment and the magnitude of human environmental exposure which they undergo. As part of its rule adding chemicals to the list, EPA must identify those uses of such chemicals which would continue "a significant new use" under Section 5(a)(2).

For all new chemicals included on the risk list as well as all significant new uses of existing chemicals included on the list, the PMN submitter must develop and provide EPA with data demonstrating that the chemical "will not present an unreasonable risk of injury to health or the environment" pursuant to Section 5(b)(2)(B). If EPA declines to initiate any action against such chemicals during the PMN period, Section

---

[21] *See, e.g.,* proposed test rules for chloromethane and chlorinated benzene, 45 *Fed. Reg.* 48524 (July 18, 1980).

5(g) requires the Agency to publish a *Federal Register* notice stating its reasons for not initiating such action.

Under Section 5(h)(1), EPA may exempt the manufacturer of a "risk list" chemical from the requirement to submit test data for that chemical in its PMN. Presumably, such an exemption would be appropriate where the risk list included a broad chemical category for which other companies had developed or were developing satisfactory test data. Where exemptions are granted, Section 5(h)(2)(B) requires the exemption holder to provide reimbursement to the companies which are bearing the expense of testing.

By designating chemical categories for inclusion on the "risk list" because they may present an unreasonable risk of injury, EPA could potentially require premanufacture testing for a broad range of new chemicals. While the Agency has not designated chemicals for inclusion on the list to date, it has announced that it "will carefully evaluate use of the Section 5(b)(4) 'risk list' to increase the reporting for new chemicals in specified classes or categories."[22]

E. The Submission and Description of Test Data Required under Section 5(d) of TSCA

Wholly apart from EPA authority to mandate additional testing on new chemicals, there exist significant questions concerning the obligation of the PMN submitter to submit existing health and safety information to EPA as part of its PMN. As noted above, Section 5(d)(1)(B) of TSCA requires the submitter of a PMN to provide EPA with any "test data" in its possession or control which are related to the effect of any manufacture, processing, distribution in commerce, use, or disposal of a new chemical substance or article containing it on health or the environment. Section 5(d)(1)(C) then requires the submitter to describe "other data" concerning health and environmental effects of the substance which are known to it or reasonably ascertainable.

In interpreting these provisions, EPA has: (1) defined "test data" to include not simply scientific studies but both nontest information and subjective evaluations of test results; (2) required the PMN to include test data relating not merely to the new chemical substance but to a broad range of other chemicals "related to" that substance as well; and (3) defined the term "other data" in Section 5(d)(1)(C) to include a broad range of nontest information. All of these EPA positions raise substantial legal and policy issues.

1. The Definition of "Test Data"

Proposed § 720.2 of the EPA PMN regulation provides that the term "test data" means:

> (a) data, including chemical identity, from a formal or informal study, test, experiment, recorded observation, monitoring, or measurement; and (2) information concerning the objectives, experimental methods and materials, protocols, results, data analyses (including risk assessments), and conclusions from a study, test, experiment, recorded observation, monitoring, or measurement.

In the preamble to the proposed regulations, EPA has stated that this definition includes "both formal and informal tests." It has also advised that the term "test data" includes "not only the raw data *per se*, but other information which is relevant to the development and analysis of the data," including "any risk assessments concerning the data, as well as protocols, results, and conclusions." 44 *Fed. Reg.* 2252. As described below, industry representatives have argued that this definition departs from the terminology of TSCA in two important ways.

a. The Difference between "Test Data" and Other Information

Several different terms are used throughout TSCA to refer to different types of data and information. The narrowest term is "data from health and safety studies" (Section 14[b]), because it refers only to one type of test (a health and safety study) and only the "data" from that particular type of test. A somewhat broader term is "test data" (Section 5[d][1][B]), which clearly includes data from health and safety studies as well as data from other types of tests and studies. A still broader term is "existing data" (Section 8 [a][2][E]), because it is not limited to "test" data, but includes all data from any sources whatever, whether accumulated by a specific "test" or otherwise. The broadest term is "information" (Sections 8[a][3][A][i] and 9[a][1]), which includes test data, nontest data, and any other nondata type of information of any kind and from any source. The term "information" thus includes evaluations, observations, and conclusions that do not fall within the meaning of "data" or "test data".

A test is ordinarily understood to include a means of examination, trial, or proof. It encompasses any

---

[22] Letter of Honorable Douglas Costle to Senator Muskie, March 18, 1980.

systematic method of obtaining data. For this reason, industry has concurred in the portion of the proposed definition that equates "test" with "a formal or informal study, test, experiment, recorded observation, monitoring, or measurement." The term "data" is ordinarily understood to mean specific numerical facts or figures, usually derived from some kind of test, from which some conclusions can then be reached. As industry has argued, however, neither the term "test" nor the term "data" include evaluations, observations, or conclusions (including risk assessments) that were not part of the test itself.

b. The Limitation of "Test Data" to Data Obtained for the Purpose of Evaluating Health and Environmental Effects

In view of the careful statutory distinction between "test data" and "other information," industry has also argued that EPA must narrow the definition of "test data" by acknowledging that it only includes data which have been developed for the specific purpose of evaluating health and environmental effects of a chemical. The broad statement by EPA that "test data" may derive from "recorded observations, monitoring and measurements" — and from "both formal and informal tests" — could potentially cover *all* information which has some bearing on the safety of a chemical, even if that information was generated for a wholly unrelated purpose. This standard could impose unmanageable burdens on the submitter of a PMN. Instead of collecting merely those studies which were undertaken to ascertain the health and environmental effects of a chemical, it would be forced to conduct a wide-ranging file search to identify all information which bears on these effects, whether developed for that purpose or not.

c. The Difference between "Test Data" and Evaluations

To conform to the careful statutory terminology, industry has argued that EPA must also redefine "test data" so the term is limited to actual studies of health and environmental effects and does not encompass evaluative and interpretative documents prepared either before or after those tests are performed. In normal parlance, the term "test" connotes a scientific procedure for observing or measuring the properties of a chemical. Thus, when a chemical is administered to laboratory animals in order to ascertain its potential toxicity, the laboratory experiment is a "test" and measurements of the animals' responses are "test data." On the other hand, the terms "tests" and "test data" are not normally used to describe opinions or subjective evaluations concerning properties and effects of a chemical. While such materials may refer to or rely on test results, they are not part of the testing process and cannot be equated with the tests themselves.

Important policy considerations support this interpretation of the term "test data". Many chemical manufacturers encourage their scientific employees to evaluate and comment on toxicological and environmental studies which the manufacturer performs. These evaluations are undertaken voluntarily and are an important part of the internal decision-making process of the manufacturer. In many respects, they resemble an internal audit or "inspector general" function within a company. They aid the company in forming judgments about suitability for various uses and applications of a chemical, its short- and long-term health and environmental effects, and the precautions which should be observed in its manufacture, transport, or use. They also assist the manufacturer in determining whether further testing is needed and, if so, in what respects.

Many in-house evaluations of test data are preliminary documents which are based on incomplete data or fragmentary information. They thus reflect views which are subject to change after further discussion, additional reflection, or the development of more data. Moreover, scientists may often have conflicting opinions concerning the validity of particular test procedures and the significance of particular test results. Accordingly, internal risk assessments by a company may vary depending on the stage of the testing process at which they were prepared and the predilections of the scientists preparing them. A "definitive" or "final" risk assessment may not exist or, if it does exist, will be the result of considerable discussion and debate.

It is important for scientists participating in the evaluation of test data to express their views candidly. Such candor is essential for reasoned and responsible marketing and safety decisions. The present EPA approach, however, would severely inhibit frank analyses of test data. Because such analyses would ultimately be divulged to EPA, industry scientists would hesitate to express their views forthrightly. This hesitancy, in turn, could lower the quality of corporate decision-making.

2. Data on Related Chemicals

Under § 720.23 of the proposed EPA PMN regulations, the submitter must provide EPA not merely with test data relating to the new chemical substance, but with test data related to certain "associated" chemicals as well. These chemicals include byproducts, co-products, degradation products, unintended reaction products, or other chemical substances relating to the manufacture, processing, distribution, use, or disposal of the new chemical substance which are not included in the Section 8(b) Inventory.

Industry has argued that this requirement is an impermissible attempt to expand the scope of premanufacture notification under Section 5. The purpose of the TSCA premanufacture notification provisions is to permit EPA to review the potential health and environmental effects of new chemical substances which the submitter of the PMN intends to manufacture for commercial purposes. For this reason, it can be argued, the test data which Section 5(d)(1)(B) requires to be submitted are data relating to the properties of the new chemical substance for which the PMN is submitted. Data concerning other chemicals which play a role in some phase of manufacture, processing, or disposal of the substance should therefore be beyond the purview of the inquiry which Section 5 directs EPA to conduct.

3. The EPA Definition of "Other Data"

Proposed § 720.23(d)(1)(i) of the EPA PMN regulations requires each PMN to be accompanied by descriptions of "any data", other than test data, that are in the possession or control of the submitter and "any data", including test data, that are not in the submitter's possession or control but are known to or reasonably ascertainable by him. The proposal does not define "any data" or specify how such data differ from "test data".

Read in sequence, it is apparent that Section 5(d)(1)(B) covers test data in the possession or control of the submitter, while Section 5(d)(1)(C) covers other test data that he does not physically possess or control but are known to him or reasonably ascertainable. Thus, construed in light of the surrounding statutory provisions, Section 5(d)(1)(C) merely would appear to insure that the submitter informs EPA of test data concerning the new chemical substance which the submitter does not actually possess but is aware of.

If the term "other data" were intended to expand the types of information which a PMN must contain, one would expect to find legislative history confirming the broad scope of Section 5(d)(1)(C). However, the legislative debate and committee reports on TSCA nowhere reflect a recognition that the Section would have this effect. Indeed, the consistent assumption of Congress was that the PMN would only contain "test data" relating to the new chemical substance. For example, the House Report States that a PMN "must include information respecting the substance, its chemical identity and molecular structure, proposed amount of production, uses, and *test data* respecting health and environmental effects." H.R. Rep. No. 94-1341, *supra*, at 23 (emphasis added). Thus, the House Committee apparently did not believe the PMN submitter would have a broad obligation to describe "any" data relating to the health and environmental effects of a new chemical substance. For the above reasons, industry has urged that "any data" should be redefined as "test data" which the submitter does not possess or control and proposed § 720.23(b)(1)(i) should therefore be deleted.

V. INFORMATION INCLUDED IN PMNS

For any prospective manufacturer of a new chemical, the type and amount of information which the PMN contains will be a matter of great concern. The collection and review of this information will require the expenditure of time and money; if the PMN submitter believes that these burdens are not justified by the economic potential of the new chemical, no PMN may be prepared. In addition, the information that a PMN contains will have a major bearing on EPA evaluation of the new chemical. Depending on the type and quantity of the information which the submitter provides, EPA may decide that no further scrutiny of the chemical is needed or, alternatively, that the chemical should be subjected to additional examination, including an extension of the PMN period under Section 5(c) or even a Section 5(e) order.

There are significant unresolved issues concerning the type and quantity of information which EPA can require PMN submitters to provide. EPA's initial proposed regulations embodied the position that the categories of information enumerated in Section 5(d) are only illustrative of the information that a PMN must contain, and that EPA is authorized to require additional information unmentioned by the statute. Industry representatives have vigorously disputed this position, arguing that Section 5(d) provides an all-inclusive description of the contents of a PMN and advocating a PMN form which calls for only the information specified in the statute. In its reproposed regulations, EPA has reduced the scope of its proposed form in response to industry objections, but continued to request information which industry considers unnecessary and unauthorized. More recently, however, there have been signs that EPA is considering adoption of streamlined PMN form that would seek only information required by statute.

Another major area of dispute between industry and EPA concerns the obligation of the PMN submitter to obtain information it does not already possess for inclusion in the PMN. The initial proposed EPA regulations would have required the PMN submitter to send prospective customers an elaborate questionnaire concerning their potential activities involving the new chemical. While EPA reproposed regulations indicated the Agency was reconsidering this approach, the matter has not yet been put to rest. In addition, the Agency has not yet clarified the obligation of the PMN submitter to obtain information which is "reasonably ascertainable." On the one hand, the Agency has asserted that this concept includes all information which can be obtained without "unreasonable" cost or effort. On the other hand, relying on portions of TSCA legislative history, industry representatives have argued that information is "reasonably ascertainable" only when a normal business in the position of the PMN submitter would be likely to possess such information already.

These matters are discussed in greater detail below.

A. The All-Inclusive Nature of Section 5(d)(1)

In the two PMN forms it proposed, EPA proceeded on the assumption that the description of a PMN in Section 5(d) is merely illustrative, and that PMN submitters can be required to provide the Agency with additional information that the statute does not specifically prescribe. Industry representatives, however, argued that such discretion is precluded by the language and legislative history of Section 5 of TSCA.

Under Section 5(d)(1)(A), a PMN "shall include" the information described in subparagraphs (A), (B), (C), (D), (F), and (G) of Section 8(a)(2). This information, in turn, consists of:

(A) The common or trade name, the chemical identity, and the molecular structure of each chemical substance or mixture for which such a report is required.

(B) The categories or proposed categories of use of each such substance or mixture.

(C) The total amount of each such substance and mixture manufactured or processed, reasonable estimates of the total amount to be manufactured or processed, the amount manufactured or processed for each of its categories of use, and reasonable estimates of the amount to be manufactured or processed for each of its categories of use or proposed categories of use.

(D) A description of the byproducts resulting from the manufacture, processing, use, or disposal of each such substance or mixture.

* * *

(F) The number of individuals exposed, and reasonable estimates of the number who will be exposed, to such substance or mixture in their places of employment and the duration of such exposure.

(G) In the initial report under paragraph (1) on such substance or mixture, the manner or method of its disposal, and in any subsequent report on such substance or mixture, any change in such manner or method.

In addition, Sections 5(d)(1)(B) and (C) provide that PMNs "shall include";

(B) in such form and manner as the Administrator may prescribe, any test data in the possession or control of the person giving such notice which are related to the effect of any manufacture, processing, distribution in commerce, use, or disposal of such substance or any article containing such substance, or of any combination of such activities, on health or the environment, and

(C) a description of any other data concerning the environmental and health effects of such substance, insofar as known to the person making the notice or insofar as reasonably ascertainable.

These provisions do not state that this detailed list is merely "illustrative" and that EPA would be free to require additional information unmentioned by Congress. Moreover, one can argue that it would have been irrational for Congress to describe the contents of a PMN in painstaking detail if it expected EPA to reshape the PMN by prescribing additional requirements of its own. Finally, Section 5(d)(1) speaks of the information that a PMN "shall include," language that suggests that the information specified by the Section is all-inclusive and cannot be enlarged at the discretion of EPA. Supporting this conclusion is the failure of Congress to grant EPA rulemaking power under Section 5. If Congress had vested EPA with broad power to define the contents of a PMN, it undoubtedly would have authorized the Agency to exercise that power by issuing rules. Numerous other regulatory statutes, for example, explicitly authorize agencies to issue rules that prescribe requirements for the submission of data.[23]

---

[23]   *See, e.g.*, the premarket approval provision of the Medical Device Amendments to the FD & C Act, 21 U.S.C. § 360e(c)(1)(G), which provides that applications for premarket approval must contain "such other information relevant to the subject matter of the application" which FDA may require.

In the Support Document for its initial proposed regulations, EPA took the position that the information listed in Sections 5(d)(1) and 8(a)(2) is merely "illustrative".[24] To support this contention, EPA pointed to language in the TSCA Conference Report which suggests that Section 8(a)(2) provides representative examples of the information which EPA may require firms to report.[25] However, in addition to being far more ambiguous than EPA suggested,[26] this language relates to the power of EPA to issue reporting rules under Section 8, not to the premanufacture notification requirements of Section 5. Congress could have considered the list of information of Section 8(a)(2) exclusive for purposes of Section 5, but illustrative for purposes of Section 8. In contrast to Section 5, for example, Section 8(a)(1) expressly grants rulemaking power to EPA and, thus, contemplates that the Agency will issue rules to define TSCA reporting requirements more fully.

More relevant evidence of Congressional intent is provided by the portions of TSCA legislative history which relate directly to Section 5. For example, in the course of discussions between EPA officials and Congressmen McCollister and Eckhardt during the 1975 House hearings on TSCA, Deputy Administrator Quarles emphasized that, for the great majority of new substances the information contained in a PMN would be minimal:

> But starting off with the assumption that the new chemicals, we are only talking about number only 600 a year, that is the chemicals used in commercial use, then it is anticipated that the notification would be very, very modest.
> One issue you might wish to explore and push us on is just how modest they will be. Hearings on H.R. 7229, H.R. 7548, and H.R. 7664, *supra*, at 226.

In response to Mr. Quarles, Congressman McCollister urged that Section 5 clearly specify the contents of a PMN in order to prevent EPA from imposing broad notification requirements of its own:

> Mr. McCollister. I would like to explore the issue of how modest your notification requirements are going to be.
> Mr. Quarles. *If they are held to a very limited amount of information, as I feel they should be —*
> Mr. McCollister. *Would you agree that should be in the bill?*
> Mr. Quarles. *In some way, I think so.*
> Mr. McCollister. You see, Mr. Quarles, let me explain the background of that. We live in an era where this Congress has been increasingly reluctant to grant Government or other agencies of Government authority which is likely to be abused. It seems to me that what you are asking for in many instances here is a very wide discretion in determining how this act shall be implemented and I am reluctant to do that. So, would you have any objection, I repeat, to letting us put into the language of the act what notification requirements there shall be?" *Ibid.* (emphasis added).

Seeking to answer this question, Mr. Schweitzer of EPA stressed that EPA envisioned that a PMN would contain *only* the identifying information required in Section 8(a)(2), perhaps augmented by test data which the manufacturer already possessed.

> *All versions of the bill which I have seen, with a couple of exceptions, set forth very modest reporting requirements.* I believe it is on page 25 of your bill, but it is simply the name of the chemical, the chemical identity, the projected uses and the estimates of amounts, and uses and the byproducts.
> Mr. McCollister. And you are saying that the premarketing notification might simply consist of that?
> Mr. Schweitzer. Well, there are two versions. One version which has been simply that. There have been other versions which also call for the company to send any test data which it might have. *But we have never anticipated that the premarket notification would go beyond that. Simply, these categories which have been spelled out essentially the same in most versions of the bill.* Then there is the question whether or not the manufacturer would submit test data which he might have generated on his own, and that is included in several versions of the bill. *In all our planning, we anticipated it to be a very modest premarket notification. Id.* at 226—227 (emphasis added).

After the EPA testimony, the Subcommittee sent a questionnaire to Deputy Administrator Quarles seeking

---

[24] Support Document for Premanufacture Notification Requirements and Review Procedures at 38.

[25] H.R. Rep. No. 94-1679, *supra* at 80.

[26] The TSCA Conference Report states that Section 8(a)(2) "provides an illustrative list of the kinds of activities for which recordkeeping and reporting may be required." H.R. Rep. No. 1679, *supra*, at 80. However, this statement would merely appear to recognize EPA power to select various items on the Section 8(a)(2) list for inclusion in particular reporting rules. Thus, the list would be "illustrative" of the different information that EPA could require industry to report when it fashioned reporting rules for individual chemicals. Had the Conference Committee intended EPA to possess broader reporting powers, it would have included explicit language to this effect in the statute or, at the very least, provided general guidance to EPA on the criteria it should use in expanding the Section 8(a)(2) list.

amplification of the Agency position. That questionnaire noted that "[d]uring the hearings before the Subcommittee. Mr. Schweitzer stated that the information listed in Section 8(a) of H.R. 7229 would be the information required in a premarket notification program." *Id.* at 242. It then inquired about the burden which collecting this information would entail. In its written response, EPA did not dispute the understanding of the subcommittee that only information listed in Section 8(a) would be included in PMNs. To illustrate the modest amount of information required in a PMN, EPA also provided the Subcommittee with a draft form for premanufacture notification. Only two pages long, this form required manufacturers of new chemical substances to provide a limited amount of information which duplicates almost precisely the present requirements of Section 8(a)(2) of the Act.

In sum, the House Subcommittee which drafted the final version of TSCA clearly understood that the contents of a PMN would include only the limited information which the statute itself required. This understanding was based on EPA assurances that premanufacture notification would entail minimal burdens, and on its preparation and circulation to Congress of a draft PMN form which closely mirrored the statutory requirements.

B. The Contents of the PMN

In its original proposed regulations under Section 5, EPA published a draft PMN form which called for nearly 40 pages of mandatory information and 20 more pages of optional information. This form was sharply critized by industry groups, particularly the Chemical Manufacturers Association (CMA) which submitted a proposed PMN form of its own which, it argued, comported more closely with statutory requirements. In its reproposed Section 5 regulations of October 16, 1979, EPA published a revised PMN form which was considerably simplified and reduced in scope but which industry submitters still believed contained a number of questions that were unnecessary and beyond the statutory requirements. In the absence of final PMN regulations, EPA has advised submitters that they are free to utilize any PMN form they wish. 44 *Fed. Reg.* 28565. In practice, most PMN submitters have used the EPA reproposed form or the CMA form, although some PMNs have utilized formats that the submitter has devised itself. While the Agency has not yet reached any conclusions on the subject, it would appear that both the CMA and reproposed EPA forms have proven adequate to evaluate most new chemicals under Section 5.

As a guide to PMN submitters, the following discussion first describes the information called for by the CMA-proposed PMN form. The discussion then focuses on the more significant differences between the CMA form and EPA reproposed form in an effort to identify areas where industry believes the EPA form still exceeds the statutory requirements. Finally, the current EPA and industry thinking on the subject of PMN forms is described in an effort to summarize the impact of the PMN experience to date.

1. The Contents of the CMA Proposed PMN Form under Section 5

Enumerated below is the information which is required in the CMA proposed form for premanufacture notification under Section 5. The PMN form proposed by CMA has two mandatory and two optional parts.

a. Part I: Mandatory General Information

Part I of the form for premanufacture notification proposed by CMA requires submitters to provide EPA with the following information:

Identity of submitter — The PMN form would be signed by an authorized official of the submitter and indicate the legal title and mailing address of the submitter. In addition, the form would designate a technical contact for future communication with EPA and provide an address and telephone number where that contact could be reached. Information on related companies and the state of incorporation of the submitter — which the EPA reproposed form calls for but are not specifically required by the statute and are normally unnecessary in any event — would be omitted.

Chemical identity — This portion of the CMA form would be used to satisfy subparagraph (A) of Section 8(a)(2), which requires the PMN submitter to provide the chemical identity and molecular structure of the new chemical. Using the present EPA format for describing chemical identity, the CMA PMN form requires submitters to categorize their substances as Class I, Class II, or polymers. Manufacturers of Class I and Class II substances would provide the CAS Registry Number of the substance, specific chemical name, synonyms, trademarks, and either its molecular formula or immediate precursors and reactants. In addition, manufacturers of polymers would provide the chemical name and CAS Registry Number of those monomers and other reactants used at greater than 2% in the manufacture of polymers. They would also describe the composition of the polymer by indicating the percentage of its total weight represented by various monomers and the maximum amounts of each monomer which may be present as a residual in the polymer as it is distributed in commerce. Finally, the submitter would estimate the degree of purity of the chemical substance, list those identified impurities which may reasonably be anticipated to be present in the commercial form of the chemical substance, and estimate the maximum percentage of the weight of the substance which each impurity represents. Thus, the Agency would have detailed information concerning the chemical identity of the substance with which to begin its evaluation of the human and environmental effects of the substance.

Production and use information — Submitters are required to indicate the total estimated production volume of the substance for the first, second, and third years of manufacture. This information would reflect the expected production volume of other manufacturers with which submitters have contractual relationships. In addition, submitters would list the actual and proposed categories of use for the substance for this 3-year period and indicate the percentage of anticipated annual production which will be devoted to each use category. Breakdowns of use into function and application — and efforts to correlate each use with population exposure — are not required because this information is not required under the Act, is often confidential, and involves a large element of guesswork. The information contained in this portion of the form would satisfy subparagraphs (B) and (C) of Section 8(a)(2), which call for, respectively, the new categories of the chemical and proposed categories of use and its expected production volume, both on an aggregate basis and for each use category.

Federal Register notice — The CMA PMN form requires the submitter to prepare a *Federal Register* notice concerning the substance. This notice would indicate the generic class of the substance (or, if the manufacturer so elected, its chemical identity) and its actual and proposed categories of use. If the manufacturer has performed testing pursuant to Sections 5(b) or 4 of TSCA, the submitter would also describe the nature of these tests and any data developed. The elaborate information concerning population exposure and environmental release required in the original proposed EPA form would be omitted, and summaries of test data would be required only where the substance is subject to testing requirements under Sections 4 and 5(b). This format for the *Federal Register* notice conforms to Section 5(d)(2) of the Act and performs the public notification function envisioned by Congress without containing excessive detail.

b. Part II: Mandatory Risk Assessment Data

Part II of the CMA PMN form requires submission of the following information for EPA risk assessment concerning the new substance:

Listing of health and environmental data — The PMN form requires submitters to provide a list of the types of available test data relating to the physical and chemical properties and health and environmental effects of the substance and to indicate whether, pursuant to Section 5(d)(1), EPA is receiving the data themselves, a description of the data, or a literature citation. The manufacturer is not required to evaluate the sufficiency of its test data or the overall risk posed by the substance, although such evaluations could be provided to EPA voluntarily.

Occupational exposure — For each site at which the submitter will manufacture, process, use, or dispose of the substance, the CMA PMN form requires it to estimate the number of employees who will be exposed to the substance and the maximum duration of exposure (in hours per day and days per year). The submitter would also disclose the route of exposure and the number of employees subject to each route. The magnitude of exposure, which is not required by the statute and is frequently extremely speculative, would be unnecessary; this information will appear in the test data accompanying the PMN if the submitter already possesses it. Detailed information about other aspects of exposure would also be unnecessary. Thus, in accordance with subparagraph (F) in Section 8(a)(2), EPA will obtain a general picture of worker exposure to the new substance without being encumbered with highly quantitative exposure information.

Disposal — In accordance with subparagraph (G) of Section 8(a)(2), for each site where the submitter will manufacture, process, or use the chemical substance, the CMA PMN form requires him to identify the proposed methods (i.e., air, water, land, or destruction) by which the substance will be disposed of at that site, including release into the environment. The submitter would indicate whether these methods of disposal will occur incidental to manufacture or processing of the substance, or whether they will follow its end-use. The submitter would also indicate whether disposal into the environment by each method will be "minimal", i.e., too small to have material environmental effects. The highly detailed information relating to disposal which is included in EPA proposed PMN form is omitted. This information is often highly speculative, will rarely be useful, is duplicative of EPA functions under other environmental laws, and is not required by statute.

Byproducts — In compliance with subparagraph (D) of Section 8(a)(2), the submitter is required to list the CAS Registry numbers of the byproducts associated with the manufacture of the chemical substance. In contrast to the EPA proposed form, information on feedstocks and intermediates would not be required, and there would be no schematic flow diagram describing the manufacturing process of the submitter.

Industrial Sites Not Controlled by the Manufacturer — Using information already obtained from potential purchasers or in-house estimates and any other information that is reasonably ascertainable, the submitter is required to describe the work place exposure and methods of disposal associated with the processing, use, or disposal of the chemical substance at industrial sites which the submitter does not control. This information could be presented either in an aggregated form or for each individual site. The submitter would give EPA the same information that it has provided for its own manufacture/use sites to the extent it already possesses this information or it is reasonably ascertainable. In addition, the form does not require information concerning exposure to the substance during transport or consumer use, since the only exposure information which the statute requires relates to exposure in the work place. Descriptions of the modes of transport of the substance and information concerning consumer exposure, to the extent relevant, will often appear in the submitter's list of the categories of use of the substance.

c. Part III: Optional Risk Assessment Information

Although the mandatory portions of the CMA form closely parallel the statutory requirements, the form contemplates that PMN submitters may decide to provide additional information to EPA on an optional basis. When a new substance is expected to be manufactured in large quantities or has a potential for adverse effects, it will be in the manufacturer's self-interest to include detailed use, exposure, and environmental information in its PMN. Such substances are likely to receive close scrutiny from EPA, and prudent manufacturers will therefore wish to assemble and submit all possible information which can allay potential EPA concerns. Failure to submit such information would expose the manufacturer to possible uncertainty and delay in manufacturing the new chemical. To guide manufacturers in selecting the additional information that will be submitted to EPA, the optional portions of the CMA form call for the principal types of risk assessment and exposure information which EPA may wish to review for chemicals that receive close scrutiny.

Risk analysis — This section permits the submitter to assess the level of risk associated with the new substance, comment on the implications of the data contained in the PMN, and explain the underlying rationale for its testing program.

Related chemicals — This section permits the submitter to discuss environmental release and human exposure data for related chemicals which are relevant to the new chemical substance, and to point out structure/activity similarities between the new substance and other chemicals which may bear on its biological activity.

General industrial hygiene program — This section allows the submitter to describe the general elements of its industrial hygiene program, thus identifying the industrial hygiene practices which will be applicable to the new chemical substance.

Specific safeguards — This section permits the submitter to describe the anticipated level of exposure to the new chemical substance at work place sites and to indicate the special precautions applicable to the new substance which will be introduced to assure employee safety. The submitter would also be able to identify contemplated environmental release safeguards and disposal procedures and to provide copies of warnings, labeling, material safety data sheets, and instructions which relate to the substance. Similarly, the submitter would be able to indicate whether it intends to control the concentrations of impurities in the new chemical substance because of their potential health and environmental effects. Finally, this section permits the submitter to indicate whether manufacture, processing, or use of the substance will require a new or revised permit governing air or water emissions or solid waste disposal under the Clean Air, Federal Water Pollution Control and Resource Conservation and Recovery Acts. This information would notify EPA if the activities of the submitter regarding the new chemical substance will receive special scrutiny under other environmental statutes that EPA administers. Thus, the Agency would be alerted to any special need for careful examination of the air and water discharges by the submitter and solid waste disposal activities either at the PMN stage or after manufacture has begun.

Industrial process restriction data — This section, which does not appear in the optional or mandatory portions of the EPA proposed form, allows the submitter to provide special information concerning new substances which will be used under controlled industrial conditions. The submitter could describe the specialized categories of industrial use to which the substance will be devoted and factors relating to that use which affect the level of risk associated with the substance. In addition, the submitter would be able to indicate whether the substance will be distributed from its site of manufacture and, if so, in what amounts.

Process chemistry — The submitter would be able to provide any information about the process chemistry for the new chemical substance which may assist the Agency in evaluating its procedures for controlling impurities, by-products, feedstocks and intermediates.

Additional production and use information — This section asks whether the new substance has been manufactured before and, if so, in what amounts. The submitter may indicate whether prior production was restricted because of governmental action or adverse health or environmental effects. The submitter also has the opportunity to explain whether its estimates of anticipated production volume are based on firm orders or forecasts and to identify other persons who will manufacture the new chemical substance by virtue of an existing or planned business arrangement.

Transport — This section calls for a description of special circumstances relating to the transport of the new chemical substance which the submitter believes EPA should consider.

Non-risk factors — This section would permit the submitter to direct EPA's attention to economic factors, such as the benefits of the substance, which EPA should consider in evaluating whether regulatory action against the substance is appropriate.

d. Part IV: Optional Additional Information on Worker Exposure and Environmental Release

The final optional Part IV of the CMA PMN form permits manufacturers of selected new chemicals to report more detailed exposure and environmental release data which they want EPA to review. This part of the form is intended for chemical substances with unusually great exposure potential and chemical substances for which more detailed, quantitative exposure information is appropriate because of physical or chemical properties, test data, or other factors which make it desirable to examine exposure to the new substance with special care. The following information is called for in this part of the form:

Worker exposure — The submitter is permitted to quantify the magnitude of worker exposure to the new chemical substance at each site where it will manufacture, process, use, or dispose of the substance. The submitter would provide these estimates on the basis of "area sampling" and the exposure experienced by a "representative employee." Exposure estimates would be presented in the form of time-weighted concentrations for an 8-hr day, 40-hr work week schedule, and peak concentrations for 15 min. If it wished, the submitter could explain how its exposure estimates were derived; estimate minimum detectable levels of the substance in the work place air; estimate the number of workers exposed to various projected levels of the substance; and list other chemicals which may be present in the workplace environment and are included on a list or in a rule or order issued or entered under Sections 4(a), 5(b)(4), 6(a) or (b), or 7 of TSCA.

Environmental release — The submitter would indicate the production volume for the new chemical substance at each manufacturing site, the hours of operation of the site, and whether manufacture will be "batch" or "continuous." In addition, the submitter would estimate the air and water discharges of the substance expected at each site. This information would be presented on an aggregate basis. The section would not call for discharge estimates for each specific emission point. Nevertheless, substantial background data concerning environmental release could be provided. The submitter would be able to: identify the receiving water bodies into which the substance will be discharged; describe its pollution control equipment, explain the basis for its discharge estimates, list potentially hazardous degradation products that may be formed as a result of the release of the substance, and estimate the minimum level of the substance detectable in air emissions and effluent streams. In addition, the submitter could identify byproducts, coproducts, feedstocks, and intermediates related to the manufacture and processing of the substance which may be present in effluent streams or air emissions and appear on a list that EPA would develop of potentially dangerous pollutants.

Exposure from processing and use at sites not controlled by the submitter — The submitter would provide the same information concerning worker exposure and environmental release for processing and use sites which it does not control. Since this information is optional, the submitter would have no obligation to contact processors or users of the substances in order to fill out the form.

Exposure from consumer use — This section of the form permits the submitter to provide information concerning consumer exposure to the new chemical substance.

2. *Differences between the CMA and Reproposed EPA Forms*

The reproposed EPA form calls for virtually all of the information required by the CMA form, although the terminology of EPA questions often differs from that of CMA questions. In addition, however, the EPA form calls for certain mandatory information which is not required by the CMA form and arguably either exceeds the statutory requirements or is unnecessary. Summarized below are the more important portions of the EPA form that fall into this category, together with a brief description of the reasons of industry for opposing their inclusion in the PMN.

a. *Part I — General Information*

1. Drinking Water Treatment

Question 2c asks whether the new chemical substance will be used to treat drinking water or the drinking water supply, or will be a component of products that will come into contact with drinking water. There is no specific basis in the statute for requiring information about contact with drinking water. Moreover, for most new chemical substances, that information will be irrelevant, either because the substance will never come into contact with drinking water or because it has no potential for adverse effects.

ii. Labels and Material Safety Data Sheets (MSDS)

Question 4 requires the submitter to attach a copy or reasonable facsimile of any material safety data sheet for the new substance, any hazard warning statements, instructions, labels, and technical data sheets concerning the substance that he will supply to customers. Sections 5(d)(1) and 8(a)(2) of TSCA do not call for the submission of labeling or instructional materials which submitters develop for their own and their customers' use. Nevertheless, there may be situations where the submitter wishes to inform EPA of the precautions which it and its customers will take to assure safe use of the substance. As the CMA form suggests, wherever appropriate, such instructional materials could be provided voluntarily.

iii. Customer Information

Question 5 asks the submitter to indicate the number of customers or potential customers for the new chemical which plan to use that chemical for a purpose that is unknown to the submitter. The submitter also must estimate the percentage of the production volume of the chemical that will be purchased by such customers during its first 3 years of manufacture. There is no specific statutory basis for requesting this information. In addition, other PMN information concerning the new uses of the chemical and its worker exposure and environmental release at sites not controlled by the submitter will normally enable EPA to evaluate the completeness of the PMN and the need for obtaining additional information about the activities of the users of the new chemical and processors.

iv. Transport

Section E requests information concerning the transport of the new chemical substance. This information includes the mode of transport which the submitter believes may be used and the Department of Transportation shipping code and hazard class of the substance. Section 5(d)(1) of TSCA does not require a PMN to contain information relating to the transport of a new chemical substance. Nevertheless, in the event that there are special circumstances relating to the transport of a new chemical substance which require explanation, this information can be provided in the optional part of the PMN form.

v. Risk Assessment

Section 5 would require the submitter to provide EPA with any evaluations it has made of the possible risks to health or the environment associated with the new chemical substance. Sections 5(d)(1)(B) and (C) require the submitter to provide EPA with all test data in its possession and control and to identify all other data which are known to it or are reasonably ascertainable. However, these Sections do not require the submitter to prepare its own evaluation of these data or to disclose such an evaluation if it has been made. There may be occasions where the submitter believes a risk assessment would be available to EPA in reviewing its PMN. On such occasions, the submitter may wish to furnish such assessments to the Agency voluntarily. This will typically occur in situations where testing raises some questions about possible toxicity or adverse environmental effects.

vi. Detection Methods

Section G requires the submitter to indicate whether there is an analytical method available to identify and quantify the presence of the new chemical in the work place, air, effluent streams, materials requiring disposal, and end-products for which the new substance is an intermediate. Information concerning detection methods is not included in the statutory description of a PMN in Sections 5(d)(1) and 8(a)(2) of TSCA. Moreover, routine submission of this information will not necessarily contribute to the PMN review process. If a chemical will not be released in large quantities and will have little environmental or human exposure, the availability of methods for detecting it is of little consequence. Similarly, if the chemical has little potential for harm and therefore can be disseminated throughout the environment safely, detection methods again have little relevance. It is possible that there will be a small number of chemicals for which the availability of adequate detection methods will be of concern to EPA — because of the substantial potential for harm, widespread use, and exposure of these chemicals or a combination of these factors. In this event, the manufacturers would describe available detection methods voluntarily in the optional portion of the PMN form.

b. Part II — Human and Environmental Release
i.Process Information

Subsection 1 of Section A requires the submitter to provide the name and location of each site within its control where the new substance will be manufactured or used. For each such site, the submitter must indicate whether the substance will be manufactured, processed, or used; whether its production operations are continuous or batch; the days per year and hours per day during which the operation is carried out; and the amounts of the substance which the submitter expects to manufacture, process, or use at the site.

TSCA does not specifically require a PMN to contain such detailed, site-by-site descriptions of the manufacturing operations of the submitter. Rather, EPA may merely obtain the total production volume for the substance, its categories of uses, and estimates of employee exposure to the substance and its disposal into the environment. Apart from statutory requirements, moreover, detailed breakdowns of the amounts of a substance manufactured or used at a particular site — and data concerning the nature of the production operations of the submitter and the hours and days when those operations are underway — will only be relevant if EPA decides that it needs to make an elaborate quantitative analysis of work place exposure or environmental release. Such analyses will be rare and will be limited to situations where, because of the potential for harm of the substance, manufacturing or processing restrictions are actively being considered.

ii. Block Diagram

The EPA proposed form (January 10, 1979) required the submitter to provide a detailed schematic flow diagram for each site where it manufactured, processed, or used the new chemical substance. The EPA reproposed form replaces this requirement with a request for a "block diagram." The submitter must (1) identify the major chemical reactions and major side reactions involved in its process, (2) provide the approximate mass of all feed materials, by-product materials, and reaction products which are entering and leaving each major operation and conversion, indicating the method of transfer and whether the operation is open or closed, and (3) identify those points in the process where the new chemical substance or by-product materials will be released into the air, land, or water environment.

In its reproposal, EPA has asserted that "this information on the manufacturing and processing operations and resultant environmental releases is critical for EPA's exposure assessments." 44 *Fed. Reg.* 59772. However, the validity of this position is dubious. A description of the production process of the submitter will, at most, make it easier for the Agency to interpret data which it obtains from other sources. Moreover, under Section 5(d)(1) of TSCA, a block diagram is not part of the information which a PMN must contain. Thus, Congress arguably mandated that EPA evaluate potential risks of a new substance not by studying a blueprint of the manufacturing operations of the submitter, but by evaluating the information on chemical identity, uses, worker exposure, and test results which other portions of the PMN will contain. Instead of looking "upstream" to the mode of manufacture of the substance, Congress thus believed that EPA would look "downstream" to its properties and effects.

iii. Occupational Exposure

In addition to requiring the number of individuals who will be exposed to the new substance at their places of employment, the probable duration and route of such exposure, and the number of employees who will be exposed to the substance by each route, EPA proposes to require substantial additional information concerning worker exposure which is arguably unnecessary and unauthorized. Specifically, the Agency wants submitters to estimate the average and peak concentration levels to which workers are exposed during manufacture, processing, use, and disposal of the substance. Section 8(a)(2)(F) does not require the submitter to quantify work place exposure. Instead, it focuses on the number of individuals exposed and the duration for which such exposure will continue. Moreover, rigorous analyses of work place air content or of potential inhalation or skin absorption levels may often be impossible until commercial production of the new chemical substance has actually begun. Without conducting monitoring studies in the actual plant environment where the substance will be produced or processed, efforts to quantify exposure will be so imprecise as to be valueless.

Subsection 3 also asks the submitter to describe those operations in which workers will be directly exposed to the new chemical substance and to identify the physical states (e.g., aerosol or mist) of the substance during such exposure. This information, too, is arguably not required by the statute. Moreover, it will only be useful to EPA in those instances where the toxic potential of the chemical or other factors necessitate a detailed evaluation of work place exposure conditions. Thus, industry has argued that to require these data on a routine basis is unjustified.

iv. Related Chemicals

For each site where the new chemical substance is manufactured, Question 3.5 of Subsection 3 requires the submitter to list other substances — including byproducts, coproducts, feedstocks, and intermediates — associated with the manufacture of the new chemical substance that may reasonably be anticipated to be present in the work place and to which workers may be exposed. Under Section 8(a)(2)(D) of TSCA, which is incorporated into Section 5(d)(1), the PMN for a new substance must describe the byproducts resulting from the manufacture of the substance. Section 8(a)(2) does not, however, refer to the submission of information about other chemicals, such as feedstocks, co-products, and intermediates, which may be associated with the manufacture of a new chemical substance. EPA is therefore arguably without authority to require the submitter of a PMN to provide information about these chemicals.

Moreover, as discussed above, industry has taken the position that requiring such information would be contrary to the intent underlying Section 5. The purpose of TSCA premanufacture notification provisions is to permit EPA to review the potential health and environmental effects of manufacturing, processing or distributing *new chemical substances* which the submitter of the PMN intends to produce for commercial purposes. Data concerning existing chemicals which play a role in some phase of the manufacture, processing, or disposal of the substance are not directly relevant to this subject. To the extent that existing chemicals used in the manufacture of new chemical substances such as feedstocks or intermediates require regulation because of their effects on work place safety, EPA can invoke the other provisions of the Act where appropriate.

v. Environmental Release and Disposal

Subsection 4 of this part of the reproposed form requires the submitter to provide information concerning environmental release and disposal of the substance at each site where the submitter will manufacture, process, use, or dispose of it. The submitter must indicate the duration of release of the substance (in hours per day and days per year) into the air and water environment. For air, land, and water releases, the submitter must then indicate the annual amount in which the chemical will be discharged. In addition, the disposition of water discharges (e.g., POTW, navigable waterway, etc.) must be identified and the submitter must estimate the effluent stream flow rate (in gallons per day) from the site. Finally, for each release point indicated in the block diagram of the submitter, it must characterize the composition of release materials and describe any pollution control equipment and disposal operations used to treat those materials.

Section 8(a)(2) of TSCA, which is incorporated into Section 5(d)(1), does not mention environmental release as such. Instead, it merely provides for a description of the ''manner or method of disposal'' of the new chemical substance. At most, this provision arguably requires the submitter to specify the mode of disposal for the substance (including the environmental media into which it will be released) at each manufacture and processing site that the submitter controls or at other sites where the substance is used or discarded. This information concerning disposal is required by the CMA proposed PMN form. The statute contains no authorization for an additional requirement that the submitter indicate the duration of release for the new substance, the quantities in which those releases will occur, the composition of release materials, or the pollution control measures of the submitter. Moreover, since commercial production of the substance has not yet begun, the environmental release data required by EPA would often be both difficult to obtain and highly speculative. Instead of analyzing existing operations, submitters would be forced to make projections about production processes that may not be fully defined, let alone implemented. The result would be to complicate the task of the submitter while providing EPA with data that may have a large margin of error. This is particularly true of EPA questions concerning the composition of release materials at each release point and the pollution control measures of the submitter.

vi. Byproduct Materials

Question 3 in Subsection 4 of Section B requires the submitter to list by-product materials containing the new chemical that are generated during manufacture, use, and processing operations and are disposed of. The submitter must indicate the method of disposal for each such material, estimate the amount of the material to be generated, and estimate the percent of the material comprised by the new chemical substance. While Section 5(d)(1) permits EPA to obtain the identities of byproducts generated during the manufacture or processing of a new substance, the additional information that this question seeks is arguably not authorized by the statute and has little relationship to basic EPA responsibilities under Section 5.

Obtaining and compiling the information that this question calls for may also require considerable effort. If the new chemical will be manufactured, processed, or used at several sites, the byproduct materials generated during these activities could be quite numerous. Moreover, determining the amounts in which these materials will be produced — and the portion of such amounts that the new chemical represents — may be difficult and time-consuming, particularly since this information must be obtained from persons other than the submitter.

These other chemicals are, of course, subject to regulation under all of the other provisions of TSCA. Thus, EPA can require their testing under Section 4, restrict or prohibit their production under Section 6, or impose reporting requirements concerning their uses and effects under Section 8. Moreover, EPA now administers an elaborate statutory and administrative scheme — including RCRA, a statute expressly aimed at waste materials — for regulating chemical discharges into the air, water, or earth. Existing mechanisms for informing the states and EPA about the disposal practices of chemical manufacturers, and permitting these authorities to initiate regulatory action where necessary, make it inappropriate for EPA to enlarge the premanufacture notification process so it encompasses all chemicals disposed of during production, processing, or use of a new substance.

vii. Consumer and Industrial User Exposure

Section C of the EPA reproposed form requires the submitter to identify consumer and commercial categories of use for the new chemical. The submitter must indicate whether these use categories are consumer or commercial in character and whether the product in question is manufactured by the submitter or someone else. For each such use category, the form also calls for the routes of population exposure, the number of persons exposed, and the frequency of exposure. Finally, for each product containing the new chemical, the submitter must explain any aspect of its construction or formulation which limits the potential for exposure to the new chemical and, in the case of products that are mixtures, the percent by weight that the new chemical comprises. Byproducts formed from all consumer and commercial categories of use must also be listed.

While Section 8(a)(2) of the Act provides that the PMN must describe use categories for a new chemical substance, including consumer and commercial uses, the Section does not require the PMN to provide quantitative data concerning population exposure to that substance. Moreover, such estimates are necessarily the most speculative of the various forms of exposure information available since they rest on projections which are subject to a large margin of error. This is particularly true when, as is often the case, the consumer or commercial product containing the substance is manufactured or sold by a company several steps removed in the distribution chain from the submitter.

For these reasons, consumer and commercial exposure information arguably cannot be included in the mandatory portion of the PMN form but should be requested, if at all, in the optional part of the PMN. Such optional information — coupled with the submission of existing data or descriptions of data and a judicious use of EPA powers under Section 5(e) — will allow EPA to make selective risk evaluations where appropriate for those new chemicals used in consumer and commercial products.

viii. Federal Register Notice

In response to the comments of CMA and others, EPA has substantially streamlined Part IV of its reproposed PMN, which calls for information that will be published in the *Federal Register* in accordance with Section 5(d)(2) of TSCA. Nevertheless, three areas of the EPA reproposed *Federal Register* notice remain objectionable to industry:

1. Section 5(d)(2) specifies that the *Federal Register* notice will identify the substance "by generic class" unless EPA determines that "more specific identification is required in the public interest." Section A on the EPA proposed notice, however, requires the submitter to enter the specific chemical name except when it is confidential; in this event, a generic name is appropriate. This approach appears to invert the statutory requirements. In accordance with Section 5(d)(2), the *Federal Register* notice prepared by the submitter must contain only the generic class of the substance. The specific chemical identity of the substance may arguably be included in the notice only if, after reviewing the PMN, EPA expressly determines that a more specific identification is required in the public interest or if the submitter consents to the use of the specific chemical identity voluntarily.

    Under proposed § 720.22(b) of its reproposed regulations, EPA would use a generic name to describe the substance in the *Federal Register* whenever the specific chemical identity of the substance is confidential. This approach, too, appears to conflict with the statutory requirements. While other provisions of the Act such as the inventory requirements of Section 8(b) contemplate the use of a generic name if the identity of the substance is confidential, Section 5(d)(2) specifically provides that the *Federal Register* notice concerning a new substance will describe its "generic class."

2. Section B of the EPA *Federal Register* notice requires the manufacturer to disclose its identity. Section 5(d)(2) does not provide that the *Federal Register* notice must contain this information. Thus, Congress would appear to have made a considered judgment that disclosure of the generic class and uses of a chemical — together with a list of test data concerning the chemical in certain limited circumstances — will provide the public with the basic identifying information about a PMN that it needs. EPA arguably should not bypass this determination by requiring disclosure of other information, i.e., identity of the manufacturer, which Congress concluded was unnecessary to inform the public of EPA receipt of a PMN.

3. Section D of the EPA *Federal Register* notice calls for a list of all data submitted, described, or cited as part of the PMN. The submitter must also provide a brief abstract of the results of tests that it has performed. However, Section 5(d)(2) expressly provides that the *Federal Register* notice will describe only data developed pursuant to Section 5(b) or a test rule under Section 4. Thus, Congress decided that a description of data would not be required in the *Federal Register* notice for every new chemical substance, but only when the substance is included in the EPA list of "suspect" chemicals under Section 5(b) or is subject to testing requirements under Section 4. EPA arguably should therefore revise its format for the *Federal Register* notice so that test data will only be described in these limited circumstances.

3. Experience of EPA and Industry with Different Forms during the PMN Process to Date

To date, most PMN submitters have used the EPA reproposed form, but a significant number of submitters have used the CMA form, and many submitters have utilized formats that they have devised themselves. In the economic analysis it prepared for CMA, the Regulatory Research Service ("RRS") found that, in accord with the earlier Arthur D. Little estimates, the CMA form has been significantly less expensive for PMN submitters than the reproposed EPA form. For example, for firms of comparable size, the pre-filing costs of submitters that used the CMA were approximately $1,000 less than those of their counterparts who used the reproposed EPA form. Significantly, these estimates were based on the costs of completing the CMA form in its entirety, including the optional sections. Moreover, while post-filing costs were somewhat greater with the CMA than the EPA form, these differences were not substantial. Thus, use of the CMA form has apparently not resulted in incomplete information or in repeated follow-up contacts between the Agency and the submitter.

In addition, regardless of the form used, many submitters have exercised discretion in selecting the information provided to the Agency, answering some questions but not others. While the Agency has not yet announced any conclusions on the subject, the absence of a standardized PMN format has apparently not undermined EPA's ability to perform its review function. Equally significant, the Agency's success in reviewing a variety of PMN submissions since July 1, 1979, demonstrates that it does not need the extensive information called for by its PMN form and that PMN notices that contain only the information prescribed by statute are adequate for risk assessment purposes.

Significantly, the history of PMN submissions to date indicates that the initial PMN submission is only a starting point for the Agency's assessment process. In reviewing PMNs, the Agency has frequently supplemented this information from a variety of other sources, including scientific literature and other federal agencies. It has also directed informal requests for additional information to the PMN submitter. In the great majority of cases, these requests have been reasonable in scope, and PMN submitters have provided the information sought. EPA's reliance on these informal information-gathering mechanisms suggests that the most cost-effective way to implement Section 5 is to prescribe minimal mandatory reporting requirements and then to use informal procedures to obtain supplemental information on a chemical-by-chemical basis.

Supporting this conclusion is the emerging evidence that, for many PMN chemicals, EPA's proposed form calls for more information than the Agency will in fact utilize in assessing potential risks of the new chemical. It is increasingly apparent that EPA will devote minimal attention to chemicals that will be manufactured in limited volume, have limited human or environmental exposure, or have physical and chemical properties associated with limited toxicity. For such chemicals, the relatively detailed information called for by EPA's proposed PMN form — or even by the statutory provisions themselves — is excessive and unnecessary. Nevertheless, fearful that EPA may decide that the information in their PMNs is incomplete, manufacturers of such chemicals may often take unnecessary pains to prepare full and complete PMNs.

The experience of EPA and industry with different forms during the PMN process to date has motivated a reevaluation of the Agency's approach to obtaining PMN information. EPA officials have recently expressed an interest in a PMN form more streamlined than the reproposed form now in use. In addition, some industry representatives have suggested that no PMN form may be necessary and that, instead, the Agency should simply provide general guidance to PMN submitters concerning its review process and information needs.

C. The Obligation of the PMN Submitter to Obtain Information It Does Not Already Possess

Section 5(d)(1)(A) of TSCA provides that a PMN must contain the information listed in Section 8(a)(2)(A) — (D) and (F) — (G) insofar as this information is "known to" the submitter or is "reasonably ascertainable." Similarly, under Section 5(d)(1)(C), the submitter must provide a description of data concerning the environmental and health effects of the new substance insofar as such data is "known to" it or is "reasonably ascertainable." Thus, under limited circumstances, the PMN submitter is required to take steps to include in its PMN information which it does not presently possess but is able to obtain.

There have been substantial differences of opinion between EPA and industry concerning the scope of this requirement. As discussed below, these differences involve: (1) the responsibility of the PMN submitter to obtain any and all information which is available with "reasonable" expense and effort, and (2) the duty of the PMN submitter to obtain information about the activities of prospective customers for the new chemical.

1. The General Definition of "Reasonably Ascertainable"
Proposed § 720.2 of the EPA proposed PMN regulations provides that:

"known to or reasonably ascertainable" means all information in a person's possession or control, plus all information that a reasonable person similarly situated might be expected to possess, control, or know, or could obtain without unreasonable burden or cost.

Under this definition, the submitter would be required to include in its PMN all information which it "could obtain without reasonable burden or cost." This information must be provided even though neither the submitter nor other businesses in its position would obtain that information in the normal course of preparing to manufacture and sell the new chemical substance.

The TSCA legislative history suggests that the EPA concept of "reasonably ascertainable" information is unduly broad and that PMN submitters have no obligation to develop new information or to obtain existing information which a normal business in the position of the submitter would not already possess. For example, responding to fears that Section 5 would impose excessive burdens on industry, the Senate Report states that:

While the EPA administrator must be given the authority to act during the premarket notification period to gather more data or to take appropriate restrictive action, the notification burden itself should not be onerous. *Unless testing has been otherwise required, notification only consists of reporting routine information which should be in the hands of the manufacturer in the first place.* Included is information as to the identity of the product, categories of use, estimates of the amount to be produced and, insofar as reasonably ascertainable, to be produced for each of the categories of use, a description of byproducts, lists of existing test data, and estimates of the number of persons who will be exposed in their places of employment. S. Rep. No. 94-698, *supra*, at 10-11 (emphasis added).

A similar understanding was expressed by EPA itself in its written response to the questionnaire of the House Subcommittee. When asked whether manufacturers would already possess the information listed in Section 8(a) and how EPA would define the term "reasonably ascertainable", Deputy Administrator Quarles responded as follows:

The information required as part of premarket notification of a new chemical should be information which is readily available to the manufacturer. *We do not anticipate that the manufacturer will be required to perform additional research and testing to determine any of the information requested.* A copy of a preliminary draft of a combined annual and premarket notification report form is attached. *"Reasonably ascertainable" information would include that information that the manufacturer has already developed or obtained.* Hearings on H.R. 7229, H.R. 7548, and H.R. 7664, *supra*, at 242 (emphasis added).

The EPA position was that a PMN would only include existing information which the manufacturer had already obtained.

In discussing the term "reasonably ascertainable", the Conference Report identifies one narrow situation where a submitter is expected to obtain information which is not already "known to" it:

The conferees intend that the "reasonably ascertainable" standard be an objective rather than a subjective one. Thus, the manufacturer or processor must provide information of which a reasonable person similarly situated might be expected to have knowledge. H.R. Rep. No. 94-1679, *supra*, at 80.

Under this approach, information will be "reasonably ascertainable" only if it is information which a normal business in the submitter's position would usually possess already. If it would not be customary for such information to be obtained as part of the regular business operations of a firm in the submitter's position TSCA would create no special obligation to obtain it merely so it can be included in a PMN.

Based on TSCA's legislative history, CMA has proposed that "known to or reasonably ascertainable" be redefined as follows to clarify that submitters need not obtain information which they would not customarily acquire in the normal course of business:

"known to or reasonably ascertainable" means all information which a business in the submitter's position would usually possess in the normal course of preparing to manufacture the new chemical substance for commercial purposes, taking into account customary business practice in light of all relevant economic and safety considerations relating to the new chemical substance.

Under this definition, the submitter would be required to obtain information it does not already possess only if it would be reasonable business practice for a firm initiating a comparable venture to acquire that information. The definition makes it clear that, in applying this test, the controlling criteria will be the normal practices of a business initiating a venture comparable to that of the submitter, taking into account all relevant safety and economic considerations relating to the particular new chemical substance involved.

For example, a submitter would not be compelled to identify the byproducts produced during its manufacturing process in small quantities unless a chemical manufacturer marketing a comparable product would usually obtain that information on the basis of normal business, safety, or environmental considerations. Similarly, if a company regularly used "Toxline" and other established data banks to identify all published toxicological studies concerning comparable new products, then information derived from these sources would be "reasonably ascertainable" for purposes of providing descriptions of data under Section 5(d)(1)(C). On the other hand, if the submitter and other companies in its position did not normally use "Toxline", or if they conducted a literature search concerning new chemicals but did not contact the original investigators and ask to examine the underlying data, these steps would not be required under Section 5 because the information in question would not be "reasonably ascertainable." Thus, the controlling standard would be one of customary business practice; submitters would not be expected to possess information inappropriate for a venture with comparable needs, objectives, and responsibilities.

A definition of "reasonably ascertainable" based on normal business practice would avoid the legal and practical pitfalls in the EPA proposed approach. Whether information could be obtained "without unreasonable burden or costs" will often be a matter of subjective judgment. In the absence of a clear standard, EPA could often assert that the efforts of the submitter to obtain information have been insufficient and that additional effort should be devoted to chemical analyses, projections of human or environmental exposure, or contacts with potential users and processors. Faced with an expansive EPA interpretation of "reasonableness," submitters would be hesitant to allocate their resources in a prudent and realistic manner for fear that EPA will later determine that their PMNs lack "reasonably ascertainable" information.

A test based on normal business practice would avoid these dangers and effectuate the Congressional intent to include in the PMN only that information which the submitter would normally have already. Moreover, in accordance with Congressional desires, such a test would be based on "an objective, rather than a subjective" standard and, thus, would provide clear guidance both to submitters and EPA. See H.R. Rep. No. 94-1679, *supra,* at 80.

2. Mandatory Customer Contact

Section 720.20 of the initial EPA-proposed PMN regulations would have created an elaborate scheme for "mandatory customer contacts" by submitters of a PMN. Under this scheme, submitters would have had to contact all persons who are parties to contracts to obtain the substance, all persons who have expressed an interest in obtaining the substance, all persons who have obtained a sample of the substance, and all others whom the submitter has either contacted or will contact and firmly believes will purchase the substance during its first 3 years of production. The submitter would have had to provide these persons with the detailed EPA "processing and consumer use form" and request that the form be completed and either be returned to it or forwarded to EPA. This form called for extensive information concerning the uses of the substance, its release into the environment, its disposal, and the disposal of associated materials and general population exposure. 44 *Fed. Reg.* at 2343.

If the submitter did not contact all of the persons specified by EPA, it would have had to demonstrate that the information which they would provide is either duplicative or will not add to the information which the submitter has included in the PMN. Moreover, in the PMN itself, the submitter would have had to certify that it has contacted all prospective users in accordance with proposed § 720.20 and list their names and addresses. EPA indicated that recipients of the "processing and consumer use form" were not legally obligated to complete it. Nevertheless, the Agency warned that, when it considered the data in a PMN concerning such uses to be incomplete, it would make "worst case assumptions about possible exposures and risks associated with" the new chemical substance. 44 *Fed. Reg.* 2244.

Industry vigorously opposed this approach. It argued that the number of persons that the submitter would have to contact could be extraordinarily large, and that merely identifying all those whom the submitter would be obligated to contact under proposed § 720.20 could itself be a very time-consuming task. It also argued that mandatory customer contact would create confusion and ill will among potential purchasers of a new substance at the very time when the submitter is attempting to create a favorable marketing climate. According to industry, there is a strong possibility that prospective customers would perceive EPA interest in the substance as evidence that it is "suspect" and might be subject to regulatory action or further reporting requirements over time. Troubled by this prospect, the customer could well decide that using the substance entails a high level of commercial risk and, for this reason, might turn to existing chemicals which are not under government scrutiny.

The sizeable drawbacks of mandatory customer contact, industry maintained, would not be offset by equivalent benefits. Because there is generally a substantial incubation period before the use patterns of a new chemical are firmly established, industry argued, most prospective customers for a new chemical substance would therefore not have firm plans for its use. Accordingly, the prospective customer would probably not yet know the quantities of the substance which it plans to use and the precise function which the substance would serve in its manufacturing or processing operations. For this reason, virtually all of the information requested on the EPA "processing and consumer use form" would be either totally unavailable or available in such a speculative form as to be essentially meaningless.

Finally, industry argued that, if "reasonably ascertainable" is defined in accordance with the Congressional intent, EPA would lack statutory authority to impose mandatory customer contact. As discussed above, Congress apparently considered information "reasonably ascertainable" if it were information that a firm in the position of the submitter would usually possess in the normal course of business operations. Under this approach, EPA would have no authority to require customer contacts which would not occur as part of normal business practice. Rather, such contacts would be required only to the extent that the submitter would normally obtain processing and use information from its prospective customers. If it is not customary to contact some or all of these customers, or if such contacts normally involve a discussion of some matters but not others, TSCA would not impose additional obligations on the submitter.

In view of these concerns, the EPA reproposal indicated that the Agency is reconsidering its scheme for mandatory customer contact. 44 *Fed. Reg.* 59770. However, the EPA reproposal also announced that the Agency is considering two alternative mechanisms for obtaining processing and use information from prospective purchasers of new chemicals.

One mechanism is requiring the PMN submitter to provide a list of the customers who have made a firm commitment to purchase the new chemical. 44 *Fed. Reg.* 59770. However, no provision of Section 5(d) or Sectin 8(a)(2) of TSCA contemplates the inclusion of this information in a PMN. Moreover, as EPA itself recognizes, the routine submission of customer lists is unnecessary, since the Agency will not normally ask purchasers of a new substance to provide processing and use information. Finally, lists of prospective customers of new chemicals represent highly confidential information.

The second approach which EPA is considering is to require the PMN to indicate the number of customers who have made firm commitments to purchase the substance for unknown categories of use and to state the percent of the estimated production volume of the new chemical that such customers are expected to purchase. Based on this information, EPA hopes to ascertain the completeness of the use information in the PMN and the desirability of obtaining additional information directly from the prospective purchasers of the new chemical.

Once again, however, the information that EPA proposes to require would not appear authorized by the statute. Moreover, the EPA objective — assessing the completeness of the processing and use portions of the PMN — can be achieved without obtaining this information. By examining other portions of the PMN, EPA can determine whether there is any aspect of the processing or use of the new chemical which warrants further investigation. In the event that EPA concludes that it needs further information about the activities of the processors or users of the chemical, it can obtain this information voluntarily or under Section 5(e).

Because of these factors, industry has taken the position that EPA does not need to expand the PMN beyond the statutory requirements in order to obtain information about the processing and use activities of the purchasers of a new chemical. Accordingly, it has argued that even the narrowed customer-related questions that EPA is considering including in its PMN form should be deleted entirely.

VI. CONFIDENTIALITY ISSUES

Almost by definition, any new chemical substance will be the result of innovative activity — either in synthesizing a hitherto unknown compound or creating a commercial application for a compound that is known but has only been used in the past for scientific purposes. Because of the time and money which this R & D effort will require, and because of the commercial potential of the new chemical, PMN submitters have an important interest in maintaining the confidentiality of PMN information. This interest is particularly great before the new chemical has actually been manufactured for commercial purposes since, at this stage, the PMN submitter has not yet obtained marketplace acceptance for the chemical. Virtually all of the information contained in a PMN could potentially be confidential. This includes not just data concerning the projected uses and production volume of the new chemical, but information concerning its physical and chemical properties, the specific identity of the chemical, and even the identity of the PMN submitter itself.

During EPA efforts to implement Section 5, a host of complex and difficult issues have emerged concerning the confidentiality of PMN information. These issues reflect the detailed requirements of the TSCA confidentiality provision, Section 14, and other applicable legal standards. They also reflect the well-intentioned but inevitably controversial effort of EPA to balance, on the one hand, the substantial interest of the PMN submitter in trade secret protection and, on the other, the pressure of public interest groups to maximize the openness of the PMN review process. The mechanisms that EPA has devised to resolve these tensions, while ingenious, are burdensome, potentially counterproductive, and, at present at least, legally untested.

Below, we discuss the major confidentiality issues raised by the EPA January 10, 1979, proposal and October 16, 1979, reproposal. These issues are (1) the disclosure of specific chemical identity as part of a health and safety study, (2) the concept of advance substantiation of confidentiality claims, (3) public disclosure of information claimed confidential in a generic form.

A. The Disclosure of Specific Chemical Identity as Part of a Health and Safety Study

1. The EPA Proposal

Section 14(a) of TSCA prohibits EPA, except in certain limited circumstances, from disclosing information that falls within Exemption 4 of the Freedom of Information Act (FOIA), 5 U.S.C. § 552(b)(4). This exemption applies to "trade secrets and commercial or financial information obtained from a person and privileged or confidential." Thus, under Section 14, the protection of confidential business information is nondiscretionary. In accordance with other recent regulatory statutes, TSCA places EPA under a mandatory duty to withhold such information from public disclosure.[27]

Section 14(b) creates an exemption from this duty for "any health and safety study" submitted to EPA under TSCA which relates to a chemical that has been offered for commercial distribution or is subject to premanufacture notification under Section 5. Under Section 14(b)(2), EPA may not invoke FOIA Exemption 4 to withhold such studies or any data obtained from them unless the information in question is among the categories of information enumerated in Section 14(b)(1). This provision, in turn, applies to data which discloses confidential processes used to manufacture or process a chemical or the portion of a mixture comprised by the component chemicals. Thus, trade secret data which meet this description cannot be disclosed even when they are contained in a "health and safety study."

EPA has taken the position that, since chemical identity is not specifically mentioned in Section 14(b)(1), Section 14(b)(2) requires the disclosure of such identity whenever it appears in a "health and safety study" for a new chemical which has entered commercial production. Reflecting this analysis, § 720.43 of the EPA-proposed PMN regulations precludes the Agency from disclosing the specific chemical identity of a confidential new chemical substance before manufacture has begun. Under the Agency proposal, however, it will disclose the specific chemical identity of a confidential new substance after the commencement of manufacture if a "health and safety study" for that substance is submitted to the Agency.[28] Moreover, when the specific chemical identity of a new substance appears in such a study, EPA proposes to use that identity to describe the substance on the Section 8(b) Inventory. *See* 44 *Fed. Reg.* 2257. The Agency proposes to follow this procedure even if there has been no FOIA request for health and safety studies where the specific chemical identity of the substance appears.

2. Industry Legal Arguments

Industry representatives have argued that EPA has no authority under Section 14 to disclose the chemical identity of a confidential new substance at *any* time, either before or after manufacture, and that, in asserting such authority, the Agency has adopted an impermissibly broad definition of "health and safety studies" which is contrary to TSCA.

Section 3(6) of the Act defines "health and safety study" as:

> any study of any effect of a chemical substance or mixture on health or the environment or on both, including underlying data and epidemiological studies, studies of occupational exposure to a chemical substance or mixture, toxicological, clinical, and ecological studies of a chemical substance or mixture, and any test performed pursuant to this Act.

The focus of this definition is on the health or environmental effects of the new chemical substance, and on underlying data in the study which relate to these effects. Thus, industry has argued, information which is not directly related to the health or environmental effects that are under investigation cannot be considered either part of a "health or safety study" or "data" underlying that study. EPA itself has recognized this principle. Under proposed § 720.43(a)(3), confidential information in a health and safety study will not be disclosed if it "is not in any way related to the effects of a substance on human health or the environment."

Section 3(6) does not state that the identity of a chemical is automatically information included in a "health or safety study". EPA has asserted, however, that "it is clearly *implicit* in the statute that chemical identity is part of a health and safety study." 44 *Fed. Reg.* at 2256 (emphasis added). The Agency apparently believes that chemical identity is *always* related to an evaluation of the health or environmental effects of

---

[27] *See, e.g.*, Section 6 of the Consumer Product Safety Act, 15 U.S.C. § 2055.

[28] It is unclear whether EPA intends to disclose chemical identity only when it actually appears in such a study, or whenever a health and safety study has been submitted, regardless of whether it refers to specific chemical identity. On the one hand, proposed § 720.41(b)(2)(ii) requires disclosure of specific chemical identity "as part of a health and safety study." On the other hand, "health and safety study" is defined in proposed § 720.2 to include the specific chemical identity of the substance to which the study pertains.

a new chemical and thus is automatically "part" of any study of such effects for purposes of Section 14(b).

Industry has challenged this reasoning, however, by pointing out that it is common practice for many health and safety studies to omit an explicit description of the chemical under investigation. The reason for this omission is usually that the investigators do not need to know the specific identity of the chemical in order to accomplish the objectives of the test. For example, studies of occupational exposure frequently will monitor the parts per million of contaminant X at a given factory site; knowledge of the specific chemical identity of the contaminant will be unnecessary to measure such exposure. Similarly, a chemical may be administered to test animals in order to evaluate its potential for short- or long-term toxicity; the specific identity of the chemical will normally play no part in monitoring and evaluating the reactions to the chemical which test animals exhibit. Indeed, commercial testing laboratories commonly process test samples under code names. This practice not only serves to safeguard confidentiality, but avoids biasing the test results. Thus, knowledge of specific chemical identity often is not merely unrelated to the purposes of testing, but may even interfere with proper test procedure as well.

Based on these considerations, industry has argued that the mere fact that the test substance is identified either in a study itself or in the data accompanying the study does not establish a connection between that identity and the scientific objectives of the study. Normally, the specific chemical identity will be mere background information which has played no role in developing test data. In this situation, industry has asserted, there is no reason why inclusion of chemical identity in the study should jeopardize its confidentiality. Had knowledge of the generic class of the substance, rather than its specific chemical identity, been provided to the investigators, their ability to generate adequate test data concerning the substance and evaluate those data intelligently would not have been impaired.

3. The Competitive Impact of Disclosing Chemical Identity

Under the EPA approach, industry has warned, the specific chemical identity of a new substance could be disclosed within a matter of days after the PMN submitter has notified EPA that manufacture has commenced. Thus, competitors would learn the identity of the chemical several weeks or months before the product is actually distributed, and years may pass before the manufacturer has established a solid marketing base. If competitors could copy and market the chemical immediately after manufacture has commenced, the advantage possessed by the innovator would be reduced, and his opportunity to reap rewards from his R & D investment would be curtailed. Similar competitive harm would be caused by disclosure of the chemical identity of a new substance which plays a vital role in manufacturing a finished product but is present in that product in undetectable amounts. If competitors learn the identity of such chemicals, their ability to copy the production process of the original manufacturer and reproduce the attributes of its finished product will be greatly enhanced.

EPA itself has recognized that there is often great commercial value in preserving the confidentiality of chemical identity. In the preamble to both the proposed and final inventory regulations, EPA stated that chemical identity can be confidential and adopted regulations specifically designed to protect that confidentiality.[29] Thus, the final inventory regulations specifically provide that a manufacturer may claim confidential treatment for the chemical identity of a chemical substance and that a generic chemical name will then be used to describe the substance on the inventory. As EPA has explained the basis for this approach:

> The inventory is a list of chemical substances manufactured (including imported) or processed for a commercial purpose. Many chemical substances have been developed and synthesized for which no commercial purpose has been found. The fact that someone has found a commercial purpose for a particular chemical substance may be a confidential trade secret. Placement of the specific chemical identity on the inventory would announce that fact to potential competitors who might be able to narrow their research activities. This problem would be further compounded if the chemical substance were newly synthesized and known only to the person reporting it to EPA or if the substance were patentable, in which case inclusion on the inventory might constitute a publication and limit the person's patent rights. 42 *Fed. Reg.* at 64590.

The Agency-proposed regulations under Section 5 continue the inventory procedure by permitting the PMN submitter to publish the generic chemical class of the new substance in the *Federal Register* under

---

[29]   *See* 42 *Fed. Reg.* 39182, 39188—39190 (August 2, 1977) and 42 *Fed. Reg.* 64572, 64573—64574 (December 23, 1977), proposing and promulgating 40 C.F.R. § 710.7.

Section 5(d)(2)(A). Moreover, in proposing to protect the confidentiality of specific chemical identity before manufacture of the new substance has begun, EPA has acknowledged that "often the first entrant into a new market has a very real competitive advantage" and "[d]isclosure of confidential identity at that time could result in competitive harm and, more broadly, in reduction of technological innovation." 44 *Fed. Reg.* at 2256. These considerations arguably apply as forcefully after manufacture as before.

EPA procedures for protecting the confidentiality of chemical identity will be undermined if, after manufacture begins, that information can be obtained by any member of the public by requesting disclosure of a health and safety study. It could easily become common practice for the trade press or commercial competitors routinely to request public disclosure of health and safety studies for all chemical substances for which PMNs have been submitted. The serious impact on innovation which this situation could have is suggested by the EPA estimate that PMN submitters will seek confidential treatment for chemical identity "in a high percentage of cases." 44 *Fed. Reg.* 2256.

4. The Public Interest Considerations Involved in Disclosing Chemical Identity

According to EPA, disclosure of chemical identity will facilitate public evaluation of health and safety studies for a new chemical substance, permit the replication of tests on that substance, and generally promote "public oversight" of the premanufacture notification process. 44 *Fed. Reg.* 2256. While these goals are not unimportant, it can be questioned whether the disclosure of chemical identity is necessary to achieve them.

As discussed above, it is only in extremely unusual cases that the chemical identity of a test substance is pivotal for the public to review test procedures or analyze test results. Even then a generic class will often suffice to provide the needed information. Thus, if chemical identity is withheld but the public has access to the remainder of a health and safety study, informed and intelligent public discussion of the safety of the substance will normally be unimpaired. Moreover, replication of test data is extremely uncommon among noncommercial observers of the regulatory process. Duplication of toxicological tests is expensive and time-consuming, and citizen groups and independent scientists therefore concentrate primarily on assessing the significance of existing data. Finally, if there are flaws in testing methodology, they will normally be detected by the EPA staff, who will necessarily review the data for a new chemical substance much more carefully than public observers and will, of course, have full knowledge of the specific identity of the substance.

In any event, if knowledge of chemical identity has any role to play in evaluating health and safety studies, that role could probably be served adequately by disclosing the generic classes (i.e., key functional groups) of new substances. Generic class descriptions of new chemicals represent the identification method which Congress chose for the *Federal Register* notice required under Section 5(d)(2). If the generic class approach is sufficient to provide public notice under this portion of Section 5, then it would seem adequate to facilitate other aspects of public oversight under Section 5 as well. It should be possible to devise generic class descriptions which are sufficiently informative to enable public observers to interpret test results and review test methodology.

*   *   *

The degree of protection that EPA affords to chemical identity will have a significant bearing on the impact of Section 5 on innovation in the chemical industry. EPA resolution of this issue will therefore be viewed by the industry with great interest.

B. Advance Substantiation of Confidentiality Claims

1. Description of the EPA Approach

In its October 16, 1979 PMN proposal, EPA substantially expanded the procedures that PMN submitters were required to observe in requesting and obtaining confidential treatment for PMN information. These expanded procedures were motivated by the desire of the Agency to deter unwarranted confidentiality claims and thereby to maximize the disclosure of PMN information to the public.

The January 10 EPA proposal would have required submitters to *assert* all claims of confidentiality when PMNs are submitted. This could be done simply by checking a box adjacent to each item on the PMN form. The January 10 proposal would have required submitters to *substantiate* their claims at the time of submission for only two types of PMN information — specific chemical identity and information contained in a health and safety study. To substantiate confidentiality claims for this information, the submitter could respond to several questions developed by EPA. Substantiation of other confidentiality claims could be required by EPA at a later date, either in response to an FOIA request or at the initiative of the Agency.

In contrast to the simple check-off method for asserting confidentiality claims under the January 10 proposal, the October 16 reproposal would require that such claims be "linked" to one of six categories. These are: (1) identity of manufacturer, (2) specific chemical identity, (3) production volume, (4) uses, (5) process information, and (6) other. To *assert* a claim of confidentiality for information that does not clearly and automatically fall within one of the categories; a submitter would then have to "link" the information to one of the categories and explain how disclosure of the information would in turn disclose information in the specified category. The submitter would then have to *substantiate* all claims of confidentiality at the time the PMN is submitted. For identity of manufacturer, the submitter would only need to certify that confidentiality is necessary. For the other categories, the submitter would have to answer various questions that are keyed to each category.

In its Revised Interim Policy concerning PMN requirements, EPA urged PMN submitters to substantiate their claims of confidentiality at the time of PMN submission. 45 *Fed. Reg.* 73478 (November 7, 1980). However, on July 2, 1982, EPA published a *Federal Register* notice announcing that it had reconsidered its substantiation policy and concluded that advance substantiation is unnecessary for most categories of PMN information. 47 *Fed. Reg.* 28969. Accordingly, EPA stated, it would not routinely request such substantiation at the time of PMN submission. The Agency made it clear, nevertheless, that it would continue to require advance substantiation of chemical identity confidentiality claims for PMNs on which the Agency receives notices of commencement of manufacture. The Agency also stressed that it would continue to consider whether an advance substantiation requirement should be included in its final PMN rule.

2. The Legal and Policy Objections to Advance Substantiation

Under the EPA advance substantiation approach, confidentiality claims for PMN information would include detailed substantiation even though there is no present or anticipated reason for disclosing the data. This approach represents a significant departure from the practices of other federal agencies, which uniformly do not require advance substantiation of confidentiality claims. In addition, such substantiation is not required by EPA-published regulations respecting the confidentiality of information submitted to the Agency under all the other statutes it administers.[30]

Industry has argued that the advance substantiation procedure proposed by EPA is impermissible under TSCA. Section 14(c)(1) of TSCA provides that PMN submitters may "designate the data" which they believe are entitled to confidential treatment, and may "submit such designated data separately from other data submitted under this Act." Throughout this subsection, Congress focused on the *mechanics of designating and identifying* that PMN information which submitters considered confidential. Accordingly,

---

[30] The EPA general confidentiality regulations provide that EPA may initiate action to determine whether business information is entitled to confidential treatment whenever EPA (1) learns of an FOIA request for the information, (2) desires to make a confidentiality determination even though no request for release was received, or (3) "[d]etermines that it is likely that EPA eventually will be requested to disclose the information at some future date and thus will have to determine whether the information is entitled to confidential treatment" 40 C.F.R. § 2.204(b)(1)—(3).

At the time this regulation was published, EPA rejected comments of interested parties urging the Agency to require detailed substantiation of every claim of confidentiality at the time information was submitted. 41 *Fed. Reg.* 36902, 36919 (1976). EPA believed then that advance substantiation would result in unnecessary burdens to both submitters and the Agency. The regulation does provide that EPA may initiate action to request substantiation of confidentiality claims if the Agency determines that certain information is likely to be the subject of an FOIA request. However, both the Agency and persons submitting comments on the proposed regulation apparently assumed that this would entail determinations on a case-by-case basis, not determinations respecting a class of information submitted under a particular statute. As EPA explained in responding to comments on the proposed regulation:

A provision has been added to § 2.204(a) which would encourage EPA offices to initiate the determination process as soon as it appears likely that requests for the information eventually will be received by EPA, even though they have not yet been received. This should produce speedier decisions. But to require detailed substantiation of a claim in every case at the time of submission of the information itself would inevitably slow the responses of business to requests for information, would in many cases cause businesses to expend needless effort, and would complicate EPA recordkeeping injustifiably.

industry has maintained, EPA authority over the initial assertion of confidentiality claims is limited to regulating the way in which submitters *identify* PMN information as confidential.

Industry has supported this conclusion by pointing to the notice provisions of Section 14(c)(2). That section requires EPA to give *30 days written notice* to the submitter when EPA intends to release data designated as confidential, except when such release is made pursuant to the special disclosure provisions of Section 14(a)(1)—(4). The most likely explanation for the unusual length of this notice period is that Congress intended that PMN submitters would prepare and submit substantiation of their confidentiality claims *after* a request for disclosure, not *before* such requests as EPA proposes.

This construction of TSCA comports with the prevailing practice at EPA, as well as other Federal agencies, when TSCA was enacted. At that time, neither EPA nor any other federal agency required submitters to *substantiate* claims of confidentiality in advance of an FOIA request for the information or some other reason for proposing release. On the other hand, many agencies, including EPA, required submitters to *assert* claims of confidentiality when submitting the data and to identify for the Agency the specific items of data claimed to be confidential. It would appear that Congress therefore assumed that this would be the practice under TSCA, since nothing in the statute or its legislative history indicates that Congress expected or authorized a different approach.

An advance substantiation requirement would force PMN submitters to undertake the time, effort, and expense of substantiation for *every* confidentiality claim, even though there never may be any occasion for disclosing the information in question. To the extent that disclosure of the information is never sought, the PMN submitter will have been burdened unnecessarily. Furthermore, if the occasion for disclosure arises long after the PMN is submitted, then the substantiation provided along with the PMN may be out-of-date. Reliance on that substantiation might therefore result in the needless withholding of information from the public. On the other hand, if the Agency decides to require new substantiation of the confidentiality claim, then the time, effort and expense of the submitter in preparing the original substantiation will have been largely wasted because the whole substantiation process will now have to be repeated.

3. The Policy Justification for Advance Substantiation

EPA originally identified three policy goals to support requiring advance substantiation of PMN confidentiality claims. 44 *Fed. Reg.* 59775. First, EPA said it expected to reduce the possibility of unwarranted claims of confidentiality, since submitters supposedly would "better understand" the nature of and relationship among their claims. Second, EPA said it hoped to enhance the ability of the Agency to respond to FOIA requests in a timely fashion. The third objective of EPA was to disclose to the public, within the 90-day PMN review period, a maximum amount of nonconfidential data.

Industry has argued that these justifications are unconvincing.

Before signing the confidentiality certification required by EPA, a PMN submitter will have to assure itself that all claims of confidentiality are made in good faith, and that, if necessary, the claims can be substantiated in accordance with the criteria set forth in the certification statement. In certifying to the truth and accuracy of the above statements, the normal submitter will not take his obligations lightly, since false statements could subject him to criminal liability under 18 U.S.C. § 1001. Furthermore, it would seem irrational to assume that PMN submitters — as distinct from all other persons submitting data to EPA and other federal agencies — will be likely to assert unwarranted confidentiality claims. As noted above, EPA-published confidentiality regulations, 40 C.F.R. § 2.204, require persons submitting data to EPA to assert *but not substantiate* confidentiality claims when the data are submitted.

The EPA second reason for wanting to enhance its ability to respond to FOIA requests in a timely fashion was the purported intention of public interest groups to file FOIA requests for all PMNs. *See* 44 *Fed. Reg.* 59775. At the inception of the PMN process, some public interest groups did, in fact, file FOIA requests for several PMNs. However, since that time, the number of FOIA requests filed by public interest groups has decreased substantially.

It is also open to question that Congress intended "strong citizen involvement" in PMN review, and that such participation is impossible unless EPA publicly discloses a "maximum amount of information" within the PMN review period. *See* 44 *Fed. Reg.* at 59774. In Section 5(d)(2) of TSCA, Congress directed EPA to publish a *Federal Register* notice consisting of the generic class of the chemical substance, the uses and intended uses of the substance, and a description of certain health and safety test data. The care and specificity with which it prescribed this information suggest that Congress considered it sufficient to permit public participation in EPA decisions under Section 5. Moreover, the publication of the Section

5(d)(2) *Federal Register* notice is expressly made subject to the confidentiality protections afforded by Section 14. Thus, Congress recognized that such disclosure was secondary in importance to the protection of trade secret information.

The balance between disclosure and confidentiality reflected in Section 5(d)(2) also suggests that Congress did not believe the public could or should review all the detailed data submitted to EPA concerning uses, chemical identity, and health and environmental effects of a new substance. Furthermore, even if public interest groups were expected to participate actively in the details of PMN review, they would be likely to focus on only major new substances, not every chemical for which a PMN is filed. In any event, even without advance substantiation, the 90-day PMN review period provides adequate time for disclosing nonconfidential information to parties requesting information under the FOIA. Since a *Federal Register* notice must be published within 5 days of the receipt of a PMN, and since by law the Agency must respond to FOIA requests within 30 days, a requester who promptly files an FOIA request for a PMN can obtain nonconfidential information in approximately 40 days from the submission of the PMN.

In recently suspending the advance substantiation approach included in its Revised Interim Policy, EPA concurred that its original rationale for the approach had not been borne out by experience:

> EPA has been operating under the Revised Interim Policy for a year and a half. In that time, EPA has not received large numbers of FOIA requests for confidential PMN information. At the same time, the substantiation requirement has not led to a reduction in the volume of information claimed confidential.
>
> Based on this experience, EPA has reconsidered its policy of automatically requesting substantiation of all confidentiality claims in each PMN. Requiring substantiation of confidentiality claims can be an additional burden on submitters and the Agency. If there are no FOIA requests for the information claimed confidential in a PMN, and EPA has no other reason to make a final confidentiality determination, EPA and the submitter have incurred that burden unnecessarily. The volume of FOIA requesters seeking confidential business information from PMNs has decreased since the substantiation requirement was imposed. Most requesters are satisfied with the sanitized copies of PMNs which submitters or EPA produce. At the same time, EPA has had little reason to initiate final confidentiality determinations for any other reason. In addition, when companies have been asked to substantiate claims, they have not withdrawn or limited their claims. Thus, requesting substantiation does not seem to have reduced the number of confidentiality claims. 47 *Fed. Reg.* 28970.

C. Specific Substantiation Questions of EPA

The original EPA substantiation approach would have required PMN submitters to "link" their claims of confidentiality to one of six categories of information: identity of manufacturer, specific chemical identity, production volume, uses, process information, or "other." The submitter would then have to substantiate his claims by answering a number of extremely detailed questions that are keyed to each category.

Regardless of when substantiation is required, the EPA scheme would call for a high level of documentation to uphold confidentiality claims. Accordingly, industry has argued that EPA should require the PMN submitter to provide a simple narrative discussion of basis for its confidentiality claims and that a presumption of confidentiality should be accorded to some categories of PMN information without any substantiation at all.

The following examination of specific EPA substantiation questions illustrates the potential burdens imposed under the approach of the Agency: The Agency's approach to confidentiality claims for chemical identity bears special scrutiny since EPA will continue to require such substantiation where a notice of commencement of manufacture has been filed.

1. Specific Chemical Identity

To substantiate a claim of confidentiality for specific chemical identity, a manufacturer must respond to eight complex questions devised by EPA.

a. Specific Information Regarding Disclosure and Harm

EPA asks the submitter to provide details about the connection between disclosure and competitive harm. This question would seem appropriate insofar as it simply requires submitters to provide a brief description of the harm that might occur if chemical identity is disclosed.

b. Length of Time for Confidentiality

EPA asks submitters to give the exact length of time for which confidentiality is required, and to explain why that period was chosen. In many instances, it is impossible for manufacturers to answer this question with any accuracy at the time the confidentiality claim is asserted. Manufacturers can only speculate as to the length of time required for them to earn a fair return on their R & D investment in a new chemical substance. Moreover, they cannot possibly know when competitors might independently discover the specific chemical identity of the substance, or whether competitors might already be developing products with similar or superior uses.

c. Patents for Chemical Substances

EPA asks submitters whether the chemical substance is patented and, if so, why it should be treated as confidential. However, disclosure of chemical identity in a patent is not the same as disclosure in a PMN. Since a patent is merely a description of an invention, a large number of chemicals are patented which are never developed commercially, and a patent may be issued before commercialization occurs. Moreover, an array of variations on a given formula will usually be patented, and competitors cannot know from the patents which of these, if any, is being commercialized. For these reasons, the existence of a patent will not necessarily mean that a specific new chemical is being, or will be, manufactured for a commercial purpose, yet this is the precise information that a PMN would disclose. To require patent information for every new chemical would therefore be unnecessary if not totally misleading.

d. Extent to Which Intent to Manufacture Has Been Disclosed, and the Precautions Taken with Respect to Such Disclosures

In the confidentiality certification, a submitter must affirm that information claimed confidential is not and has not been reasonably obtainable or publicly available, and that the company has taken and will continue to take measures to protect that confidentiality. EPA should presume that these representations, which are made under penalty of perjury (*see* 18 U.S.C. § 1001), constitute sufficient substantiation for claims of confidentiality for specific chemical identity. To require manufacturers to set forth in detail all of the facts regarding prior access to the chemical identity of a substance, and the specific measures taken to prevent security leaks or other disclosures, is unnecessarily burdensome.

e. Extent to Which the Substance Leaves the Manufacturing Site in Any Form, and the Measures Taken to Safeguard Its Identity

EPA asks whether a new substance has left or will leave the manufacturing site in any form, either as product, effluent, or emission. If so, the manufacturer must detail the measures that have been and will be taken "to guard against discovery of its identity."

This question indicates that EPA expects manufacturers to take extraordinary efforts to prevent any industrial espionage that might lead to the discovery of the identity of a chemical. However, it is highly unlikely that competitors will be monitoring and analyzing the effluents and emissions of every other manufacturer in the hope of identifying a marketable new substance. Furthermore, even if a substance leaves the site of a submitter as part of a product before manufacturing begins, competitors will not necessarily be aware that the chemical is new, or of the intent of the PMN submitter to manufacture the chemical for a commercial use. Consequently, competitors might have no reason to believe they should analyze a given product to detect the presence of the substance. Because of the many obvious difficulties in attempting to isolate and identify a new substance, this question focuses on an issue which will rarely, if ever, be relevant to a determination of confidentiality.

f. Identification by Analysis of a Product

EPA asks whether the identity of a new substance can be ascertained by analyzing a product in which that substance is contained. This question is misleading, since almost any product conceivably could be analyzed to identify its constituent chemical substances if enough time and resources were devoted to that task. The real issue is not whether such analysis is theoretically possible, but whether competitors will in fact attempt it when they have no reason to believe that a product contains a new chemical and no idea of the possible composition of the chemical and when the cost of such "reverse engineering" is high.

g. Confidentiality Determinations by Other Agencies or Courts

EPA asks submitters whether any other agency or court has made a confidentiality determination concerning specific chemical identity. If so, the submitter must attach copies of such determinations. This question is arguably unnecessary. If an agency or court had already denied a confidentiality claim, a PMN submitter would have no basis for re-asserting a confidentiality claim before EPA. On the other hand, if another agency or a court had granted confidential treatment to the identity of the chemical, EPA would still have to make an independent determination, particularly since the prior determinations might be out-of-date.

h. Link between Identity of the Manufacturer and Chemical Identity

In order to obtain protection for chemical identity, a submitter must show that competitive harm would occur if only chemical identity were disclosed without linking it in any way to the identity of the manufacturer. One problem with this question is that competitors may be able to infer the identity of the manufacturer if given the chemical identity, thus compromising the protection otherwise accorded the identity of the manufacturer. For example, if a chemical is particularly complex, competitors may know that it could have been developed by only one or a very few manufacturers. A second problem is that the actual intent by *any* company to manufacture the given chemical substance may itself be a commercially sensitive trade secret. Mere knowledge of the fact that a competitive product is entering the market can be of significant value to competitors seeking to minimize adverse effects from the new entrant.

2. Production Volume, Process Information, and Use Data

In order to substantiate a claim of confidentiality for production volume, process data, or use information, a manufacturer would be required to answer two questions. The first question is why and how disclosure of the information would result in competitive harm if the identity of the manufacturer remains confidential. The second is why and how such disclosure would result in competitive harm if the specific chemical identity remains confidential. According to EPA, "[t]he two questions are designed to lessen the need for multiple confidentiality claims." 44 *Fed. Reg.* at 59774.

Production volume is typically considered confidential business information. *See, e.g.,* 18 U.S.C. § 1905; 16 C.F.R. § 4.10(a)(2) (FTC confidentiality regulations); 1 R. Milgrim, *Trade Secrets* § 2.09[8][a], at 2-146 and n.447 (1978). FDA regulations accord a presumption of confidentiality to such information. *See* 21 F.F.R. §§ 71.15, 171.1(h), 314.14, 431.71.

Process information is also expressly considered a trade secret under 18 U.S.C. § 1905. Moreover, Congress accorded special status to this type of trade secret in Section 14(b)(1) of TSCA, in which it provided that data from health and safety studies cannot be disclosed to the extent it would reveal processes used in manufacturing or processing a chemical substance. Again, FDA regulations accord a presumption of confidentiality to manufacturing methods or processes, including quality control procedures. *See, e.g.,* 21 C.F.R. § 171.1(h) (food additive petitions). Like specific chemical identity, process information is normally carefully protected against disclosure in order to assure a sufficient return on the R & D investment of the innovator before the onset of competition.

Use data, as EPA itself has acknowledged, "may be the most commercially sensitive type of information included in the premanufacture notice." 44 *Fed. Reg.* at 2253. If EPA discloses information concerning the specific use of a new chemical substance, other manufacturers can learn of imminent competition in a particular product market. Such advance warning would allow them to engage in defensive marketing tactics, such as favorable long-term contracts with relevant purchasers, that could effectively shut an innovator out of the market for a substantial period of time.

Thus, the commercially sensitive character of production volume, use, and process data is so well established that this information should be presumed confidential. Accordingly, substantiation of confidentiality claims should be unnecessary except when EPA may have some specific reason to question the validity of a claim.

3. Other Information

Items of information that do not fall within one of the five categories identified by EPA are within the category "other infomation." To substantiate claims of confidentiality for "other" information, a submitter must respond to five questions. The first question is whether the item is confidential in and of itself, and if so, exactly what it reveals. The second question is whether the item would reveal other information that is confidential, and if so, exactly what that information is. The third question is whether the item would reveal other confidential information when it is disclosed in association with other PMN items, and if so, what the other PMN items are. Two substantiation questions for the "other" category relate to the need for confidentiality when EPA grants confidential treatment to either identity of the manufacturer or specific chemical identity.

The effect of these questions is to require submitters to detail exact links between various pieces of information. Accordingly, although a submitter may generally know that one kind of information may reveal other confidential information, it may need to do extensive research and paperwork in order to explain to the satisfaction of the EPA the causal connection between the two categories of data. Moreover, some data might be claimed confidential because they might reveal competitively sensitive information if combined with other data potentially held by or available to a competitor. In such circumstances, it may be extremely difficult to establish with certainty what confidential information is likely to be revealed. Instead of these detailed linkage questions, EPA should only require submitters to certify that the information claimed confidential is neither customarily disclosed to others not publicly available and that disclosure of the information is likely to cause substantial competitive harm.

d. EPA Proposal Requirements Respecting Generic Descriptions of Confidential Business Information

Under the reproposal, a submitter who asserts a claim of confidentiality for (1) identity of manufacturer, (2) specific chemical identity, (3) use data, or (4) physical and chemical properties data, must also provide a *generic description* of the confidential information, or explain why a generic description cannot be provided. If the submitter does not provide a generic description, or if EPA deems the description "more generic than necessary", EPA will devise a description and notify the submitter. EPA will publish (30 days after such notification) in the *Federal Register* its generic descriptions of the identity of the manufacturer and use, and will place in the public docket the EPA-assigned ranges for physical and chemical properties data. In addition, EPA has proposed a special procedure for settling disputes between the agency and the submitter over an appropriate generic name when specific identity is claimed to be confidential. Under this procedure, the PMN submitter must submit three proposed generic names, and EPA would reserve the right to publish a second name in the *Federal Register* if the first name is too "generic".

1. The Legal Objections to the EPA Generic Information Approach

There is reason to question the legality of the EPA generic information approach. Section 14(c) of TSCA permits EPA to prescribe the *manner* in which confidential data is *designated*. The clear focus of this Section, however, is on the mechanics of identifying PMN data as "confidential." EPA would appear to lack authority under Section 14(c) to require the submission of *additional* information not otherwise included in a PMN because *other* information in that PMN has been claimed confidential.

Similarly, in Section 5(d)(2), Congress has set forth the specific information that must be included in the *Federal Register* notice publicly disclosing receipt of a PMN. The notice must identify the generic class of the chemical, list the uses of the chemical, and describe the nature and results of any health and safety tests required by EPA under Sections 4 or 5(b). However, publication of this information is *explicitly made subject to the confidentiality protections of Section 14*. Nothing in Section 5(d)(2) requires the submitter to provide *additional* information for inclusion in the *Federal Register* notice as a "trade off" for obtaining confidential treatment.

The key premise underlying the EPA decision to require and disclose generic descriptions of confidential data is that EPA can thereby provide the public with information useful in PMN review. However, the basis for this premise is open to challenge. In fact, it may be impossible to create generic descriptions of confidential PMN data that at once will promote meaningful public scrutiny of new chemicals yet avoid revealing the sensitive data about the business of the submitter that motivated its original confidentiality claim. For this reason, EPA and manufacturers will inevitably become embroiled in frequent disputes over the adequacy and confidentiality of particular generic descriptions.

For example, if EPA decides that the manufacturer has submitted a description that is "more generic than necessary" to protect the confidentiality of the underlying data, the Agency will devise, and then publish or disclose, its own generic description. On the other hand, companies may frequently conclude that EPA-proposed "generic" descriptions are too specific and would reveal sensitive data that is entitled to confidential treatment. Except for procedures respecting generic names of chemicals, no mechanism has been established by EPA for resolving disputes over generic descriptions. Nevertheless, due process considerations, coupled with the inevitability of frequent disputes, may require EPA ultimately to create such a mechanism. Moreover, in order for manufacturers to protect their property rights in confidential information, they could be forced to seek judicial review of EPA determinations with which they disagree. Thus, both on the administrative and judicial level, conflicts over generic descriptions could create a class of litigation distinct from and in addition to both FOIA and "reverse-FOIA" litigation, with attendant costs to EPA and submitters.

2. Specific Categories of Generic Information Proposed by EPA

Some of the legal and practical shortcomings in the EPA scheme for requiring generic descriptions are revealed by an examination of the categories of generic information that EPA proposes to require:

a. Identity of Manufacturer

The reproposed EPA regulations would require manufacturers to submit generic descriptions of their company when claiming identity of manufacturer confidential. If EPA does not approve of the submitted description, or if no generic description is submitted, EPA will devise a description and publish it in the *Federal Register* 30 days after notifying the submitter.

Section 5(d)(2) of the Act expressly requires *Federal Register* publication of generic chemical class, use data, and certain categories of test data; at least to the extent such information is not protected under Section 14. In contrast, the Section does *not* provide for inclusion of identity of manufacture in the *Federal Register*. Since Congress did not contemplate *Federal Register* publication of identity of manufacturer even when it was not confidential, it should not be deemed to have authorized EPA to identify the company generically when the precise identity of the manufacturer is protected as a trade secret.

b. Use Data

Where the uses or intended uses of a substance are confidential, EPA proposes to require and to publish a description of those uses that is "only as generic as necessary to protect the confidential business information." In lieu of a list of specific uses, this description may include information concerning population exposure and environmental release of a substance — including the degree of containment, the level and mode of environmental release, and the type and average frequency of human contact.

Under Section 5(d)(2), human exposure and environmental release data are not among the categories of information which the *Federal Register* notice is to contain. Moreover, there is no reason why the *Federal Register* should describe the substance in such elaborate detail, whether or not other information in the notice is claimed to be confidential. Indeed, much of the "generic" information which EPA seeks to require need not be included in the PMN form — much less in the *Federal Register* notice summarizing that form — under Sections 5(d)(1) and 8(a)(2) of the Act.

c. Chemical Identity

Under the EPA-reproposed regulations, submitters asserting a claim of confidentiality for specific chemical identity would be required to submit three proposed generic names for the new substance, unless a generic name had been agreed upon in pre-notice consultations with EPA. EPA would then publish one of the three names in the *Federal Register* notice. However, if EPA later determines that the published name was "more generic than necessary" to protect the confidentiality of the specific identity, EPA will publish an amended *Federal Register* notice containing one of the other names proposed by the submitter or a name devised by EPA.

As discussed above, Section 5(d)(2) provides that the *Federal Register* notice must identify the new chemical substance by generic *class*, not by specific chemical identity. Section 5(d)(2) does state that EPA can give a "more specific identification" when "required in the public interest." However, Congress apparently intended that this would be the exception, not the rule. Against this backdrop, the EPA proposal to publish a *second* generic name in an amended *Federal Register* notice when the first name is deemed too "generic" would appear unreasonable. EPA authority to publish *any* generic name in the *Federal Register*, absent a specific public interest determination, is questionable. Therefore, it is arguably unnecessary and excessive to publish in the *Federal Register* two different generic names for a new chemical.

d. Physical and Chemical Data

The EPA-reproposed regulations would require PMN submitters to "report physical and chemical properties in ranges if disclosure of the specific value of the property would reveal confidential business information." 44 *Fed. Reg.* 59777. If a submitter does not use the EPA-devised ranges, it must explain why use of the ranges would reveal confidential data and must also provide an alternative generic description. *See Id.* at 59777, 59823.

Taken out of the context of detailed information about uses and effects of a chemical, range data for physical and chemical properties will often have little meaning to the general public. Moreover, these range data may prove valuable to competitors of a submitter. Information on the five key properties of a substance — vapor pressure, density, solubility, melting point, and boiling point/sublimation point — itself discloses significant information about chemical structure, class, and composition. Thus, especially when combined with generic information on chemical identity, uses, or the identity of the manufacturer, such data may reveal to competitors the specific identity of the new chemical.

*   *   *

Confidentiality issues are among the most complex and time-consuming matters which PMN submitters must address. Because of the commercial sensitivity of information on new chemicals, these issues are also extremely important. EPA has complicated the task of the PMN submitter, as well as created barriers to the protection of legitimate trade secret information, by proposing stringent procedures for seeking and obtaining confidential treatment for PMN data. Industry has expressed serious legal and practical reservations about these procedures as they relate to the protection of confidential chemical identity, advance substantiation of confidentiality claims, the level of evidence necessary to sustain those claims, and the generic disclosure of confidential information. How EPA responds to the industry comments on its expanded confidentiality procedures has substantial long-term implications for PMN submitters. It is encouraging that, with respect to advance substantiation, EPA has recently expressed considerable sympathy with the industry position.

VII. SIGNIFICANT PROCEDURAL AND SUBSTANTIVE ISSUES UNDER SECTION 5

The proposed and reproposed EPA regulations for implementing Section 5 raise a number of important procedural and substantive issues whose resolution will have a major impact on how EPA implements its premanufacture notification program. Equally important issues have emerged from other EPA activities under Section 5. These issues, which are discussed below, include: (1) statutory authority of EPA to promulgate binding rules implementing Section 5; (2) EPA power to reject PMNs, upon their receipt, because they lack the information prescribed by Section 5(d) or otherwise fail to comply with applicable legal requirements; (3) the EPA proposal to use Section 8 of TSCA to require follow-up reporting on new chemicals while their PMNs are still under review; (4) the use of Section 8 reports to monitor the commercial development of new chemicals after the completion of PMN review; (5) the EPA proposal to require the submission of PMNs for new chemicals manufactured solely for export; (6) the scope of EPA's Section 5(e) power to prohibit or limit manufacture of a new chemical; (7) the potential legal issues associated with the "significant new use" provisions of Section 5(1)(2); and (8) recent EPA efforts to implement the exemption provisions of Section 5(h)(4).

A. EPA's Power to Promulgate Rules Implementing Section 5

As the proposal and reproposal by EPA of January 10 and October 16, 1979, demonstrate, the Agency plainly contemplates that the promulgation of substantive regulations will play a major role in defining the obligations of PMN submitters under Section 5. Such regulations will be used to establish a uniform form for the PMN, for prescribing the information which the PMN must contain, for defining statutory terms like "reasonably ascertainable", and for creating procedures to govern such matters as the Agency treatment of PMNs which contain insufficient information. There is, however, significant doubt whether Congress permitted EPA to use substantive rules to implement Section 5. Indeed, the most plausible construction of TSCA is that Congress specifically withheld such substantive rulemaking power from EPA, and instead expected the Agency to implement the premanufacture notification provisions of the Act on the basis of the statutory requirements alone.

It is universally recognized that administrative agencies derive their authority from Congress and, hence, may exercise only those powers granted them by statute. *E.g., Federal Trade Commission* v. *National Lead Co.*, 352 U.S. 419, 428 (1957); *Civil Aeronautics Board* v. *Delta Airlines, Inc.*, 367 U.S. 316, 322 (1961). No provision of TSCA grants EPA the general power to issue substantive regulations to carry out the requirements of the Act. Nor does TSCA, either in Section 5 or elsewhere, authorize the promulgation of regulations which define the obligations of manufacturers under the provisions of the Act for premanufacture notification.

Several decisions have upheld the authority of particular agencies to develop, by rule, general standards which will then apply in particular proceedings. *E.g.*, *Mourning* v. *Family Publication Service, Inc.*, 411 U.S. 356 (1973); *United States* v. *Storer Broadcasting Co.*, 351 U.S. 192 (1956); *American Trucking Associations* v. *United States*, 344 U.S. 298 (1953); *National Broadcasting Co.* v. *United States*, 319 U.S. 190 (1943); *National Nutritional Foods Association* v. *Weinberger*, 512 F. 2d 688 (2d Cir. 1975); *Bell Aerospace Div. of Textron, Inc.* v. *NLRB*, 475 F.2d 485 (2d Cir. 1973); *National Petroleum Refiners Association* v. *FTC*, 482 F.2d 672 (D.C. Cir. 1973); *Public Service Commission of State of New York* v. *FTC*, 327 F.2d 893 (D.C. Cir. 1964). In each of these cases, however, the statute creating the agency contained a specific or general grant of rulemaking power.[31] There is no reported decision upholding substantive regulations that were promulgated by an agency which lacked both general and specific rulemaking authority. Yet this is the case with Section 5 of TSCA.

The provisions of TSCA which empower EPA to issue substantive rules to implement other sections of the Act suggest that the absence of rulemaking authority in Section 5 was clearly intended by Congress. For example, Section 4(a) of the Act empowers EPA to issue regulations which require testing. Section 4(c) permits EPA to issue regulations which allow those who conduct such tests to obtain reimbursement from other manufacturers and processors. Section 6(a) of the Act empowers EPA to promulgate regulations which prohibit or restrict the manufacture or processing of substances or mixtures. Section 8(a) authorizes the issuance of regulations which require manufacturers or processors to report information concerning substances and mixtures to EPA. Analogous grants of substantive rulemaking authority can be found throughout TSCA.[32] Elsewhere in Section 5, in fact, Congress provided for the promulgation of regulations governing the submission of notices for "significant new uses,"[33] and authorized EPA to develop, by regulation, a list of "suspect" chemical substances and mixtures for which premanufacture testing would be required.[34] Thus, Congress was certainly aware of the availability of substantive rulemaking when it enacted Section 5.

When Congress has included highly specific grants of authority in a regulatory statute, the courts will be reluctant to imply broad powers which Congress did not convey to the Agency expressly. *E.g.*, *Addison* v. *Holly Hill Fruit Products, Inc.*, 322 U.S. 607, 616-617 (1944). This principle directly applies to TSCA, which contains numerous specific grants of rulemaking power yet no general authority for the issuance of regulations to carry out the requirements of the Act or specific authority for the issuance of regulations to implement the premanufacture notification provisions of Section 5.

Indeed, the legislative history of TSCA specifically confirms that Congress intended to deny general rulemaking authority to EPA:

> [C]ertain regulatory statutes contain provisions granting general rulemaking authority to the agencies administering the statutes. See, e.g., section 701 of the Federal Food, Drug, and Cosmetic Act. However, such provisions have been construed to grant such agencies substantive rulemaking authority. The bill contains specific grants of substantive rulemaking authority to the Administrator (see, e.g. sections 4, 5, 6, and 8) and the Committee does not intend that the Administrator have any substantive rulemaking authority which is not specifically granted. A general rulemaking authority is not needed to authorized the issue of procedural, interpretative, or similar administrative rules, consequently, such a provision is not included in the bill. H.R. Rep. No. 94-1341, *supra*, at 62 (footnotes omitted).

As an unambiguous statement of the intent of Congress to withhold substantive rulemaking power from EPA, this passage could hardly be written more plainly.

Since EPA has not yet promulgated final regulations under Section 5, it has not responded to industry arguments that it lacks substantive rulemaking authority under this provision of TSCA. Nevertheless, in the absence of substantive regulations, EPA has been implementing its premanufacture notification program on the basis of the statutory requirements alone. As explained in the Agency interim statement of policy under Section 5, the only obligation of manufacturers of new chemicals is to submit a PMN which complies with the requirements of Section 5(d)(1). 44 *Fed. Reg.* 28565. Beyond satisfying the statutory requirements, EPA has indicated, the PMN submitter has substantial freedom to determine the contents of its PMN, including the type of form it uses to present PMN information.

---

[31] For example, Section 701(a) of the FD & C Act, 21 U.S.C. § 371(a), contains a broad grant or rulemaking power to FDA, and the main issue in National Nutritional Foods Association v. Weinberger, *supra*, was the scope of this power. The same situation existed in National Petroleum Refiners Association v. FTC, *supra*, which involved the rulemaking power of the FTC. At the time the case was decided, Section 6(g) of the FTC Act, 15 U.S.C. § 46(g), also contained a broad grant of rulemaking power.

[32] *See, e.g.*, Sections 5(h)(2)(B), 5(h)(4), 6(c)(4)(A), 6(e), 8(a)(3)(i), 8(d), 18(b) and 26(b).

[33] Section 5(a)(2).

[34] Section 5(b)(4).

The experience of EPA under Section 5 to date demonstrates that TSCA premanufacture notification provisions can be implemented in an efficient and orderly manner without the aid of binding substantive regulations. Based on this experience, EPA could decide that its final regulations under Section 5 will have the status of a statement of policy, and thus would not be binding on PMN submitters. This approach would probably not undermine EPA authority to enforce Section 5 requirements and would arguably comport with the plain intent of Congress.

B. EPA's Power to Declare PMNs "Invalid"

Inevitably, there will be situations where the PMN submitted to EPA fails to contain the information prescribed by Section 5(d)(1). Normally, the absence of this information will not be intentional; rather, the PMN submitter will have misconstrued the statutory requirements or simply omitted required data from the PMN inadvertently. In these situations, it is plainly proper for EPA to inform the submitter of the deficiencies in its PMN and afford it an informal opportunity to submit a revised PMN containing the omitted data. It is less clear, however, whether EPA can or should reject a PMN which the submitter believes complies with the statutory requirements. An open-ended EPA power to declare PMNs "invalid" when it believes they omit necessary information could expose PMN submitters to unfair and arbitrary treatment; in the guise of rejecting a PMN because of allegedly inadequate information, EPA could indefinitely delay review of the new chemical, thereby effectively barring its commercial production.

Since Section 5 took effect, there have been a number of PMNs which clearly lacked the information described by the statute and which EPA therefore refused to review on their merits. In all of these cases, the manufacturer submitted an expanded PMN which contained the information that the original PMN lacked. EPA thereafter proceeded to review the PMN on its merits and manufacture of the chemicals involved ultimately commenced. While such informal mechanisms for correcting incomplete PMNs have apparently satisfied both EPA and industry, the PMN proposal of the Agency contains elaborate and highly formalized procedures for declaring PMNs "invalid" when EPA believes they lack adequate information. The effect of this finding of "invalidity" would be to reject the PMN for filing, thereby suspending the required 90-day review period. Industry has questioned the need for and validity of these procedures on a number of grounds.

1. Details of EPA's "Invalidity" Provisions

Under the "invalidity" provisions of its proposal, if EPA believes that a premanufacture notice contains a "minor or technical" deficiency, it must request correction of that deficiency within 30 days after the notice has been received. Requests for correction will be transmitted to the submitter by certified mail and will explain why EPA believes the notice is deficient and what steps are necessary to remedy the deficiency. The premanufacture review period for the notice will be suspended on the date when a certified letter has been sent by EPA and will resume when EPA has received the corrections requested. In proposed § 720.34(a)(1), EPA has identified the following deficiencies in a PMN that will be considered "minor or technical": (i) failure to date the notice forms; (ii) typographical errors which render answers to any questions unclear or ambiguous; (iii) confusing responses; and (iv) answers which do not conform to premanufacture notice instructions.

Deficiencies in a PMN that EPA considers more "substantive" will result in a determination that the notice is "invalid". This determination can be made at any time during the notification period, not merely in the first 30 days. Moreover, if EPA makes a determination of invalidity, the premanufacture review period will be deemed never to have begun. EPA determination that a notice is "invalid" must be conveyed to the submitter in a certified letter which states the basis for that determination and explains how the deficiencies in the notice can be corrected. The submitter then has the option of providing EPA with a new notice or withdrawing the substance from premanufacture review. Proposed EPA regulations contain no mechanism by which the submitter can dispute the determination by the Agency of "invalidity" and no procedure by which the Agency can be forced to consider assertedly "invalid" notices on their merits.

Proposed § 720.34(b)(1) provides that EPA may declare a premanufacture notice "invalid" for the following reasons: (i) failure to sign the PMN; (ii) failure to comply with procedures for obtaining information from other persons or to certify that these procedures have been observed; (iii) failure by an importer to comply with procedures for obtaining information from foreign manufacturers or suppliers; (iv) failure to provide any information requested on the PMN form unless that information is optional; (v) failure to remedy any "minor or technical" deficiency within 30 days after corrections have been requested; (vi) submittal of intentionally false or misleading responses; (vii) submittal of a PMN by a person who neither intends to manufacture or import a chemical substance nor is the designated agent of such a person; (viii) failure to provide any information required by Sections 5(d)(1)(B) and (C) of TSCA; (ix) failure to include test data or other information required pursuant to a rule promulgated under Section 4 of TSCA; (x) failure to include the specific chemical identity of the new chemical substance in question; and (xi) failure to submit data which show that the chemical substance will not present an unreasonable risk of injury if that substance appears on the list issued by EPA under Section 5(b)(4) of TSCA.

It should be noted that these grounds for a determination of "invalidity" are not exclusive. By stating that the grounds for invalidity in proposed § 720.34(b) "include" those listed, EPA has apparently proposed to reserve the right to make a determination of "invalidity" on other grounds unspecified in its regulations.

2. Industry Arguments against the EPA Proposal

Industry has challenged the legality of the EPA procedures for declaring PMNs "invalid" unlawful on a number of grounds.

First, industry has argued that, except as authorized by Sections 5(c) and 5(e), Section 5 nowhere empowers EPA to extend the initial 90-day review period, or to take any other administrative action within that period or after it has expired, on the ground that a PMN is technically or substantively incomplete and therefore "invalid." The absence of any such provision is entitled to great weight, industry has maintained, in view of the general statutory presumption that manufacture of a new chemical substance may begin after 90 days and the great particularity with which Congress has defined the few situations where manufacture can be postponed beyond that period.

Second, industry has argued that, where EPA's believes that a PMN is incomplete and the submitter declines to supplement that PMN, EPA has several adequate statutory remedies: (1) EPA's "inherent" power to promulgate purely procedural rules would authorize it to prescribe ministerial requirements for the submission of PMNs and then to reject PMNs for filing when these requirements have not been observed; (2) under Sections 5(e) or 5(f), if the submitter declines to make corrections in a PMN when given the informal opportunity to do so, EPA may issue a proposed order or rule to prohibit or limit the manufacture of a chemical substance upon a determination that sufficient information about the substance is unavailable and the other criteria prescribed by these Sections are met; and (3) under the False Reports to the Government Act, 18 U.S.C. § 1001, the failure to include any of the statutorily prescribed information in a PMN, if knowing and willful, would expose the submitter to possible prosecution. Thus, EPA has a full range of explicit statutory remedies against the submitter of a PMN and the chemical substance that it covers when the PMN is technically or substantively incomplete.

Third, industry has argued that Congress assigned great value to prompt premanufacture review by EPA under Section 5 and intended that EPA expeditiously inform manufacturers if production could not proceed. During the debate on TSCA in the 92d Congress, for example, Senator Baker expressed the following reservations about extending the period for premanufacture review by even 90 additional days:

> The extra 90-day period in section 104(c) can, on the other hand, have several serious disadvantages:
> First, it may encourage delay and procrastination by the Agency. To stall and delay is a natural human tendency, and the presence of authority for an additional 90 days would seem to encourage such an approach.
> Second, for the manufacturer of a new chemical substance, an extra 90 days' delay can be a most serious matter. For the producer of fertilizers, for example, an additional 90-day delay can mean a whole year's delay if he misses the growing season for that year. For the producer of textile dyes, or the manufacturer of flame-retardant chemicals, the delay can make him miss the season of heavy production. And in the vast majority of cases, the extra 90 days of unforeseen delay can impose extra costs which will burden the producer, discourage his efforts to produce improved products, and needlessly increase his costs. 118 Cong. Rec. 19163 (May 30, 1972).

Under the EPA-proposed "invalidity" procedures, however, those Congressional policies could be frustrated.

Finally, industry has argued that, under the EPA-proposed regulations, even if EPA's grounds for declaring the notice "invalid" are open to question, the manufacturer would lack a mechanism for forcing EPA to consider its PMN on its merits. In the absence of such a mechanism, manufacturers could consume substantial amounts of time and money endeavoring to respond to demands for information by EPA that are unreasonable and unauthorized. In addition, the absence of a procedure for contesting "invalidity" determinations would effectively eliminate the procedural rights guaranteed to the submitter under Section 5(c), 5(e), or 5(f), which provide that an EPA decision to delay or prevent manufacture of a new chemical substance is subject to legal challenge by the affected manufacturer. Equally troubling, unless manufacturers have some recourse when EPA has made unwarranted demands for additional information, the Agency could use its power to declare a PMN "invalid" as a vehicle for expanding the scope of premanufacture review beyond the boundaries set by Congress.

3. The Industry Proposal

To eliminate the serious unfairness and legal defects in the EPA approach, industry has proposed a procedure for supplementing a PMN that seeks (1) to minimize delay by EPA in screening and taking action concerning new chemical substances and (2) to provide submitters with recourse against requests for additional information that are unnecessary or unreasonable. This procedure would contain the following elements:

a. PMNs which fail to conform to ministerial requirements relating to form would be rejected for filing upon receipt. EPA would immediately advise the submitter that the PMN had been rejected and indicate what corrections must be made. Grounds for rejecting the PMN for filing would include failure to sign or date the PMN, and failure to utilize the PMN format prescribed by EPA.

b. If EPA believes required information is omitted from a PMN, it would inform the submitter that the notice is "incomplete" within 30 days of the receipt of the notice. EPA's action would be set forth in a certified letter to the submitter which identifies the areas where the notice is deemed incomplete and specifies the additional information required. There is no reason why deficiencies in a PMN cannot be identified promptly. A 30-day deadline for completing that task would insure that the initial screening of a PMN for possible omissions is not deferred until the last few weeks of the review period.

c. The grounds for determining that a PMN is "incomplete" would be limited to the specific grounds stated in EPA regulations. The Agency should not have unbridled discretion to reject a notice as incomplete on any ground it chooses. Comprehensive criteria for "incompleteness" should be articulated in advance and must relate directly to the statutory requirements for a PMN.

d. The grounds for determining a PMN "incomplete" would be divided into technical and substantive deficiencies. The category of "technical deficiencies" would include: (1) terminology which is unclear or difficult to understand, and (2) failure to provide information required by the mandatory portions of the form and "known to or reasonably ascertainable by" the submitter if, in the absence of that information, EPA review of the new substance is still able to proceed. The category of "substantive" deficiencies would include the following: (1) failure to provide test data within the "possession or control" of the submitter as required by Section 5(d)(1)(B), (2) failure to provide other information in the mandatory portion of the PMN form which is "known to or reasonably ascertainable by" the submitter when the absence of this information has prevented EPA from reviewing the new substance, (3) failure to include test data or other information required pursuant to a rule promulgated under Section 4 of TSCA, and (4) failure to submit data which show that the chemical substance will not present an unreasonable risk of injury if that substance appears on the "risk" list issued by EPA under Section 5(b)(4) of the Act.

e. Upon being informed that its notice is "incomplete," the submitter would elect to (1) withdraw the notice without prejudice to its resubmission at a later date, (2) resubmit the notice with the supplemental information requested by EPA, or (3) dispute the EPA claim of "incompleteness" and instruct the Agency to consider the notice on its merits. In the case of a substantive deficiency where the submitter supplements the notice with additional information, the period for premanufacture review would be deemed to commence on the date when EPA has received the revised PMN. However, in the case of a technical correction, or an alleged substantive deficiency where the manufacturer instructs EPA to consider the original notice on its merits, the premanufacture review period would be tolled from the date when the EPA letter is received by the submitter to the date when the response of the submitter is received by EPA. Any time which elapses after the initial notice is filed and before the EPA letter is sent would be included in the 90 days allotted EPA for premanufacture review.

*   *   *

Adoption of industry-proposed procedures for correcting incomplete PMNs would provide an orderly and efficient procedure for remedying legitimate deficiencies in a PMN while affording PMN submitters recourse against unjustified demands for additional information. Accordingly, the PMN process will be significantly enhanced if, in lieu of proposed "invalidity" procedures, EPA adopts the industry approach.

C. EPA Authority to Require Supplemental Reporting during the PMN Period

A major issue raised by the EPA proposal and reproposal is the authority of the Agency to require PMN submitters to report additional information, which was omitted from their PMNs, during the PMN review period. Recognition of such authority on the part of EPA would provide the Agency with a powerful tool for expanding PMN reporting requirements beyond the categories of information specified by Congress.

1. The EPA Supplemental Reporting Proposal

Section 720.50 of the original EPA-proposed PMN regulations would have granted the Agency broad power to require supplemental reporting on new chemical substances. Such reports could have been mandated either while the PMN for the substance was still under review or after manufacture had begun. Under the proposal, supplemental reports could be extremely broad in scope; the proposed regulations provided only a very general description of the information that EPA could require such reports to contain. Moreover, EPA could require reporting not merely by the manufacturer of a new chemical, but by its processors as well. Reporting requirements could be imposed unilaterally, and manufacturers and processors would have had no opportunity to challenge the underlying basis or scope of information demands of the EPA.

The EPA reproposal contains a somewhat more limited supplemental reporting procedure. As before, reporting requirements may be imposed on both manufacturers and processors of new chemicals. These requirements may relate to the new chemical itself or to related substance, including impurities and by-products. Under reproposed Section 720.50, however, EPA can only require supplemental reporting while the premanufacture review period is still in progress. In addition, under the reproposal, EPA can base reporting requirements on one of two alternative sets of findings. First, the Agency must find that the substance or articles or mixtures containing it (1) may present a significant hazard to human health or the environment and (2) have a potential for release resulting in human or environmental exposure. Alternatively, the Agency must find that (1) insufficient data on the toxicity of the substance are available to determine whether it may be harmful to humans or the environment and (2) the substance has a potential for significant release resulting in significant human or environmental exposure.

The reproposal also attempts to identify the categories of information which must be submitted once these findings are made. The Agency must first identify the stage of the life cycle of the substance where exposure might occur — e.g., manufacturing, processing and industrial use, distribution in commerce, disposal, or end-use. For each such stage, EPA has specified various items of information which it can ask to receive. This information encompasses such diverse matters as engineering safeguards, process chemistry, packaging and labeling information, waste handling procedures, product formulation data, identifying information concerning affiliated companies of the submitter or prospective purchasers of the new chemical, social benefits of the new chemical, and the economic consequences of possible regulatory measures.

When EPA has decided to require supplemental reporting, it will send a letter, signed by the Assistant Administrator or a Deputy Assistant Administrator for Toxic Substances, to the persons subject to reporting requirements. This letter will, among other things, state the findings on which reporting requirements are based and describe the information which must be submitted. The recipient of a reporting requirement will have 10 days to file written objections with EPA; the Agency must answer these objections in writing. If the Agency does not withdraw its proposed reporting requirement, that requirement will be final and effective upon receipt of the response by the Agency to the objections of the recipient.

2. Legal Problems Raised by the EPA Approach

Utilizing the supplemental reporting authority contained in its reproposal, EPA could require PMN submitters to provide extensive information which the PMN form does not call for and which is nowhere authorized in Sections 5(d)(1) or 8(a)(2) of TSCA. Moreover, EPA could require the submission of equally detailed information from processors and users of new chemicals, who are not assigned *any* role in premanufacture notification under Section 5. Based on these factors, industry has argued that this supplemental reporting scheme would circumvent the detailed statutory provisions which govern the submission and evaluation of information on new chemicals under Section 5.

Section 5(d)(1) describes the contents of a PMN in great detail. Moreover, as discussed above, the legislative history of TSCA suggests that the requirements of these Sections were intended to be all-inclusive, and thus represent the *only* information which Congress believed EPA would need to evaluate a new chemical substance under the premanufacture notification provisions of Section 5. Since Congress withheld from EPA the authority to impose additional information requirements under Section 5, it would seem highly improbable that it intended to grant EPA such power under Section 8.

If the PMN form specified by the statute lacks sufficient information to evaluate a particular new chemical substance, Section 5 itself provides EPA with an adequate remedy. Under Section 5(e), the Agency is empowered to prohibit or limit the manufacture of the substance if it may present an unreasonable risk or will have substantial human and environmental exposure. When issuing an order under Section 5(e), EPA will undoubtedly identify the information which it believes it needs to complete its evaluation of the new chemical substance. Thus, the submitter will be put on notice of the information gaps which EPA perceives in the PMN. At this point, the submitter may either provide additional information to the Agency or file objections to the EPA Section 5(e) order, thereby testing the determination by the Agency that more information about the new substance is needed.

That is not to say that EPA is precluded from making requests for the *voluntary* submission of additional information at any time during the PMN period. Section 5(e) contemplates such a procedure by requiring EPA to inform PMN submitters of the substance of its determination together with or before the issuance of a proposed order postponing or limiting manufacture. Moreover, it is likely that manufacturers will frequently comply voluntarily with requests for additional information which are not unduly burdensome. It is, after all, in the interest of the manufacturer to avoid the potential uncertainty and delay that could result from an order under Section 5(e). Indeed, the Agency has asked some submitters of the PMNs it has already received to provide additional information voluntarily and these requests generally have been complied with. However, if the submitter and EPA cannot agree on the scope of an adequate supplemental report and the Agency remains convinced that it needs additional information, there are strong arguments that the only remedy available to EPA is the remedy designated by Congress — an order under Section 5(e).

To date, EPA has implemented Section 5 without the aid of any supplemental reporting authority. Under the threat of a Section 5(e) order if they fail to cooperate with the Agency, PMN submitters have responded voluntarily to EPA requests for additional information on the PMN chemical. Thus, experience would appear to refute the EPA belief that a supplemental reporting authority is essential to the effective implementation of Section 5. Hopefully, EPA will therefore decide to sidestep the difficult legal and policy issues raised by a supplemental reporting scheme and omit such a scheme from its final regulations under Section 5.

D. Follow-up Reporting

While EPA may not be authorized to use Section 8 to obtain additional information during the PMN review period, the Agency is clearly entitled to require follow-up reporting under Section 8 on new chemicals which have been added to the TSCA Inventory. Once commercial manufacture of a new chemical has begun, it is no longer subject to the premanufacture notification provisions of TSCA and, instead, will be treated like any existing chemical. Thus, EPA would be entitled to invoke all of the provisions of TSCA which authorize it to engage in regulatory, testing, and information-gathering activities for substances and mixtures distributed in commerce.

In comments on the initial EPA PMN proposal, industry pointed to the EPA power to require follow-up reporting on new chemicals under section 8 and argued that this power lessens the need for detailed, intensive scrutiny of all chemicals at the PMN stage. To demonstrate how the EPA follow-up reporting authority might be used, CMA recommended that the Agency adopt appropriate Section 8(a) reporting requirements for new chemicals.

Based on these reports, the Agency may find that production of a chemical has ceased and it is no longer a matter for Agency concern. Conversely, it may find that production of the chemical has increased beyond the original expectations of the manufacturer or that it is being produced in large quantities by other manufacturers. Where the statutory criteria are satisfied, this information would provide the foundation for regulation under the testing provisions of Section 4, the "significant new use" provisions of Section 5, the rulemaking provisions of Section 6, or the "imminent hazard" provisions of Section 7. Follow-up reports may often be more meaningful to EPA than the earlier PMN since they would reflect the experience of the manufacturer during an extended period of actual production and use.

In its PMN reproposal and on subsequent occasions, EPA has announced that it intends to develop a system of follow-up reporting requirements for new chemicals which warrant continued monitoring. EPA is expected to take major steps in this direction in the near future.

E. The Application of Section 5 to New Chemicals Intended Solely for Export

Section 720.10(a)(1) of initial EPA-proposed regulations applies the premanufacture notification requirements of Section 5 to new chemical substances manufactured in the U.S. but intended solely for export. While the EPA reproposal contains an abbreviated PMN form for export chemicals, it is the intent of the Agency to subject export chemicals to the full spectrum of regulation under Section 5, including postponements in manufacture pursuant to Sections 5(c), 5(e), and 5(f). Industry has argued that the EPA approach is in direct conflict with the requirements of Section 12.

Section 12(a) provides that "this Act (other than Section 8) shall not apply to any chemical substance" which is "manufactured, processed or distributed in commerce for export from the United States." Thus, except for Section 8, the provisions of the Act are inapplicable to new chemical substances manufactured for export. Section 12(a)(2) creates a limited exception to this principle. It provides that chemical substances manufactured for export will be covered by the Act if EPA "finds" that they "will present an unreasonable risk of injury to health within the United States or to the environment of the United States."

Under this provision, EPA would appear to lack authority to apply Section 5 to chemicals manufactured for export merely because it believes that information about these chemicals may be useful. Rather, the Act contemplates a specific determination that particular chemicals intended for export may pose an unreasonable risk within the U.S. EPA cannot make this determination for exports as a general class. Rather, the statutory scheme requires EPA to select an individual export chemical for regulation under the Act, based on particular evidence relating to its safety and a specific showing that it may have adverse effects within the U.S.

In drafting Section 12, Congress recognized that EPA might need to monitor export chemicals in order to determine whether particular substances present an unreasonable risk within the U.S. and therefore can be subjected to TSCA requirements pursuant to Section 12(a)(2). It is apparently for this reason that, under Section 12(a), export chemicals are exempt from all provisions of the Act except Section 8. Thus, Congress intended EPA to use its reporting authority under Section 8 — and not the premanufacture review provisions of Section 5 — to keep apprised of chemicals manufactured in the U.S. for export to other countries.

In its interim statement of policies and procedures for premanufacture notification, EPA specifically excluded export chemicals from PMN submission requirements. 44 *Fed. Reg.* 28564, 28566 (May 15, 1979). Hopefully, EPA will incorporate this approach in its final Section 5 regulations.

F. EPA Actions under Section 5(e)

To date, EPA has issued nine proposed orders under Section 5(e) prohibiting the manufacture of new chemical substances pending the development of additional information concerning their health and environmental effects. The fullest explanation of EPA's approach to Section 5(e) is provided in its first Section 5(e) order which was issued by EPA on April 23, 1980, and involves 6 phthalate ester compounds proposed to be used as plasticizers.[35] The reasoning supporting the EPA Section 5(e) order for these compounds deserves careful scrutiny, since it provides important insights into the EPA approach to implementing this provision of TSCA.

The EPA-proposed Section 5(e) order was based on Section 5(e)(1)(A)(ii)(I), under which EPA may prohibit the manufacture of a new chemical if it determines that (1) the available information on the chemical is "insufficient to permit a reasoned evaluation of [its] health and environmental effects" and (2) manufacture of the chemical "may present an unreasonable risk of injury to health or the environment."

In determining that the PMN submitter met the first element of this test, EPA pointed out that the PMNs did not contain any test data on the new chemical, no data were available in the published literature, and chemicals of similar structure have been recommended for further testing by the ITC under Section 4(e). While acknowledging that data were available on structurally related compounds, EPA determined that these data were "not adequate to assess the risk or control options." According to EPA, data on a substance itself "are certainly the best" basis for evaluating the effects of the substance. While recognizing that "in many cases it should be possible to reasonably evaluate a substance . . . on the basis of data on structurally-

---

[35] The six PMNs were assigned EPA Numbers 5AHQ-1279-0079 through 5AHQ-1279-0084.

analogous compounds," EPA indicated that it was "not prepared to do so with respect to these PMN substances."[36]

Equally significant is the EPA explanation of why it believed the PMN substances "may present an unreasonable risk" to health and the environment. In construing this statutory test, EPA emphasized that definitive evidence of a health or environmental hazard is unnecessary:

> The conclusion that a chemical "may present" a hazard will not be based on definitive scientific data; if EPA already knew in detail, prior to production, the type and degree of hazard that a new chemical would be likely to present, the Agency would not need to seek additional data to perform a reasoned evaluation of the effects of the substance. Thus, determination that a substance "may present a hazard" under §5(e) must involve reasonable scientific assumptions, extrapolations, and interpolations. Proposed Order, at 14—15.

Evidence of a possible "unreasonable risk", EPA further indicated, need not derive from test data on the PMN chemical, but could be based on other types of evidence, including physical and chemical properties or structure/activity relationships:

> EPA considers a variety of factors to be suggestive of potential health or environmental hazards. Among other things, knowledge of a chemical's physical and chemical properties can be very helpful; such information can indicate, for example, whether a chemical is likely to be excreted from the body or accumulated in fatty tissue of humans or other organisms, thus increasing the potential for long-term adverse effects. Another major factor is whether the chemical is structurally related to another chemical with known adverse health or environmental effects. *Id.*, at 16.

Most disturbingly, EPA took the position that, once a possible hazard were found to exist, EPA could determine that this hazard entailed a "risk" within the meaning of Section 5(e) even though exposure to the new chemical is "low or moderate." EPA explained the basis for this conclusion as follows:

> Because §5(e)(1)(A)(ii)(II) authorizes EPA to regulate a chemical which is produced in substantial quantities and results in substantial or significant exposure, even in the absence of any indication that a chemical may be toxic, it is apparent that § 5(e)(1)(A)(ii)(I) establishes a lower exposure threshold for Agency action. EPA uses the term "risk" to describe the conclusion drawn from combining the information on hazard with the information on the degrees and likelihood of exposure to the substance. Potential human exposure is the primary concern in health risk assessment; for environmental risks, the exposure side of the assessment consists of two factors: release to the environment, and environmental fate and transport. Moreover, since little or no actual monitoring data on releases of the new chemical substances to the environment are likely to be available to EPA prior to manufacture, and the exact use conditions may not be known, it will not generally be possible for the Agency to perform a comprehensive or highly detailed analysis of exposure prior to proposing an order pursuant to §5(e). *Thus a chemical "may present an unreasonable risk" for purposes of §5(e) if EPA finds that there is a reasonable likelihood that exposure may arise because of activities associated with the manufacture, processing, distribution, use, or disposal of the substance, and that, if exposure occurs, it will not be clearly so insignificant, when considered with potential toxicity, as to render the risk of harm inconsequential. Id.*, at 15—16 (emphasis added).

Thus, in the view of EPA, once there is some evidence suggesting that a chemical is hazardous, that chemical will be deemed to present a "risk" for Section 5(e) purposes if it will have sufficient exposure so that the possibility of harm cannot be considered "inconsequential". In practice, this standard could be satisfied in any situation where human or environmental exposure to a chemical is more than negligible.

In determining that the particular PMN chemicals before it presented a "risk", EPA relied on analogies between those chemicals and a category of chemicals known as dialkyl phthalates. In utilizing this approach, EPA maintained that "structural analogy, as a predictive principle for biological activity, has been the subject of considerable study in recent years, and has been found to be a highly useful tool for estimating the environmental properties and behavior . . . of chemical agents." *Id.*, at 17. The Agency then referred to preliminary data which it had received concerning a National Cancer Institute (NCI) study of a related dialkyl phthalate compound. *Id.*, at 27—29. Because these data contained significant evidence of possible carcinogenicity, and because substantial dermal and inhalation exposure to the PMN substances was possible during various processing operations, EPA determined that the substances presented a possible cancer risk to workers. *Id.*, at 30—31. In addition, again relying on data for analogous compounds, EPA found that the PMN chemicals could bioaccumulate to a high degree and, even at low levels of exposure, might harm aquatic life. Because it was possible that large quantities of the PMN substances could be released to the environment and persist for a considerable time, EPA determined that a possible risk to the environment was present as well. *Id.*, at 38—39.

---

[36] Proposed Order and Supporting Determination ("Proposed Order") at 12—13.

Finally, EPA addressed the question of whether the potential "risk" posed by the PMN chemicals was "unreasonable", as Section 5(e) requires. In considering this issue, EPA speculated that a Section 5(e) order "could be based solely or primarily on the existence of information indicating that a chemical presented a reasonable likelihood of causing an adverse health or environmental effect." *Id.*, at 43. Nevertheless, the Agency pointed out that, according to the TSCA legislative history, the term "unreasonable risk" involves balancing adverse effects of a chemical against the impact of regulatory action "on the availability to society of the benefits of the substance, taking into account the availability of substitutes." In applying this test, the Agency explained, a less stringent standard would be applicable under Section 5(e) than under Section 6, since a Section 5(e) order "generally would at most delay or restrict availability of a substance until adequate data on its effects are developed and evaluated." *Id.*, at 42—43.

Turning to an examination of the economic aspects of its Section 5(e) order, EPA determined that "nothing before the Agency regarding the benefits of these new substances indicates that they outweigh the potential risks presented by the substances." *Id.*, at 47. In addition, EPA concluded that there was "no basis to expect that the PMN substances would be clearly less hazardous to humans or the environment than the possible substitutes" identified by the Agency. *Id.*, at 49. EPA recognized that the high cost of carcinogenicity testing "may result in the Company deciding to keep the chemicals off the market permanently", but nevertheless concluded that it did "not believe that the cost of testing alone should compel the Agency to allow potentially dangerous new chemicals on the market without adequate assessment of risk. . . ." *Id.*, at 46. Based on this analysis, EPA concluded that the potential risk posed by the PMN substances was "unreasonable."

In framing its Section 5(e) order, EPA sought to preserve the flexibility of the manufacturer to the greatest extent possible. The proposed order merely required the submitter provide EPA with information sufficient to permit a reasoned evaluation of the human health and environmental concerns raised by EPA. *Id.*, at 50. EPA did not specify particular tests which it considered essential to achieve this goal. Instead, it explicitly left open the possibility that short-term testing might obviate the need for a long-term bioassay, and that one rather than all six of the PMN substances would need to be tested. *Id.*, at 51. In addition, the Agency declined to prescribe particular test procedures but merely required the data in question to be "developed according to good laboratory practices and through the use of methodologies generally accepted at the time the study is initiated." *Id.*, at 61. The Agency explicitly urged the PMN submitter "to consult with EPA before undertaking extensive information development." *Id.*, at 51.

A number of significant points emerge from the first proposed EPA Section 5(e) order:

Of paramount importance is the relatively low standard which EPA believes it must meet to justify action under Section 5(e). According to the Agency, it is entitled to find that a chemical "may present an unreasonable risk" in the absence of data relating to the chemical itself when there is evidence indicating that structurally analogous chemicals may have adverse health or environmental effects. Moreover, once the potential for harm is demonstrated, EPA believes that substantial exposure to the PMN chemical need not be shown. Instead, the position of the Agency is that it can proceed if there is sufficient exposure so that any risk of harm cannot be dismissed as "inconsequential." In effect, this standard shifts the burden to the PMN submitter to disprove the existence of a potential risk; there will be few new chemicals which have such low exposure that this showing could be made. Finally, in the EPA view, for the normal new chemical, a Section 5(e) order will be justified even if the PMN submitter will abandon the chemical rather than perform the testing required by the Agency. Thus, the only occasions on which the risk associated with a new chemical would be considered "reasonable" are where the chemical possesses substantial economic benefits and has no substitutes which are equally safe.

Another important aspect of the EPA Section 5(e) order is that the Agency is prepared to display considerable flexibility in devising an acceptable information-development program for the PMN chemical. Thus, PMN submitters subject to such orders can expect EPA to be prepared to negotiate a testing program which will satisfy Agency needs yet avoid unnecessary costs. In addition, the Agency hopes to use informal consultation, rather than legal coercion, to structure an acceptable data-development procedure. Whether the EPA's flexibility will be sufficient to persuade most manufacturers to continue developing new chemicals subject to Section 5(e) orders, of course, remains to be seen.

Finally, even after a Section 5(e) order has been issued, EPA is prepared to allow manufacturers to withdraw their PMNs if they are prepared to abandon their plans to commercialize the new chemical. Thus, the formal requirements of a Section 5(e) order will not take effect, thereby according the manufacturer a measure of flexibility if its interest in the PMN chemical revives at a later date and a new PMN is submitted.

Nevertheless, although a formal Section 5(e) order may not be in effect, the PMN submitter and the other manufacturers of the same compound must expect that subsequent PMNs will receive close scrutiny from the Agency and, unless new information is presented, trigger an additional Section 5(e) order.

G. Significant New Use Rules (SNURs)

EPA has recently begun to devote its attention to the Section 5(a) authorization for EPA rules which define "significant new uses" of existing chemicals and provide for the submission of notices to the Agency at least 90 days before manufacture or processing of such chemicals for these uses may begin. While the EPA plans for developing SNURs have not fully crystallized, the Agency has indicated that it intends to use such SNURs as a device for monitoring the subsequent development of a large number of new chemicals that have completed PMN review. The Agency has already proposed a SNUR for one such chemical, and SNURs for other new chemicals are in preparation. EPA has made less progress in developing SNURs for existing chemicals included in the TSCA Inventory, but has stated that it will eventually turn its attention to this task as well.

1. Text and Legislative History of the SNUR Provisions

The provisions of Section 5 governing a notice submitted as a result of a SNUR are in many respects identical to those governing other PMNs. The review period for a SNUR notice is 90 days, but can be extended for another 90 days for good cause. A SNUR notice must contain the information described in Section 5(d)(1), which incorporates by reference the list of information contained in Section 8(a)(2). Normally, the SNUR notice need only contain the test data that the submitter has generated voluntarily, but additional test data may be required if the SNUR chemical is covered by a Section 4 test rule or included on the Section 5(b)(2) "risk list." If EPA wishes to prohibit or limit manufacture or processing of the SNUR chemical, it may invoke Section 5(e), which authorizes EPA to take control measures when the available data concerning the SNUR chemical is insufficient and evidence indicates that it may present an unreasonable risk or other enumerated criteria are met. Section 5(f) also permits summary action respecting a SNUR chemical in appropriate cases. Finally, SNUR requirements do not apply to chemicals manufactured or processed in small quantities for research and development purposes, and an exemption from SNUR requirements may be obtained for test marketing activities.

There are, however, major differences between the SNUR and the regular PMN provisions of Section 5. While regular PMN requirements apply only to manufacturers and importers, SNUR requirements can be applied to manufacturers, importers, and processors. In addition, regular PMN requirements are self-executing, while SNUR requirements only apply to those particular uses of specific chemicals that EPA has identified by rule. Because of this, the scope and timing of SNUR requirements are matters that Congress has committed in certain respects to EPA's discretion.

Moreover, while regular PMN requirements apply to all new chemicals, a SNUR notice can be required only in limited circumstances. EPA may exercise its SNUR power to require the submission of a PMN for an existing chemical substance only if that substance will be devoted to "a use which the Administrator has determined . . . is *a significant new* use." Section 5(a)(1)(B) (emphasis added). As provided by Section 5(a)(2), "[a] determination by the Administrator that a *use* of *a* chemical substance is *a significant new* use with respect to which notification is required. . . shall be made by a rule promulgated after a consideration of all relevant factors . . . ." (Emphasis added.) The provision lists several such factors, including production volume, changes in the manner, magnitude or duration of exposure, and the reasonably anticipated manner and methods of the chemical's manufacture, processing, distribution or disposal.

The three words of the statute, italicized above, contain the essential requirements that must be satisfied before a SNUR may be promulgated:

1.   EPA must make a determination with reference to each use.
2.   The use must be new.
3.   The new use must be significant within the meaning of the statute.

Any SNUR requirements that EPA imposes must be limited to situations that meet all three of these criteria.

These statutory principles have important implications. First, since Section 5(a)(1)(B) applies only to a "new" use, the Agency's rule must, at a minimum, be limited to those uses of the chemical that are initiated after the rule is proposed. Moreover, even if a particularized use is new, EPA can subject the use to a SNUR requirement only if it can make a further showing that the use is significant. While the statute does not prescribe detailed criteria for identifying a significant new use, this concept necessarily must be defined in terms of TSCA's principal objective of preventing harm to human health or the environment. Thus, EPA must show that the alteration in use may change the nature or increase the magnitude of the chemical's adverse effects and that the resulting magnification of the chemical's risk potential could cause significant human or environmental harm.

2. EPA's SNUR for NMPT

EPA's first SNUR proposal applies to *N*-Methanesulfonyl-*P*-Toluenesulfonamide (NMPT). 45 *Fed. Reg.* 78970 (November 26, 1980). The proposal adopts alternative qualitative and quantitative tests of what constitutes a significant new use. In general, the proposed SNUR defines a significant new use of NMPT as (1) any use different in function or particular commercial or technical application from the use identified in the original PMN submitted by the National Starch and Chemical Company; and (2) the production by any person of more than 1000 lb/year of NMPT for the use identified in the National Starch PMN.

In its comments on the proposed NMPT SNUR, industry argued that promulgation of that proposal in its current form would exceed EPA's SNUR authority in at least four critical respects:

1.   EPA's proposal contained no recognition that EPA can require the submission of a SNUR notice only for a use of a chemical which it has determined to be both *new* and *significant*.
2.   EPA's proposed NMPT SNUR failed to identify each use or category of uses that it determined to present a significant health or environmental concern.
3.   EPA's proposal improperly assumed that production volume alone can be used to define a new use.
4.   EPA presented no data or other information to indicate that NMPT may have toxic effects on either man or the environment. Under such circumstances, a meaningful finding of health or environmental significance is impossible.

EPA has not yet taken final action on its proposed SNUR for NMPT. However, Agency officials have informally concurred with many of the industry criticisms of EPA's proposal.

3. SNURs for Existing Chemicals

More uncertain is EPA's thinking concerning the development of SNURs for existing chemicals on the TSCA Inventory. Before developing a SNUR for such existing chemicals, it would seem essential for EPA to develop a full profile of present uses, production volume, exposure patterns, and adverse health or environmental effects. Without such a data base, EPA would have no foundation either for defining new uses of the chemicals involved or ascertaining which of those new uses are significant.

The Agency is also undecided whether to define significance in terms of specific uses, exposure levels or production volume. One approach suggested by EPA is to develop qualitative SNURs which, for the chemicals involved, define significant new uses in the form of increases in overall exposure levels. Another approach which EPA is considering is "population focused SNURs", under which any new exposure of the chemical by particular subsets of the total population would constitute a significant new use. Yet another approach under consideration is a quantitative SNUR, under which specified increases in the aggregate production volume of a chemical would trigger SNUR notice submission. All of these approaches have a host of legal, practical, and conceptual problems which EPA has apparently not yet resolved.

A final open issue concerning EPA's SNURs for existing substances is whether they will apply to individual chemicals, entire categories, or even to all chemicals on a generic basis. Environmental groups have been pressing EPA to adopt the broadest SNURs possible. On the other hand, to the extent that SNURs apply to a broad and amorphous grouping of chemicals, the Agency will find it difficult to develop the necessary factual support to justify its findings that particular uses are new and significant. Any EPA effort to develop all-encompassing SNURs would therefore be very likely to meet with considerable industry resistance and legal challenge.

H. Exemptions under Section 5(h)(4)

In view of the mounting evidence of the adverse economic impact of Section 5, industry and the Reagan Administration have supported mechanisms for reducing the costs and delays imposed by PMN requirements without jeopardizing the basic objectives of Section 5. One such mechanism is EPA's exemption authority under Section 5(h)(4). Under this provision, EPA may, upon application, promulgate rules that exempt manufacturers of specified chemicals or chemical categories from all or part of Section 5's requirements. In granting such exemptions, EPA must find that, under their projected conditions of manufacture and use, the chemicals involved will not present an unreasonable risk of injury to health or the environment.

Recently, the subject of Section 5(h)(4) exemptions has received substantial attention from EPA and industry. EPA has published a final rule granting exemptions to certain chemicals used in the photographis industry. It has also proposed an exemption rule for certain site-limited intermediates, low-volume chemicals and polymers.

The following discussion reviews certain general aspects of Section 5(h)(4) and then describes EPA's two specific initiatives under that Section.

1. General Aspects of Section 5(h)(4)

A creative, flexible use of the Section 5(h)(4) exemption authority can afford major benefits to EPA, industry and the public. Based on the statutory concept of no unreasonable risk, EPA can focus its efforts under Section 5 on those new chemicals whose health or environmental effects require the closest scrutiny and devote correspondingly less attention to new chemicals that have little or no potential for harm. In addition, EPA can protect innovation in the chemical industry by reducing regulatory burdens on those new chemicals that are least able to withstand the delays and costs of full PMN review. Finally, by excluding low-risk chemicals from full PMN review under Section 5(h)(4), EPA can conserve its own funds and manpower, thereby using its limited resources in a more cost-effective manner.

Section 5(h)(4) permits granting exemptions from PMN requirements based on a finding that the chemicals in question do not present an unreasonable risk. The criteria for evaluating the reasonableness of a risk are essentially the same under Section 5(h)(4) as under other provisions of TSCA. EPA must first evaluate the likelihood that the chemical in question will harm human health or the environment and then assess the potential adverse economic and social impact of subjecting the chemical to notice and review under Section 5. If the Agency determines that the potential for harm to health or the environment associated with the chemical is small and the burdens associated with PMN requirements are excessive in light of the chemical's limited risks, it would have to conclude that those risks are reasonable and the chemical should be exempted from PMN requirements. On the other hand, where the Agency determines that the PMN process may afford significant protection to health or the environment and the benefits of that protection will outweigh its economic costs to manufacturers and others, the Agency would conclude that the risks presented by the new chemical could be unreasonable and an exemption from the PMN process would be unauthorized.

2. EPA's Specific Exemption Initiatives

The photographic exemption — On June 4, 1982, EPA promulgated a final exemption rule for chemicals used in or for the manufacture or processing of instant photographic and peelapart film articles. 47 *Fed. Reg.* 24308. Under that rule, which was developed in response to a petition by the Polaroid Corporation, manufacture or processing of eligible chemicals may begin immediately upon submitting an exemption notice to EPA. Certain manufacturing, processing, and use operations involving these chemicals must take place in a demarcated special production area where exposure is limited to specified levels by engineering controls and worker safeguards. The rule also requires manufacturers to implement certain controls for releases to land, water, and air. Manufacturers must maintain records of each chemical substance manufactured and processed under the exemption, and EPA may prohibit use of the exemption if the manufacturer's activities may present an unreasonable risk to health or the environment.

Generic exemption rulemaking — On August 4, 1982, EPA proposed two rules under Section 5(h)(4) with generic application to the chemical industry. These rules, if promulgated, could apply to as many as 50% of the chemicals presently subject to PMN requirements. Because of their broad coverage, the EPA proposals have received substantial attention from industry and environmentalists. CMA, which submitted a petition seeking the commencement of rule-making proceedings under Section 5(h)(4), has supported EPA's proposals as a major step in reducing the burdens imposed by PMN requirements and stimulating innovation. Environmentalists, on the other hand, have asserted that the EPA proposals could undermine the intent of Section 5.

The major provisions of EPA's exemption proposals are summarized below:

a. The Site-Limited Intermediate and Low-Volume Proposal

This proposal applies to (1) chemicals that are consumed at their sites of manufacture and (2) chemicals with an annual production volume of 10,000 kg or less.

To be eligible for an exemption, these chemicals would have to meet the following criteria for low risk: (1) site-limited intermediates and low-volume chemicals manufactured in more than 1000 kg/year could not be known or reasonably suspected to have carcinogenic or teratogenic potential or to have produced positive results in a mutagenicity test; (2) low-volume chemicals manufactured in more than 1000 kg/year could not be capable of acute toxic effects; and (3) both low-volume chemicals and site-limited intermediates could not cause serious acute or chronic effects, or significant environmental effects, under their anticipated conditions of manufacture, use, or disposal.

To ensure that EPA's detailed criteria for low toxicity are applied in a careful and thorough manner, site-limited intermediates and low-volume chemicals produced in more than 1000 kg must be reviewed by a "qualified expert" equipped by training, education or experience to make an informed evaluation of the chemical's potential risks to human health and the environment. The expert must compile and review the available information concerning the chemical's human health and environmental effects. A chemical will be eligible for an exemption only if the expert determines, and documents in the manufacturer's files, that the chemical meets EPA criteria for low toxicity. In filing an exemption notice, the firm's management must certify in writing that the expert has performed the required analysis.

The final responsibility for reviewing exemption candidates rests with EPA. To facilitate the EPA review, the manufacturer must submit a written notice to EPA at least 14 days before the commencement of manufacture. This notice would have to provide certain information about composition, intended uses, and production volume of the chemical. Based on its review of this notice, EPA could declare a chemical ineligible for an exemption if it failed to meet the Agency's exemption criteria. EPA could also postpone manufacture if significant questions about the chemical's risks arose that the Agency could not resolve in the 14-day review period.

As a condition of the exemption, manufacturers would be required to observe certain recordkeeping and reporting requirements. EPA would have the right to revoke an exemption at any time if it obtains information that the chemical does not meet applicable exemption requirements or that the manufacturer has willfully or negligently failed to comply with the terms of the exemption.

Exempt chemicals would not be added to the TSCA Inventory. Thus, if an exemption were revoked, or if the new chemical were manufactured outside the terms of the exemption, a PMN would have to be filed with EPA or manufacture would cease.

Failure to comply with any provision of EPA's exemption rule would constitute a violation of TSCA. Firms that commit such violations would be subject to civil and criminal penalities.

b. The Polymer Exemption Proposal

Certain classes of polymers, for which there is a strong base of experience which demonstrates low risk, could be manufactured upon the submission of a notice to EPA. These polymers include: (1) polyesters made from a specified list of monomers; (2) polymers with a number-average molecular weight of 20,000 or greater; and (3) polymers with certain number-average molecular weight and polydispersity values. The notice filed with EPA would identify the polymer in sufficient detail to demonstrate its eligibility for an exemption. Exempt polymers would not be added to the TSCA Inventory and, thus, would be subject to PMN review if manufactured outside the terms of the exemption.

Polymers with a number-average molecular weight above 1000 would undergo a shortened PMN review. Manufacturers of these polymers would be required to submit a limited PMN describing the polymer's identity, production volume and projected use. Such a PMN would have to be filed at least 14 days before the start of production. If serious unresolved questions concerning toxicity or exposure remained at the end of this period, EPA could postpone manufacture for up to 90 days and, thereafter, could take regulatory action in accordance with Sections 5(e) and 5(f). Polymers that complete limited PMN review would be added to the TSCA Inventory.

Certain polymers would be automatically excluded from all provisions of EPA's exemption rule. Covered by this exclusion would be (1) polymers intended or reasonably anticipated to be water soluble, (2) polymers made from living organisms, (3) polymers with certain reactive functional groups, (4) polymers designed to degrade, decompose, or depolymerize, and (5) polymers containing other than a limited list of chemical elements. All such polymers would be required to undergo full PMN review before the commencement of manufacture. EPA has recognized that its proposed exclusions are very conservative and has invited comment.

Chapter 6

SUBSTANTIAL RISK REPORTING — INDUSTRIAL REQUIREMENTS

George Dominguez

TABLE OF CONTENTS

I. INTRODUCTION

Substantial Risk Section 8[e] reporting requirements became automatically effective on January 1, 1977, when the Toxic Substances Control Act (TSCA) officially became law. In the intervening years, several hundred notices have been submitted to EPA, while untold numbers of potential reporting situations have undoubtedly been reviewed within who knows how many companies.

Since industry has been faced with this obligation for some time, experience has demonstrated certain internal administrative and organizational needs. Visualized from this perspective, these intervening years have served to highlight such needs as industry has attempted to comply with the spirit and letter of the law.

This has not been, nor will it be, an easy task since neither the statute itself nor the legislative history of this Section provides specific guidance as to what is actually required or what Congress itself actually intended. As we approach the question of practical compliance and consider industrial needs, the most logical place to start is with an examination of the statutory requirement itself and then to explore the guidance which EPA has issued as it proceeds to implement the law.

II. SECTION 8(e)

The text of Section 8(e) is very brief:

NOTICE TO THE ADMINISTRATOR OF SUBSTANTIAL RISKS — Any person who manufactures, processes, or distributes in commerce a chemical substance or mixture and who obtains information which reasonably supports the conclusion that such substance or mixture presents a substantial risk of injury to health or the environment shall immediately inform the Administrator of such information unless such person has actual knowledge that the Administrator has been adequately informed of such information.

Attempting to understand and comply with Section 8(e) immediately gives rise to several pertinent questions:

1.   Who is a person under this law?
2.   What does the phrase "obtains information" mean?
3.   What does "reasonably support the conclusion" mean?
4.   What constitutes a "substantial risk"?
5.   What constitutes "actual knowledge" that the Administrator has been informed mean?
7.   What does it mean that the Administrator has to be "adequately informed"?
8.   What need not be reported?
9.   What are the penalties for failing to report?

Critical as all of these basic questions are, there are others regarding the scientific basis for such determinations, the level and extent of verification involved, and a considerable number of policy-related issues, such as "can valid company proprietary information that might be involved in such notices be protected from public disclosure?" that must also be answered.

As previously mentioned, recourse to the legislative history of this section of the law does not provide any interpretative insights. This is admittedly a problem not just for industry but for EPA, since it is charged with implementing this imprecise charge. In order to do so, EPA issued a "Statement of Interpretation and Enforcement Policy: Notification of Substantial Risk" on March 16, 1978. This interpretative document is central to our consideration of industrial organizational, analytical, interpretational, and compliance requirements since it remains the only agency "policy" statement. It stands as the basic interpretative and guidance document insofar as an official EPA position is concerned. In it, the Agency in effect attempts to answer, from its viewpoint, the questions just posed.

III. THE EPA GUIDANCE DOCUMENTS ANSWERS TO BASIC INDUSTRIAL QUESTIONS

A. Who Is a Person under the Law?

EPA has broadly interpreted this to mean essentially all employees within the company who are "capable of appreciating the significance" of the information obtained. Since it has equally broadly approached the criteria for "capable of appreciating the significance" of the information, the number and level of responsible company employees is quite large. In order to limit this, EPA did provide a mechanism by which if an internal review and reporting procedure is established, then reporting responsibility is limited to certain clearly specified individuals within the company. Later, we will examine these requirements in greater detail since the limitation applies only when the internal system satisfies stated EPA criteria.

The first requirement of industry, then, is to determine if it wants to limit such review and reporting responsibility. If so, it must then develop a system which satisfies EPA requirements.

B. What Does "Obtains Information" Mean?

Here again, the Agency has placed a broad interpretation on this phrase, having established the criterion that the company will be deemed to have obtained such information at the time that any company employee "capable of appreciating its significance" obtains it.

This brings us to a second industrial requirement, namely, that following this premise we must internally inform our employees of this interpretation and must critically advise those who would be deemed "capable of appreciating" of their direct obligations.

C. What Does "Reasonably Support the Conclusion" Mean?

In this instance, the EPA approach has been a definition by exception rather than by explicit statement. For example, it has stated that it "does not mean conclusive proof"; however, this statement is also important in the context of the EPA interpretation of "immediately notifiable" since it goes on to say "and one cannot wait for conclusive proof" before notifying EPA of "substantial risk information". EPA has also said "reasonably supports" does not mean that merely suggestive uncorroborated information must be reported; however, between these two statements there is little other practical guidance, and this interpretative situation is further clouded in that the Agency also says that in making an assessment of this provision the emphasis is essentially on the reasonableness of the conclusion that the substance or mixture presents a substantial risk, not necessarily on the risk itself, and clearly not on the reasonableness of the risk.

Again, in the framework of providing operational definitions, EPA cited several examples of the type of information that could be considered to "reasonably support" a conclusion of substantial risk:

Information from the following sources concerning the effects described in Part V will often "reasonably support" a conclusion of substantial risk. Consideration of corroborative information before reporting can only occur where it is indicated below.

(1) *Designed, controlled studies.* In assessing the quality of information, the respondent is to consider whether it contains reliable evidence ascribing the effect to the chemical. Not only should final results from such studies be reported, but also preliminary results from incomplete studies where appropriate. Designed controlled studies include:
(i)   In vivo experiments and tests.
(ii) In vitro experiments and tests. Consideration may be given to the existence of corroborative information, if necessary to reasonably support the conclusion that a chemical presents a substantial risk.
(iii) Epidemiological studies.
(iv) Environmental monitoring studies.

(2) *Reports concerning and studies of undesigned, uncontrolled circumstances.* It is anticipated here that reportable effects will generally occur in a pattern, where a significant common feature is exposure to the chemical. However, a single instance of cancer, birth defects, mutation, death, or serious incapacitation in a human would be reportable if one (or a few) chemical(s) was strongly implicated. In addition, it is possible that effects less serious than those described in Part V(a) may be preliminary manifestations of the more serious effects and, together with another triggering piece of information, constitute reportable information; an example would be a group of exposed workers experiencing dizziness together with preliminary experimental results demonstrating neurological dysfunctions.

Reports and studies of undesigned circumstances include:
(i)   Medical and health surveys.
(ii) Clinical studies.
(iii) Reports concerning and evidence of effects in consumers, workers, or the environment.

This introduces the third industrial need, namely, the ability in those instances where the information does not clearly fall within one of these "defined" areas to provide for expert review and interpretation within the framework of a reasonable judgment regarding the reasonableness of the data supporting any substantial risk conclusion and hence any reporting obligation. Since this involves discharging an important legal obligation on the part of the company and its employees, it is a critical requirement and one which we will see is often undertaken not by the actions of a single individual but through the establishment of an interdisciplinary review "team" or "committee" with legal evaluation, advice, and support.

D. What Constitutes a "Substantial Risk"?

Here again EPA has elected to provide operational definitions to this question. First, it has stated that such "risks" can be divided into three categories:

1.   Human health effects
2.   Environmental effects
3.   Emergency incidents of environmental contamination

In addition, the Agency provides two criteria that can be employed in making substantiality judgments:

1.   The seriousness of the threat to health or the environment
2.   The probability of the occurance of the adverse effect

As we examine and reflect on these two criteria, it is also important to note that economics, social value, benefits, etc. are not to be taken into consideration insofar as the agency interpretation is concerned. Furthermore, EPA goes on to say that these criteria themselves are differently weighted when considering different types of effects. What this means from the interpretational and decision-making viewpoint is that, according to EPA, certain human health effects, e.g., carcinogenicity, are so serious that relatively little, if any weight, should be given to the probability of occurrence such as might be demonstrated by exposure opportunity or limitation. In that situation, the mere fact that the chemical is in commerce would be sufficient to satisfy any exposure consideration. From the standpoint of environmental effects or those arising from emergency incidents, however, a different standard would apply. In those instances, the situation must involve or be accompanied with significant exposure levels.

As a practical matter, EPA went even further in providing several specific examples of information in all three categories that would be considered to meet the threshold of "substantial risk" and therefore be reportable:

(a) *Human health effects* —

(1) Any instance of cancer, birth defects, mutagenicity, death, or serious or prolonged incapacitation, including the loss of or inability to use a normal bodily function with a consequent relatively serious impairment of normal activities, if one (or a few) chemical(s) is strongly implicated.

(2) Any pattern of effects or evidence which reasonably supports the conclusion that the chemical substance or mixture can produce cancer, mutation, birth defects or toxic effects resulting in death, or serious or prolonged incapacitation.

(b) *Environmental effects* —

(1) Widespread and previously unsuspected distribution in environmental media, as indicated in studies (excluding materials contained within appropriate disposal facilities).

(2) Pronounced bioaccumulation. Measurements and indicators of pronounced bioaccumulation heretofore unknown to the Administrator (including bioaccumulation in fish beyond 5,000 times water concentration in a 30-day exposure or having an n-octanol/water partition coefficient greater than 25,000) should be reported when coupled with potential for widespread exposure and any non-trivial adverse effect.

(3) Any non-trivial adverse effect, heretofore unknown to the Administrator, associated with a chemical known to have bioaccumulated to a pronounced degree or to be widespread in environmental media.

(4) Ecologically significant changes in species' interrelationships; that is, changes in population behavior, growth, survival, etc., that in turn affect other species' behavior, growth, or survival.

(5) Facile transformation or degradation to a chemical having an unacceptable risk as defined above.

(c) *Emergency incidents of environmental contamination* — Any environmental contamination by a chemical substance or mixture to which any of the above adverse effects has been ascribed and which because of the pattern, extent, and amount of contamination

(1) seriously threatens humans with cancer, birth defects, mutation, death, or serious or prolonged incapacitation, or

(2) seriously threatens non-human organisms with large-scale or ecologically significant population destruction.

E. What Constitutes "Immediately"?

The EPA Guidance Document has been very precise as to what is considered to constitute an immediate notification.

1.   Where the "substantial risk" does not involve an emergency incident, i.e., environment contamination, "immediately" is construed to mean notification within 15 working days. As we will see, given the practical problems of analysis, review, interpretation, submission preparation, or even assembling the proper parties to make the reporting decision, this is in fact an extremely short period of time.

2. Where emergency incidents of environmental contamination are involved, the response time is considerably shorter, since the Agency requires as a minimum oral reports within 24 hr — acknowledging that written reports can be submitted later (within the 15-working day reporting period). The EPA guideline (as provided in the appendixes) provides emergency reporting telephone numbers for each of its ten regions.

F. What Constitutes "Actual Knowledge" That the Administrator Has Been Informed?

In this case, there are actually two bases for making the determination that the Administrator has "actual knowledge". In the first case, the potential notifier has first-hand documentation that EPA has been informed, e.g., copies of notices that the Agency may have received dealing with the same substance and effect.

Second, the notifier can rely upon publication of the information in certain journals stipulated by EPA. These are limited to:

1. *Agricola*
2. *Biological Abstracts*
3. *Chemical Abstracts*
4. *Dissertation Abstracts*
5. *Index Medicus*
6. National Technical Information Service

Additionally, it is not necessary to report under Section 8(e) if this would be duplicative of other reporting obligations to the Agency and:

1. Such reports have been filed
2. The report contains those elements of information that EPA requires under Section 8(e) reporting
3. The reports are submitted within the 8(e) reporting time requirements

Examples of such exclusions would be reports submitted to EPA under authority of:

1. The Federal Insecticide, Fungicide and Rodenticide Act
2. The Clean Air Act
3. The Clean Water Act
4. The Marine Protection, Research and Sanctuaries Act
5. The Safe Drinking Water Act
6. The Resource Conservation and Recovery Act

G. Exclusion of Corroborative Data

EPA also excludes the need to report information which is merely corroborative of well-established adverse effects which are already in the published literature. This exclusion, however, is limited to the same six publications previously mentioned.

IV. CONTENT OF AN 8(e) NOTICE

Once the decision to report has been made, EPA, while not specifying that any exact format must be used, has indicated that the notice should contain the following information:

1. The job title, name, address, telephone number, and signature of the person making the report
2. The name and address of the manufacturer, processor, or distributor with which he or she is associated
3. The chemical identification of the chemical substance or mixture, including, where known, the CAS Registry Number
4. A summary of the adverse effect being reported
5. A summary of the nature and extent of the risk involved
6. A statement regarding the specific source of the information with a summary of any available supporting technical data

This information, when conveyed to the Agency, must also state that it is being submitted in accordance with Section 8(e) of TSCA and sent by certified mail or in any other fashion that will allow verification of its receipt by the Agency.

This outline of contents, while based on written reporting procedures, also constitutes the basis for oral reporting in emergency situations. As previously indicated, such oral reports are to be followed up within the review and reporting period by written notices.

V. PENALTIES

Before proceeding to examine company needs, options, and internal response it is also important to appreciate the legal penalties involved in nonreporting. Failing to submit an appropriate 8(e) notice could subject the company and individual employees to civil and/or criminal penalties. Section 15(3) of TSCA makes it unlawful for any person "to fail or refuse to submit reports, notices or other information . . . as required by this Act." Violations will subject the company and persons within it to possible civil penalties of up to $25,000 per day of violation. Under Section 16(b) of the Act, "any person who knowingly or willfully violates any provision of Section 15 shall, in addition to or in lieu of any civil penalties . . . be subject, upon conviction, to a fine of not more than $25,000 per day of violation or to imprisonment of no more than one year, or both."

In short, this means that companies and their employees (limited to the degree previously described), can face potential civil and criminal liabilities of up to $25,000 per day of violation and in the criminal prosecution situation up to one years imprisonment.

VI. MANAGEMENT AND ORGANIZATIONAL CONSIDERATIONS

Section 8(e) reporting obligations place new demands upon the organization. The company must now develop not only an awareness of these requirements, but the capability to have such potential reporting situations internally reported, reviewed, analyzed, interpreted, and finally, where warranted, submitted. In addition, there are several other considerations demanding company attention such as follow-up capability; possible internal, technical, or administrative alterations based on the "substantial risk information" itself; and the possibility of informing other government agencies, customers, or co-manufacturers.

VII. AUTHORITY AND RESPONSIBILITY

The first step in developing a Section 8(e) compliance program is to determine who will have the responsibility for its development and implementation and who will have the final authority for making 8(e) submissions. These may well, in fact, not be the same persons, to the extent that senior management may want to reserve for itself the ultimate authority for making the reporting decision while delegating the responsibility for development and implementation to middle managers. It is more likely, however, that baseline responsibility and authority for development, implementation, and reporting will be centralized in one individual. Even in those instances, however, it is more than likely that some form of review team will be established in order to provide him or her with the many skills and disciplines obviously required to evaluate potential 8(e) reports and to make not only the final determination but assist in the preparation of the 8(e) notice itself and any internal or external follow-ups required.

VIII. INTERNAL REPORTING PROCEDURES

A. General

While the company is free to develop its own internal reporting procedures, we must keep in mind that in order to limit personnel liability and to provide delineation the system must essentially conform to the EPA March 16, 1978, Guidance Document. In reviewing the following, bear in mind that these are the minimum requirements set forth by EPA:

1. The system must specify the nature of the information that company officers and employees must submit
2. Indicate how such reports are to be internally prepared and the company official to whom they should be directed

3. Stipulate the Federal penalties for failure to report
4. Provide the means by which to advise of the disposition of the report
5. Inform employees of their right to report directly to EPA in the event that the company decided it was not necessary to submit a report to EPA

While not a component of the reporting system itself, the Agency also requires that the company establish procedures to publicize internally the system and to make certain that it is positively implemented. This leads to the need for considering not only development of a program consistent with processing review and notification requirements but internal administrative and employee notification provisions as well.

In this later context it is also necessary to recognize that with job transfers, alteration in responsibilities, and turnover, there is an on-going need for employee training and education. This is critical if an effective 8(e) program is to be developed and maintained.

While there are several ways in which such programs can be developed and implemented, two elements to consider are

1. The desirability of developing an internal policy statement regarding 8(e) compliance implementation — this may or may not contain detailed "reporting up" procedures and reporting forms
2. System audit provisions to ensure periodic review not only of the internal publication and employee training aspects but to verify that the system is actually working and that review, documentation, etc. is being appropriately undertaken

B. Reporting Procedures and Forms

When developing the internal reporting system, it may be desirable to also provide specific internal reporting forms. Several companies have done so and others, as they become more familiar with these requirements, will also find them helpful. While once more the specific content and format will vary considerably from company to company, the basic elements that should be considered for inclusion are:

- Company name, address, and location (the one initiating the report)
- Source and nature of the information
- Overall assessment of information
- Designated internal company contact
- Statement of 8(e) and penalties for failure to report
- Disposition (including employee notification and statement of employees' right to notify EPA directly in event company does not report)

After this initial report, an individual file on the situation will probably be established containing not only this form but additional data developed during the assessment period. These might include toxicology or other studies, results of literature searches, meeting minutes, etc.

Insofar as specific notification to the Agency is concerned, as mentioned, there is no form stipulated. However, the basic data elements previously listed must be included and the company may therefore wish to consider establishing its own standard format. This could be useful not only in providing an element of consistency but serve to assure that all required information is provided.

IX. CLAIMING CONFIDENTIALITY

It is entirely conceivable that information contained in an 8(e) notice could be proprietary or that for other reasons the company might wish to claim part or all of the submission to be confidential. EPA has not been insensitive to this need, and the March 16, 1978, Guidance Document contains instructions regarding confidentiality claims. In addition, the Agency has published regulations governing its overall treatment of confidential business information. These are set forth in 40 C.F.R., Part 2 (41 F.R. 36902, September 1, 1976) as amended in 40 C.F.R. Part 2 (43 F.R. 39997, September 8, 1978). All of these documents should be carefully studied and the stipulated requirements followed in order to assure that confidential information will be protected to the extent that it can be. These requirements are highly technical, and legal consultation should be sought in order to be sure of compliance.

From the business viewpoint, it is critical to bear in mind that all of the 8(e) notice or portions of it can be claimed as confidential. However, it is also necessary to recognize that:

1. There are statuatory limitations on what the Agency can and cannot consider as confidential. In this connection, it is important to review Section 14 of the Statute which establishes this overall framework.
2. Even if the company claims information to be confidential, this is no guarantee that the Agency will agree, but it does establish the claim of the company. This in turn means that the Agency will follow certain review procedures and notify the company before divulging the information in the event that it does not agree with a confidentiality claim. In that event, the company also has certain legal rights. While none of this should be construed as encouraging the filing of frivolous confidentiality claims, it is indicative of the need for carefully considering the possibility of exercising company rights.

From the viewpoint of basic procedural requirements, whenever a confidentiality claim is made it will be necessary to furnish the Agency with two copies of the submission, one complete and the other a "sanitized" version omitting the information claimed to be confidential. The first version should, of course, clearly be marked confidential on each page while the second does not require such marking. If the Agency agrees with the classification, it will restrict access to the first copy in accordance with its internal security procedures while placing the second in a publicly available file where it can be seen by anyone upon request.

X. NOTIFYING OTHERS

While there is no direct requirement to notify anyone else based on either a statuatory obligation or Agency interpretation and guidance, there may nonetheless be reasons why a company might want voluntarily to consider notifying:

1. Other federal agencies
2. Customers
3. Other manufacturers, processors, or distributors

Examples of other agencies that the company might want to inform are OSHA, NIOSH, CPSC, or FDA. This, of course, would be done within the spirit of bringing the attention of these agencies to health, safety or other aspects of public information that might be of value to such agencies in protecting human health and the environment.

From the viewpoint of customer or other notifications, there are several reasons why this may be desirable:

1. By providing information, the company could further health and environmental protection by informing them of possible hazards
2. Informing others assures that they receive the information directly and reliably
3. Customers or others may have additional information that could have a bearing on the assessment of the risk. They, in turn, might want to bring this information to the attention of EPA or let the original notifier know of it
4. New handling, storage, use, or other precautions may be necessary. These can be recommended by the notifier or developed by the user. Bringing the 8(e) notice to customers' attention could assist in developing such additional precautions

Any internal 8(e) compliance program should include consideration, on a case-by-case basis, of these additional voluntary notices.

XI. ADDITIONAL INTERNAL ACTIONS/ACTIVITIES

An 8(e) review, regardless of the eventual reporting outcome, can give rise to several additional internal assessments and activities. Again, from the viewpoint of developing a comprehensive and effective program, these should also be built into the review, analysis, and action system. Among the more important considerations in this regard are:

Identification of the true causative agent — Initial results are often based on sales grade material or mixtures. Given the time constraints of the 8(e) reporting obligation, it may not be possible to determine precisely what is causing the effect. It may, in the case of an individual substance, be some impurity which through process change might be eliminated. In the case of a mixture it might be a replaceable component. Further study to ascertain the cause of the reportable effect may therefore be highly desirable.

Additional testing — Again, initial findings are often the basis for reporting. More extensive studies may be warranted. These may verify, contradict, or provide needed additional information useful in making a more accurate risk determination.

Identification of impurities, by-products, etc. — This could be important in connection with the preceeding point regarding the determination of causative agents.

Changes in material safety data sheets — The findings of an 8(e) review may well indicate that any product safety literature should be changed in order to provide information about new hazards.

Changes in labeling — Where new information regarding hazards is developed, existing labels should be reviewed in order to assure that they are still adequate and, if not, changed accordingly.

Employee notification — In some cases it may be necessary to undertake specific employee notification of a newly discovered hazard.

Work practice changes — New finding may necessitate changes in internal work practices including handling precautions and alterations in industrial hygiene requirements.

Storage and handling procedures — Once more, new information may result in the need to alter storage and handling procedures.

Manufacturing — Depending on the nature of the information and the identification of the causative agent, process R & D, process modification, or formulation changes may be called for.

Process withdrawal, modification — In some cases, if the risk is sufficiently great or uncontrollable, it may be necessary to modify the process or in the extreme event withdraw it completely.

Insurance — In reviewing and making 8(e) reporting decisions, personal and company liability should be considered. In addition, however, other litigative situations should be considered, i.e., private suits.

Public notice, media coverage, etc. — There are two aspects inherent in this consideration. A Section 8(e) notice, with the exception of confidential information, is publicly available. This means that it could become the subject of media or other coverage. On the other hand, depending upon the information and the circumstances, the company might want to call media attention to its own findings and inform them directly.

Marketing aspects — In many cases, the company will want to review the marketing implications of 8(e) notices. This could involve an examination of sales, customers, uses, distribution, continue/discontinue decisions, etc.

Product aspects — From the product viewpoint, reevaluation of packaging, use, processing, or physical form may be necessary or desirable.

All of the foregoing, while not mandatory under the law or EPA guidance, are a logical outgrowth of analyzing 8(e) reporting requirements. The extent to which any company actually undertakes them will naturally vary, not only from company to company, but within a given company on a case-by-case basis. The main point however, is that these additional considerations should not be overlooked.

XII. FOLLOW-UP CONSIDERATIONS

While follow-up by EPA or others is not automatic, it is not unusual for them to provide the company with an initial evaluation of the notice and to subsequently request additional information or clarification. To the extent that this occurs, it will also be necessary to provide follow-up capability within the company. From a slightly different viewpoint, if the company in its notification to the Agency indicates that it intends to undertake its own additional assessments, tests, labeling changes, or other actions, it should in turn inform the Agency when these have taken place.

In regard to this latter point, it should be clear that providing this additional information is not mandatory. However, there is a reasonable consensus that all relevant information should be given to the Agency initially. This could well include plans for internal review, additional testing, work practice changes, etc. To the extent that these can be identified or specific examples provided, these could be sent to the Agency. Sending this information in, initially or subsequently, can assist the Agency by assuring them that the company is either undertaking further analysis or taking actions to control the risk. This can be very helpful to the Agency when making its determinations and deciding what actions, if any, it will take pursuant to notice by the company.

XIII. AN 8(e) ACTION WORKSHEET

A. General Considerations

Because there are a number of elements to consider in any 8(e) reporting situation and because the time for internal review and decision making is short, it may be beneficial to use an internal worksheet. This should cover the major actions/options open to the company and following it during deliberations and checking against it before final submission will assure that no important aspects have been inadvertently overlooked. While specific content and format will vary, the following suggested contents should capture all of the most important elements that ought to be contained in such a worksheet.

B. Elements of a Section 8(e) Worksheet

1.   Date and time information first reported or received
2.   Identity of person reporting or source of information
3.   Written statement of individual reporting or copy of data from written or published sources
4.   Determination of activity of company — manufacturer, importer, processor, distributor
5.   Determination of substantial risk
6.   Nature of the effect or hazard
     - Probability of occurrence
     - Exposure data
     - Use data
     - Production data
     - Route(s) of exposure
     - Source of information
     - Nature and quality of information
7.   Scientific review of information — effects and hazard
8.   Legal review of information and reporting obligation
9.   Identity of chemical substance or mixture — including determination of impurities, by-products, etc.
10. Literature review — if conducted on compound or related compounds
11. Determination of reporting obligation — substance is or is not subject to TSCA, EPA has been previously informed
12. If EPA must be informed within 24 hr — date and time of determination, date and time of reporting, person to whom reported, date for written follow-up report
13. Notification to employee who reported information of company disposition
14. Confidentiality determination — complete report, part of report, compliance with EPA requirements, review of final submission
15. Determination of other voluntary actions — additional notification, testing, process changes, work practice changes, new labeling, etc.
16. Identification of personnel involved in review and determination
17. Review and final submission — date and documentation

XIV. MAKING A SUBSTANTIAL RISK DETERMINATION AND REPORTING OPTIONS

In the final analysis, each company will have to make its own reporting decisions. While the EPA March 16, 1978, "Guidance", other publications such as this one, and a review of EPA experience and actions with substantial risk notices received to date are instructive and helpful, individual decisions still have to be made.

Aside from the organizational and managerial suggestions made here, there are two other aspects that might be of assistance in making such determinations: consideration of the factors to evaluate in the potential 8(e) reporting situation and the reporting options open to the company.

XV. FACTORS TO CONSIDER IN 8(e) REPORTING

While the facts will vary in every case, as will the relative weight to be ascribed to each of the following factors, they must all be considered in any potential 8(e) reporting situation. Arraying and fully analyzing them will provide a basis for complete review and evaluation:

1. Nature of the hazard involved
2. Number of employees exposed (actual or potential)
3. Number of users exposed (actual or potential)
4. Environmental exposure
5. Volume of the substance or mixture involved
6. Opportunities for adverse effect(s)
7. Reliability of the basis for the determination
   a. Quality of data
   b. Source of data
   c. Age of data
   d. Application to man or the environment
8. Other supporting or contradicting information or experience
9. Previous company decisions and actions

XVI. REPORTING OPTIONS

There are essentially four reporting/nonreporting options open to the company:

1. Based on complete assessment and the determination that a substantial risk does in fact exist, there is the clear option of submitting the notice pursuant to the statute and the EPA guidance document.
2. Where there is doubt that a substantial risk exists, the company has two choices:
   a. A report can be filed in accordance with EPA guidance, relying on them to make the determination, or
   b. The company may submit but include in their submission a written disclaimer stating that in its judgment the situation does not constitute a substantial risk. This is an important option because given this disclaimer, the Agency will review the information and make a decision. If it does not agree with the company, it will classify, log in, and process the notification as an 8(e) submission. However, if EPA agrees with the company, then it will not be logged in, processed and treated as an 8(e) submission. This has the practical effect that the notice will not appear in EPA records of 8(e) submissions nor will the notice be retained in the EPA 8(e) files and hence will not be publicly accessible.
3. Where the company feels certain that the situation does not represent a substantial risk, then there is no need for any EPA notification. It is however advisable in this, as in all of the other situations mentioned, to develop and maintain complete records of the review process and the basis upon which such decisions were made.
4. Again, assuming that the company does not feel that the situation represents a substantial risk and therefore is not reportable under Section 8(e), it may still elect to make a voluntary submission. In such cases, the option is to write to the Agency pointing out the nature of the risk, and indicating that while not within the ambit of 8(e) reporting obligations, the company is providing this information because it deems it to be of potential interest to the Agency in the execution of its responsibilities and in the interest of protecting health and the environment.

In all of these deliberations and in making a final selection from the options available, it is well to keep in mind that the standard for determining the substantiality of risk is not an absolute one. In order for a situation to meet the definition of a substantial risk, it is necessary only for the information to reasonably suport the conclusion that a substantial risk is involved. This is an important distinction and one which should be fully appreciated since it is the basis upon which EPA makes its determination as to substantiality and the standard that it would impose on potential notifiers.

From the business/compliance viewpoint, there are a number of important decisions to be made, ranging from organizational to specific reporting determinations in individual situations. Those companies which have developed programs may well want to review them in the light of recent developments and Agency actions on 8(e) reports submitted. Those companies which may not yet have instituted formalized procedures may well consider their needs and the desirability of doing so.

Chapter 7

CONFIDENTIALITY — AN OVERVIEW

George S. Dominguez

TABLE OF CONTENTS

I. GENERAL

What has become known as the "confidentiality" problem is unquestionably one of the most, if not the most, complex and contentious issues to emerge from the Toxic Substances Control Act (TSCA). Given the specific disclosure requirements of the Statute, a valid concern for a concerned public's "right to know", and an equally valid industrial need for confidential treatment of sensitive and valuable private property, a seemingly irresolvable tension appears to have been created. Complicated as this is, it only characterizes the U.S. situation. However, because TSCA and foreign national or supranational laws with TSCA-like objectives have international application, we must also be concerned with their approach to "confidentiality". When we do so we immediately compound the complexity of the issue and its eventual resolution because of widely disparate international approaches and practices insofar as classification and treatment of confidential industrial information is concerned.

While we cannot hope to resolve all of the issues inherent in or deriving from these differing systems and national and international approaches to treating confidential data, we can and will examine the following six critical areas:

1. The overall legal framework
2. The overall industrial framework
3. The present TSCA situation
4. International considerations
5. Policy issues
6. Domestic and international approaches to solving the problem

Approaching the subject in this way we can: (1) gain insight into the conceptual and pragmatic legal, social, and economic foundations of "confidentiality" while, (2) separating considerations of immediate and near-term compliance and protection of private property rights under existing requirements from possible longer-term legislative, regulatory, or policy changes. By fully examining the present situation while exploring possible policy options, we enable those who are concerned with effecting future changes to better understand, elect, and work toward attaining their chosen options. The assessment of present requirements, on the other hand, provides those who are charged with day-to-day compliance responsibility with the information they need to protect the interests of their company.

Before proceeding with each of these areas, it is helpful to step back and try to place the overall situation and its evolution into perspective. This can be done by briefly reviewing the past and present social, political, legal, and economic environments, which in fact, gave rise to the "confidentiality' problem. Historically, as we will discuss in more detail later, government has long recognized the necessity for the legitimacy of trade secrecy and the need to protect certain business information from broad public disclosure. This right has been recognized in both the broad abstract context of intellectual property and in more finite terms as it applies to specific data considered proprietary based on demonstrable cost or value. This concept, however, developed and prevailed in an era of relative trust. Where the public once trusted government and industry, today it is difficult to say which they trust least. This distrust has resulted in great social and political pressure for government and industry to make their operations and actions and, more important, the basis upon which their decisions are made, more open, the theme and the objective being increased public participation based on increased public awareness and concern. An ever politically sensitive Congress has responded by "opening" government to public inspection and participation through various measures, the most notable of which is the movement toward "sunshine" in government. This approach is demonstrable in several operational procedures, and most explicitly in the 1966 Freedom of Information Act (FOIA) which, reduced to its simplest terms, provides the public with a basis for legal access to virtually all government documents. But even FOIA has its limitations and does provide for classification and disclosure protection of confidential data. However, this protection is limited and since many government documents are in reality obtained from industry, it is a direct means for public access to industrial data and property — hence, the criticalness of providing for confidentiality, particularly on a basis consistent with FOIA disclosure limitations.

From the viewpoint of forcing industrial "openness", government, through various laws, rules, regulations, and policy statements as well as court decisions, has been gradually and in some cases (TSCA, for example) aggressively changing conditions and concepts of what can be considered confidential business information. This, in turn, has a severe impact on restricting the nature and extent of data not subject to actual or potential public disclosure. In this content, TSCA is not unrealistically the most visible and compelling example of a conscious government effort to limit industrial confidentiality claims while vastly expanding public disclosure/participation opportunities. However, it should be emphasized that this is only the culmination of a growing change in public and government objective. While industry is rightfully concerned, we must not lose sight of the fact that underlying this development is an equally, if not more drastically, changed public perception of government and the governmental decision-making process. One has to consider objectively which institution, from the viewpoint of public perception, is the more suspect — government or industry. Furthermore, the Congress itself, through its legislative enactments, while acknowledging and acceding to this public demand, is (ironically) castigating itself, for it establishes the very agencies so mistrusted.

In any event, the net result is the sociopolitical and attitudinal situation that gave rise to TSCA and in this instance to serious, sometimes conflicting, objectives of preserving industrial property interests on a limited scale while satisfying a perceived public demand for disclosure on the other. To the extent that those attitudes and conditions differ around the world, they, too, lead to national and international policy differences, which exacerbate the "confidentiality" problem. But, because of the importance of the concept and its practical business/economic effects, so, too, do they accentuate the need for eventual policy resolution.

With this admittedly sketchy sociological and political overview in mind, we can now consider the legal, procedural, and operational specifics central to our immediate and long-term needs. After examining these, we will consider various policy opinions and several specific alternatives to classification, treatment, and disclosure of industrial trade secret or proprietary data.

II. THE OVERALL LEGAL FRAMEWORK

A. Background

While it is impossible to provide an exhaustive presentation of the genesis of "confidentiality" at law here, it is important for our purposes to appreciate that from the legal viewpoint it is one of long tradition and standing, to the extent that the concept of protecting data which have some actual or potential material or intellectual value for their owner derives from a more general theory of protecting private property rights. The origins of classification and protection of certain information as having proprietary value entitled to protection is related to these earlier concepts of property rights. This is also in a very real sense related to early patent practices which serve to disclose while protecting property rights.

As a further demonstration of the importance that the law attaches to protecting proprietary data, one has only to examine the theory of trade secrecy which is well established in statutory and common law. From a constitutional viewpoint, protection of private property is also recognized in that the 5th Amendment to the Constitution guarantees against deprivation of private property rights without due process, an important concept not only in the abstract but on a far more pragmatic level, since several legal experts have indicated the possibility that the TSCA confidentiality provisions may be challengable on such constitutional grounds. However, it is not our intention to pursue this intriguing possibility here but to mention it in passing only as an example of how far-reaching the TSCA confidentiality provisions are and how fundamentally the concept of private property protection is established in our legal system.

B. The Concept of Trade Secrecy

As we have seen, the basic concept of property rights is an old one, and yet there is often a need for the government and the public to have access to such information. In order to provide this legitimate access while simultaneously providing protection to the owner of the property, in this case trade secret or proprietary information, different systems have evolved. These in effect represent different attempts to satisfy these sometimes conflicting objectives through specific statutes and court interpretation and decisions which are in turn founded upon our basic legal, social, political, and economic systems. It has been and continues to be an evolutionary process with TSCA representing for the moment the most far-reaching, significant, and liberal manifestation of that progression. As we shall see later, TSCA is of critical importance not only because it so severely limits the nature and extent of data which can be claimed to be confidential and hence of limited disclosure potential (which is of immediate concern to those segments of industry affected by it), but because it establishes a broad disclosure principle that could ultimately affect others.

This principle can by relatively simple extension be applied elsewhere or provide a foundation for even less recognition of private property rights. Looked at another way, this could establish an extended disclosure requirement in the future.

C. Other Laws and Considerations

As we examine the general background of the "confidentiality" issue from the legal perspective, it is important to recognize that in addition to TSCA itself, there are several other important considerations. These involve some of the basic common law and constitutional principles that we have already mentioned and several specific statutes. It is beyond our purpose to examine each of them in detail here, but the following highlights of major additional considerations should be helpful in putting the issue into perspective.

1. Common Law

The common law does recognize proprietary data rights and confidentiality. This is clearly expressed in tort law. The question that often arises when discussing trade secrecy is, just what information does the law consider to be trade secret and what are the legal standards employed in determining whether specific information does or does not fall within the ambit of trade secrecy. While this is a somewhat controversial question without an absolute answer, tort law gives us the best working definition available. The common law standard which is applied is provided in the *Restatement of Torts* as follows:

A trade secret consists of "any formula, pattern, device or compilation of information which is used in one's business, and which gives him an opportunity to obtain an advantage with competitors who do not know or use it."

Insofar as the legal tests by which "trade secret" information shall be judged as to its actually meeting the definition, the following standards have been applied in tort law.

In determining whether given matter constitutes a trade secret, consideration shall be given to:

a.   The extent to which the data are independently known to outsiders or are used by outsiders for similar purposes
b.   The extent to which they are known by insiders
c.   The extent of the measures taken by an owner to guard their secrecy
d.   Value of the data to the owner and others, including the extent to which, if used in conduct of the business, they would confer a competitive advantage on said owner
e.   The amount of effort or money expended on developing the data; and
f.   The ease or difficulty with which the data would properly be acquired or duplicated by others

2. Patent Law

Patent laws are designed to provide exclusive rights to the invention in exchange for disclosure of the innovative technology and are a clear demonstration of the recognition of the economic and intellectual value of industrial properties. In this context, one of the questions frequently raised when discussing confidentiality and TSCA is the relationship of this patent protection and the type of data developed and submitted under TSCA, the sometime hidden question being, "doesn't patent law provide the protection that's required?" The answer to this question is an emphatic "no", for the following reasons. The Constitution (Art. I Section 8) provides the Congress with power to "Promote the Progress of Science and useful Arts, by securing for limited times to authors and inventors the exclusive right to their respective writings and discoveries."

1.   Patents do not cover all of the information subject to development and disclosure pursuant to satisfying TSCA obligations, e.g., marketing data, production data, process data (in some cases), toxicology, ecotoxicity, and risk information.
2.   Patents are not always sought for various commercial, technical, or economic reasons.
3.   Patents, while applied for, are often not granted.
4.   Patents take time to obtain and there are therefore very real concerns regarding the timing of patent applications, TSCA disclosure provisions, and the granting of the patent.
5.   TSCA disclosure provisions could in fact jeopardize patent applications and ironically create rather than solve confidentiality and patenting problems.
6.   The international patent situation is even more complicated with TSCA disclosure problems again creating international patent problems.

3. Other Laws

TSCA is not the only statute to deal with the specifics of confidentiality. The oldest statute that is of direct relevance to us is the Federal Food Drug and Cosmetic Art (FDCA), with the Freedom of Information Act (FOIA) and the Federal Insecticide, Rodenticide and Fungicide Act (FIFRA) following. Again, circumstances do not permit reviewing each of these in detail here; however, suffice it to say that all of these laws and legislation that derives from them do recognize the importance of confidentiality and trade secrecy and provide for the protection of valid trade secrets. In this connection, even the much-publicized Freedom of Information Act does not ignore nor abrogate such rights. In fact, while it may be surprising to some, there are specific provisions in FOIA regarding confidentiality determinations which in turn relate to determinations of confidentiality and trade secrecy closely paralleling those principles examined under tort law considerations. So, while FOIA is important and did create an opportunity for disclosure of considerable information once held to be confidential by government, it does not mandate wholesale divulgence of trade secret information.

D. General Agreement of Trade and Tariff (GATT)

GATT negotiations resulted in the Trade Agreements Act of 1979 which among its many provisions prohibits federal agencies from engaging in standards-setting activities that could create unnecessary obstacles to trade — so-called non-tariff trade barriers (NTBs). We will have more to say about this law and its confidentiality implications later. For the moment, we should consider it only as another element in our overall concern with proprietary data protection, since policies and practices regarding classification and treatment of data could constitute an NTB. In addition, the Act also provides a potential basis for approaching resolution of the problem which we will also consider in greater detail subsequently.

III. THE OVERALL INDUSTRIAL FRAMEWORK

From what we have seen so far, the basic matter at issue is what information shall or shall not be considered as confidential, and once such a determination has been made, to what extent will such data be disclosed to government and in turn by the government to the public. From the viewpoint of governmental disclosure, the answer is clear: government shall have access to all such data. From the standpoint of subsequent disclosure, there are at least two additional considerations: (1) what information will the government allow the company to declare confidential; (2) how will it treat such information once it has satisfied itself that in fact the data should be accorded confidential treatment? From the strictly legal viewpoint, we have already examined the background for these determinations within the common law and in the context of several specific statutes. Before we examine them within the context of specific confidentiality provisions of the TSCA, we should review why industry is so concerned with trade secrecy and confidentiality.

In broad terms, this concern stems from the investment which has been made in developing the information, product, or process and in the future profits which are to be derived from it. They also involve considerations of competitive advantage or disadvantage and, realistically, the value associated with such advantages or disadvantages.

A. What Industry Has Traditionally Considered Proprietary Information

Business and industry have traditionally considered various categories of technical, financial, and marketing information as confidential. Gradually, there has been a shift in public attitudes, laws, and circumstances which have caused business to change some of its perceptions and concerns and hence some data that were formerly considered confidential such as certain financial information (either because of changed laws or changed business attitudes) are being disclosed freely. However, there still remains a considerable body of information from what was previously considered priority which, despite these alterations, business still feels should be considered as confidential. Given increasing demands for development of new and different kinds of information, e.g., safety and health studies, and process data, the extent and nature of the information which business is being called upon to develop and disclose to government is increasing. This has led to a host of new items of information which in the proprietary sense can and should be considered trade secret. Therefore, just as we have seen an evolution of attitudes and laws regarding confidentiality, we have simultaneously seen an exacerbation of the confidentiality problem created not by new limitations on confidentiality alone, but by an extension of the nature of data developed and new disclosure demands themselves. There is an irony in this situation in that increased pressures for broader public disclosure are concurrent with broader information demands per se. In any event, all of this has led to previously unexpected information demands and disclosure problems. In the present climate and based on TSCA, the following types of information can be called for and may be either automatically disclosable or subject to potential disclosure:

- Product identity
- Process data
- Marketing data
- Production data
- Use data
- Exposure data
- Safety and health effects studies
- Environmental release data
- Labeling information
- Financial data

In short, these represent a virtually complete set of product and market data of the most sensitive nature, information which is not only now potentially available discretely (that is, element by element), but collectively, which means that a competitor gains the additional potential advantage of gathering this valuable information simultaneously and by virtue of little effort. This collective aspect not only accentuates the problem but enhances the value of the data making the total package now potentially more valuable from a commercial viewpoint than the sum of the individual components. As we can see from this list then, not only are all of the traditional elements of commercial and economic sensitivity involved, but additional scientific, R & D, and technical ones as well. It is of little wonder that the new confidentiality limitations inherent in TSCA are of such concern.

B. Industrial Concerns

The very nature of the information that we have reviewed implicitly delineates the nature and extent of industrial concern. To make this even more explicit, however, we can see that this information has a direct economic value to the company in that it can cost literally millions of dollars to develop and insofar as it reveals processes, markets, and R & D activities, has additional immediate and long-term value to competitors. It is very apparent from the nature of the data that their disclosure can meet the three classic standards for considering information to be confidential:

1. Competitors can gain advantage from the data.
2. The company can be injured competitively.
3. The ability of the company to do business can be adversely affected.

In addition to these three somewhat narrow criteria, there are several other short- and long-term considerations that might have a bearing on the concern of industry. The following tabulation highlights the more important of these reasons for industrial concern based on the potential effect that TSCA disclosure provisions has on them:

- Markets
- R & D (product & process)
- Innovation
- Investment
- Capital formation
- Voluntary programs
- (Testing, labeling, etc.)
- Licensing
- Patents
- Industrial development
- Industrial transfer

- Competition
- Industrial configuration

As we reflect on each of these, we must also recognize that they have not only national but international implications. For example, as disclosure requirements are more or less stringent in other countries, it will affect the potential for new chemical R & D and the marketing of new and/or existing substances.

Given the investment in R & D, testing, product development, commercial development, test marketing, market research, marketing, advertising and promotion, and the numerous other expenses incurred when translating a product concept to a commercially successful reality, it comes as no surprise that the "confidentiality issue" takes on such importance to the industrial community. Because of the public's desire for information and the role of government in not only protecting health and the environment, and its informational needs in this role, as well as its obligations to satisfy the public's quest for additional data, it is equally unsurprising to encounter increasing disclosure requirements. How we can better reconcile the resultant tension will be the subject of subsequent consideration.

IV. THE PRESENT TSCA SITUATION

A. Section 14

Now that we have examined the broader background issues, it is time to examine specific provisions of TSCA and industrial requirements in greater detail. The basic TSCA confidential provisions are found in Section 14, which, because of its importance, is provided verbatim:

SEC. 14. DISCLOSURE OF DATA.

15 USC 2613.   (a) IN GENERAL. — Except as provided by subsection (b), any information reported to, or otherwise obtained by, the Administrator (or any representative of the Administrator) under this Act, which is exempt from disclosure pursuant to subsection (a) of section 552 of title 5, United States Code, by reason of subsection (b)(4) of such section, shall, not withstanding the provisions of any other section of this Act, not be disclosed by the Administrator or by any officer or employee of the United States, except that such information —

(1) shall be disclosed to any officer or employee of the United States —

(A) in connection with the official duties of such officer or employee under any law for the protection of health or the environment, or

(B) for specific law enforcement purposes;

(2) shall be disclosed to contractors with the United States and employees of such contractors if in the opinion of the Administrator such disclosure is necessary for the satisfactory performance by the contractor of a contract with the United States entered into on or after the date of enactment of this Act and under such conditions as the Administrator may specify;

(3) shall be disclosed if the Administrator determines it necessary to protect health or the environment against an unreasonable risk of injury to health or the environment; or

(4) may be disclosed when relevant in any proceeding under this Act except that disclosure in such a proceeding shall be made in such a manner as to preserve confidentiality to the extent practicable without impairing the proceeding.

In any proceeding under section 552(a) of title 5, United States Code, to obtain information the disclosure of which has been denied because of the provisions of this subsection, the Administrator may not rely on section 552(b)(3) of such title to sustain the Administrator's action.

(b) DATA FROM HEALTH AND SAFETY STUDIES. — (1) Subsection (a) does not prohibit the disclosure of —

(A) any health and safety study which is submitted under this Act with respect to —

(i) any chemical substance or mixture which, on the date on which such study is to be disclosed has been offered for commercial distribution, or

(ii) any chemical substance or mixture for which testing is required under section 4 or for which notification is required under section 5, and

(B) any data reported to, or otherwise obtained by, the Administrator from a health and safety study which relates to a chemical substance or mixture described in clause (i) or (ii) of subparagraph (A).

This paragraph does not authorize the release of any data which discloses processes used in the manufacturing or processing of a chemical substance or mixture or, in the case of a mixture, the release of data disclosing the portion of the mixture comprised by any of the chemical substances in the mixture.

(2) If a request is made to the Administrator under subsection (a) of section 552 of title 5, United States Code, for information which is described in the first sentence of paragraph (1) and which is not information described in the second sentence of such paragraph, the Administrator may not deny such request on the basis of subsection (b)(4) of such section.

(c) Designation and Release of Confidential Data. — (1) In submitting data under this Act, a manufacturer, processor, or distributor in commerce may (A) designate the data which such person believes is entitled to confidential treatment under subsection (a), and (B) submit such designated data separately from other data submitted under this Act. A designation under this paragraph shall be made in writing and in such manner as the Administrator may prescribe.

(2)(A) Except as provided by subparagraph (B), if the Administrator proposes to release for inspection data which has been designated under paragraph (1)(A), the Administrator shall notify, in writing and by certified mail, the manufacturer, processor, or distributor in commerce who submitted such data of the intent to release such data. If the release of such data is to be made pursuant to a request made under section 552(a) of title 5, United States Code, such notice shall be given immediately upon approval of such request by the Administrator. The Administrator may not release such data until the expiration of 30 days after the manufacturer, processor, or distributor in commerce submitting such data has received the notice required by this subparagraph.

Notification.

(B)(i) Subparagraph (A) shall not apply to the release of information under paragraph (1),(2),(3), or (4) of subsection (a), except that the Administrator may not release data under paragraph (3) of subsection (a) unless the Administrator has notified each manufacturer, processor, and distributor in commerce who submitted such data of such release. Such notice shall be made in writing by certified mail at least 15 days before the release of such data, except that if the Administrator determines that the release of such data is necessary to protect against an imminent, unreasonable risk of injury to health or the environment, such notice may be made by such means as the Administrator determines will provide notice at least 24 hours before such release is made.

(ii) Subparagraph (A) shall not apply to the release of information described in subsection (b)(1) other than information described in the second sentence of such subsection.

(d) CRIMINAL PENALTY FOR WRONGFUL DISCLOSURE. — (1) Any officer or employee of the United States or former officer or employee of the United States, who by virtue of such employment or official position has obtained possession of, or has access to, material the disclosure of which is prohibited by subsection (a), and who knowing that disclosure of such material is prohibited by such subsection, willfully discloses the material in any manner to any person not entitled to receive it, shall be guilty of a misdemeanor and fined not more than $5,000 or imprisoned for not more than one year, or both. Section 1905 of title 18, United States Code, does not apply with respect to the publishing, divulging, disclosure, or making known of, or making available, information reported or otherwise obtained under this Act.

(2) For the purposes of paragraph (1), any contractor with the United States who is furnished information as authorized by subsection (a)(2), and any employee of any such contractor, shall be considered to be an employee of the United States.

(c) ACCESS BY CONGRESS. — Not withstanding any limitation contained in this section or any other provision of law, all information reported to or otherwise obtained by the Administrator (or any representative of the Administrator) under this Act shall be made available, upon written request of any duly authorized committee of the Congress, to such committee.

As can be seen for this section of the law,[1] EPA is essentially compelled to make a considerable amount of the information it receives public — either by initial agency publication, or upon public request. Looked at another way, the agency can be visualized as being severely constrained as to what it can classify and retain as confidential business information (CBI).

From the industrial perspective, there are other important procedural and policy questions, however. For example, from the policy viewpoint, how will EPA make confidentiality decisions? What criteria will it employ? What documentation will be required when confidentiality claims are put forward? And how will the agency protect CBI once they have classified it as confidential? From the procedural viewpoint, there are other important considerations such as how should a company go about claiming that information submitted is confidential? How should it go about submitting confidential information in order to preserve its claims? What actions are open to the company if the Agency rejects a confidentiality claim? And what substantiation will the company need in order to assert and maintain a confidentiality claim?

B. Policy Aspects

EPA has published statements regarding its overall policy on confidentiality. These are detailed statements and therefore cannot be quoted at length here; however, full text statements of the relevant documents have been published by and are available from EPA. These policy statements provide information answering all of the questions previously posed.

In addition, EPA has published a TSCA Confidential Business Information Security Manual which provides a full description of all of the precautions taken by the agency to secure CBI.

C. Procedural Aspects

Here again, the full text of the procedural aspects of filling confidential claims are too lengthy to be

---

[1]   TSCA 14(a) generally precludes disclosure of information specified in FOIA 5 U.S.C. 552(b)(4), i.e., "Trade Secret and Commercial or Financial Information obtained from a Person and Privileged or Confidential" but is significantly qualified by the § 14(b) mandate to disclose "Health and Safety Studies."

reviewed here. However, the Agency has published highly detailed and specific instructions on how to file confidentiality claims.

D. Substantiating Confidentiality Claims

One of the most important aspects of claiming confidentiality relates to the information the Agency requires in order to support a confidentiality claim and to assess the claim when it is received. The approach that EPA has taken to satisfy these needs is to request that the submitter of data which are claimed to be CBI provide the answers to several questions. While there is considerable argument regarding the Agency demands that this information be provided, in most instances to date — ''up front'' — that is, at the time that the claim is made rather than later when any request is made for the data, this so far has been their practice.

Special procedures may be required to claim confidentiality. Since these may change from time to time, they are not being delineated here. It is preferable to refer to EPA or to Counsel at the time that a specific confidentiality claim is made in order to determine and comply with current requirements. This is particularly important since it possible (1) to have waived the possibility of asserting the claim if it is not made at the time of the original submission of the data, and (2) to invalidate a claim by failing to follow the required procedures at the appropriate time or upon request for additional information relative to the confidentiality claim.

E. Transmittal of Confidential Business Information (CBI) within Government

An aspect of the confidentiality problem that has received comparatively little attention is the transmittal of CBI within government. As indicated in Section 14, limitations on disclosure of CBI do not apply in many cases when it comes to transmittal of such data within government, e.g., employees of government, government contractors, the Congress, and pursuant to Section 14, others (including the general public) if such release is deemed by the Administrator of EPA to be ''necessary to protect health or the environment against an unreasonable risk of injury . . . .''

The derivative and, for the moment unanswerable, question is what limitations if any, will be imposed to limit subsequent disclosure once these data are given to outside the Agency. When it comes to other federal agencies or federal employees, they are governed by the same limitations as EPA, their own enabling legislation, and FOIA. When it comes to the Congress or the judiciary, however, we enter comparatively uncharted territory. While we can anticipate that the judicial branch will respect the basic precepts of common and statutory law and, therefore, restrict disclosure in accordance with these tenet and commensurate with a balancing of public and private need, the congressional situation will in all probability differ, the differences in treatment being related to the perceptions and inclinations of the members involved in any given instance and the uses to which they are putting or might put the information. Since EPA has already been requested on more than one occasion to provide extensive TSCA information to individual members of Congress, this is far from an academic question.

V. POLICY ISSUES

Confidentiality is a complex issue. It is therefore, not surprisingly, one of the most important areas requiring immediate consideration in order to resolve divergent viewpoints and to accommodate the legitimate concerns previously described. To do so, it is apparent that there are a number of policy issues and options that must be addressed and ultimately resolved. Not unexpectedly, there are substantial differences in the industrial, public interest, and governmental sectors as to what the truly relevant issues are and what options exist for their resolution. However, if the basic problems attending the confidentiality issue are to be remedied, it is essential that these fundamental policy aspects and options are not only identified but, after thoughtful examination, resolved.

As we have already seen, there is a clear ''industrial'' vis-à-vis ''public interest'' viewpoint regarding what should or should not be accorded confidential treatment and, equally important, what protection should be accorded this information once its confidential status has been established. Naturally, the government also has opinions and responsibilities. Inherent in this debate is the central theme of what is in fact in the public's best interest. As we examine the public policy issues involved in this controversial issue, it is critical that we too keep this ultimate objective in mind. One of the more important considerations in the context of ultimate public good must be economics and it is here we encounter another major controversy. While legitimate business, economic, and financial interests are recognized, and while it is generally agreed that much of the information that is now disclosable under TSCA has economic value, there is no definite indication that the full negative economic impacts of present domestic policy and regulations, direct or indirect, have been fully examined or totally appreciated.

Since the various policy issues that are involved in the "confidentiality issue" are themselves complex and in turn give rise to many subset policy issues, we cannot examine them all exhaustively. However, the following summarizes the principal policy issues involved:

Determination — What data will or will not be accorded confidential status? This is not only a legal but a very basic public policy issue.

Legislative and/or regulatory change — Since not only TSCA but other statutes, rules, and regulations in effect define "confidentiality", it is necessary to evaluate the need, as well as the mechanisms for legislative or regulatory change.

Policy changes — Inherent in any consideration of policy issues is open recognition of the potential need for policy changes and their subsequent implementation.

TSCA amendment — While there is a consideration under the broader topic of legislative change, specific attention to TSCA is a valid concern since this law brings into focus the essence of the "confidentiality issue".

Health and safety data — Do health and safety data constitute a special situation and should such data automatically receive special consideration outside of the norm for establishing standards of justifiable confidential data claims or classifications? Since it is generally recognized that special treatment is deserved, the operative issue relates more to the nature, extent, or timing of disclosure rather than to the principal.

Licensing Notification/registration — Among the issues/options that are relevant in the context of determining what is or is not confidential, there is the related question of how this information might be protected (from the investment viewpoint). One approach that warrants attention is some form of licensing arrangement which may or may not be linked to some form of notification registration procedure.

Data exchange — Aside from the outright disclosure of data, confidential or otherwise, there is the important issue of exchange of such information between: (a) agencies, (b) departments, (c) federal and state officials, (d) governmental bodies such as the Executive Office or the Congress, (e) between national governments, and (f) between international or supranational legislative or quasi-legislative institutions such as the EEC, OECD, or the United Nations. Inherent in this issue is the government need to know and the equally pressing question of how far should such need be limited and what is the nature of the information and the ultimate test of public interest and responsibility in any given instance.

Universal public disclosure — While perhaps extreme, there is, nevertheless, advocacy in some areas for total, unrestricted, and complete public disclosure of all information submitted to government.

Restricted disclosure — Conceptually at least, we have a system which is based on restricted disclosure (restricted in the sense that some information can still be claimed confidential and that there are limitations), once it is so classified, on subsequent disclosure. The controversy, however, still remains when the specific application of the system is examined; e.g., TSCA provisions, FOIA. This is particularly so when we contrast the U.S. approach with that employed in most other nations.

Relationship to other laws, court decisions, and overall public policy — Inherent in all of these public policy issues is the need to frame the problem in the broader context of our overall legal system, other existing laws, court decisions, rule regulations, and public policy.

Penalties — Assuming that some data will be accorded confidential status, with protection and limitation on subsequent disclosure, this then raises the issue of penalties for abuse (a) by those making such claims and (b) by those disclosing the information. In the U.S. approach, these are clearly defined penalties applicable to violators in either category.

Compensation for damages — Again, assuming demonstrable damage results from the disclosure of confidential information, the issues involved are: should damage be allowed (our legal system does afford some relief now), and how should they be determined?

International aspects — All of the issues mentioned here are not restricted to the U.S. Since not only legal systems, but cultural, economic, social, and attitudinal differences dictate different actions, priorities, and finally decisions in this area, it is necessary to consider domestic policy as it affects and is affected by international policies. These in turn have to be considered either collectively, to the extent that they exist through various multinational or supranational organizations (such as the Organization for Economic Cooperation and Development or the U.N.) or based on the actions of individual nations.

Harmonization — Related to the above consideration of international aspects but separable is the issue of harmonization: how far should efforts be made to have compatible, if not identical, "confidentiality" policies, practices, and requirements?

Important as all of these policy areas are, they cannot be addressed in isolation. They must be examined within the overall framework of the systems we have described and in the context of the various approaches that can be considered as solutions to the "confidentiality problem" that we will be examining next.

VI. DOMESTIC AND INTERNATIONAL APPROACHES TO SOLVING THE CONFIDENTIALITY PROBLEM

No consideration of the "confidentiality problem" would be complete without a review of the various approaches which have been suggested both domestically and internationally for its solution. It is important to consider both the domestic and international scene in this context because the problems that various legal requirements and systems have given rise to are truly international. While alterations in domestic laws, rules, regulations, and policies are undoubtedly required if any change is to be made (because some of the problems we have examined are international in scope), it is also necessary to consider a method for international problem resolution. It is in fact ironic that, in the opinion of many who are knowledgeable in this area, relief and resolution ultimately will only come through international mechanisms such as bilateral or multilateral agreements or on a more broadly based international convention resulting in changes in domestic laws. Accordingly, there are not only a number of domestic activities being pursued with the objective of changing our system and their requirements, with an eye to alleviating the "problem", but there are comparable international ones treating of the problems on the more comprehensive scale of international concerns, leading to international recommendations for harmonization and conflict resolution. In this connection, one of the most important of these efforts, and the one which holds the most promise for eventually developing some acceptable recommendations is the effort now underway in the Organization for Economic Operation and Development (OECD). This was approached by OECD through the establishment of an "expert group" on confidentiality. The group consisted of "experts" from the business, industrial, public interest, and government sectors. Their charge was to consider all of the ramifications and implications of all "confidentiality problems" and develop recommendations for international harmonization in this critical area.

As in the preceding section, we cannot detail all of the approaches that have from time to time been suggested nor can we provide any extensive discussion of those major ones which we will be outlining here. However, there have been several important suggestions made and these are tabulated below with a brief description of the principal features involved. No attempt has been made to array all of the pros and cons of each since this is beyond our scope which is primarily to array the alternatives.

Before we examine them, it is also necessary to bear in mind the ultimate objective. This is to provide the government and the public with that information which they rightfully need to make sensible decisions or to participate effectively in the processes of government and society, while at the same time, to recognize that there are corresponding legitimate business industrial rights and privileges. There is in this sense no absolute right or wrong involved in this issue, nor in the suggestions that have been made to solve this complicated problem. What is needed in the final analysis is some sensible accommodation that takes these "interests" and "rights" into balanced consideration, a system which provides that which is necessary, when it is necessary, but which also provides the protections and restrictions that are essential to the protection of private property and the preservation of the principle of private property rights.

VII. SPECIFIC APPROACHES TO SOLVING THE CONFIDENTIALITY PROBLEM

The following is a brief summary of the major approaches that have been suggested and, at one time or another, actively considered for solving the "confidentiality problem." Some of them are more complicated, far-reaching, and involved than others. Some require not only changes in domestic laws, but complex international agreements that might take years to realize. Accordingly, aside from the logic or intrinsic merits or demerits of each, they have to be examined in the light of practical reality — the potential for attainment. In this sense, while they are listed here for the sake of reviewing the most important approaches proposed, several have already in effect been rejected for various reasons, while others are still under active consideration.

International convention — This involves the convening of a formal international body with appropriate national representation to review, discuss, and ultimately develop a solution to the problem. This would take the form of a specific agreement which, upon acceptance by the individual governments of those participating nations, would result in a binding agreement. In some instances, this would require and result in changes in the domestic laws involved in order to bring them into conformity with the final agreement. It is a highly formal procedure which by its nature would not only take a considerable amount of time to organize and in turn to reach decisions, but would take an even longer period to be accepted by the respective participating governments and to be translated into effective domestic actions. Despite these difficulties, it remains a viable option.

Period of exclusivity — Under some notification systems, TSCA in particular, there is insufficient protection afforded the notifier insofar as limiting the public disclosure (with severe limitations) or the information submitted once this has been provided the Agency. This is, in fact, the essence of the TSCA "confidentiality problem". Further, under a "First Notification Only" system (again, TSCA in particular), subsequent entrances into the market have no notification requirements. An "exclusivity" approach is based on a system wherein the original notifier would be granted a period of exclusive use of the data submitted. A second or third (or whatever number) notifier thereafter would not be able to rely on the information submitted by the initial notifier for some fixed period of time. In addition, the data would have to be used with the consent of the original notifier. Some variants of this approach involve paying an agreed upon fee to the original notifier, thereby providing some compensation for the original data developed and submitted.

Data summaries — Under this approach, only summaries of relevant data would be submitted. The objective is to provide as much information as might be required to make any requisite assessments, but using summaries only minimizes the potential for disclosure of "confidential" data, thereby protecting proprietary rights.

Limitations on data disclosure — In this system, complete information would be submitted to the government. However, any disclosure by them would in turn be limited to that necessary to protect health and the environment.

Limitations to use by owner — This is a variant of the exclusivity approach involving limitations of the use of the information to the owner thereof. In this approach, the owner would determine who would have access to what information and under what conditions. It avoids any potential government intervention and the establishment of any formal periods of exclusivity such as might be involved in the other closely related option.

Bi-lateral or multi-lateral agreements — Under specific provisions, accords, or agreements reached between two or more nations, various bilateral or multilateral approaches could be developed which would involve some application of one or more of the specific approaches mentioned here. The relevance of the agreements is that they could be used to make such determinations as: (1) the selection of enforcement approaches; (2) a basis for modeling a more universally accepted approach to solving the confidentiality problem; or (3) a basis for establishing other bi- or multi-lateral agreements.

Compensation — Since some of the major business concerns over protection of confidential information are the preservation of private property rights, economic disadvantage, and related financial considerations, this approach would provide for some compensation to the data owner if this information were disclosed and relied upon by others, e.g., PMN data used by subsequent notifier — a protection which does not exist under our system. In the context of the European 6th Amendment system, this would mean that instead of requiring subsequent notifiers to develop their own data there would be specific provision for compensation if the information of the initial submitter were relied upon. While TSCA does provide for limited data compensation (Sec. 4 and 5 testing costs), this is extremely narrow in scope and does not cover all of the costs of the company, such as R & D and commercial development. There are various sub-approaches that could be considered within this system ranging from letting the "market" prevail (that is establishment of charges without any form of government intervention) to the government's determining and establishing any compensation.

The non-confidential data package — This system is based on examining the problem and its solution from another viewpoint. Instead of trying to determine what is confidential, this solution is based on the concept of agreeing upon a "package" of data which government, business, and public interest groups would agree is not confidential and hence disclosable without restriction. The remaining data would then have to be made available only with sufficient justification. This latter would be approached on a case-by-case basis. What limitation might be made on subsequent disclosure of this information remains to be fully considered; however, it could take the form of one of the approaches described here. We will refer to this system in greater detail later.

GATT agreements/codes of conduct — As we saw before, the international dimensions of this problem have to be visualized not only within the context of specific national laws and international approaches such as the EEC 6th Amendment, but also in the broader dimensions of international trade agreements. The most important of these is the General Agreement on Trade and Tariff (GATT). This treaty not only establishes many trade-specific requirements such as duty rates, but also addresses broader questions of those factors which might effect international trade flows including such aspects as so-called nontariff trade barriers. These are nationally imposed requirements which, while not direct duties or tarrifs or other direct forms of trade restrictions, quotas, etc., nevertheless have the effect of imposing trade barriers. There are many who correctly assess inequitable confidentiality and disclosure conditions as potentially creating nontariff trade barriers. While it is too soon to tell exactly what effect the new GATT agreements and the Codes of Conduct (these are specific provisions regarding the conduct of individual countries in the nontariff trade barrier context) will have in this area, they are relevant to considering options since their provisions could be brought into play.

Redefining safety and health studies — One source of considerable concern within the overall proprietary data question involves TSCA provisions for "automatic" disclosability of safety and health data. Safety and health studies can be extremely costly and, depending upon how they are defined, can contain not only safety and health data per se, but product-specific technical and scientific information, e.g., product identity, R & D leads, etc. Section 14 of TOSCA provides only very limited protection of safety and health data and this, coupled with an extremely broad definition of what constitutes a safety and health study, means that virtually all information that in any way can be considered as a safety and health study is potentially discloseable under the law. One solution to this particular problem that has been proposed is to redefine what constitutes a valid safety and health study under the law; thereby limiting disclosure to that which is really necessary for a safety and health determination without unnecessarily or prematurely divulging valuable proprietary information.

Limited disclosure — Inherent in the confidentiality problem is subsequent disclosure by government to other parties. In this approach, the government would be limited in that it could only release such information if the receiving party could justify its "need to know". While the criteria for such justification might vary depending upon the potential recipient, this would in all instances provide a basis for some form of review and substantiated need, thereby not only restricting disclosure but discouraging frivolous inquiries.

Review panels — Since one of the objectives of disclosure is to insure full and adequate public participation, as well as to afford the "public" an opportunity to review Agency decisions and the data upon which they are based, it is clear that there is at least a modicum of distrust inherent in disclosure advocacy. Under this proposal, an independent panel would be established which could, upon request by a party, obtain and review all of the information without having full public disclosure. In this way, those who have questions and concerns could satisfy themselves as to the adequacy of the data and actions of the company or the government. Again, depending upon the specifics of its application, this system could also provide mechanisms for more complete disclosure if there were validated circumstances demonstrating the need.

VIII. THE NONCONFIDENTIAL DATA PACKAGE

Of all the approaches that have been or are being considered, the one which seems to have the most potential of ultimately being adopted, at least as a recommendation by appropriate experts engaged in international dialog (such as the OECD "expert group" previously described), is the nonconfidential data package. Therefore, as mentioned earlier, we must consider this in greater detail.

As indicated in Section VIII above, the essence of this approach is the predetermination of a "package" of information which "all" can agree is by definition not confidential. These data would then be automatically subject to disclosure when and as such disclosure might be called for by national or international law, requested by legitimate third parties, or deemed to be in the public interest by the respective authorities. Since such data are by mutual agreement nonconfidential, there would be no need to notify the submitter of their actual disclosure (although this could be done on a voluntary basis by the government) nor would there be any basis for claims against the government if and when such data were divulged.

Since this approach eliminates a number of problems for the public, the government, and industry and there is a substantial possibility that it will be advocated, the critical question that remains and the area of most debate, then, is not the principle but the specific data elements which would be included in the nonconfidential package. The lead for this approach so far is being taken by the Expert Group on Confidentiality within the OECD and they have accordingly attempted to define this package. The proposed nonconfidential data package consisted of the following information:

Nonconfidential Data Package Being Considered by OECD Expert Group on Confidentiality

Trade name

General data on uses — the uses must be described only broadly; for instance, closed or open system, agriculture, domestic use. Specific uses may be highly confidential

Methods of disposal and transport

Precautions as to use and measures to be taken in the event of an accident

Physical/chemical data, excluding spectra. If the rest of the physical/chemical data make it possible to find the chemical identity easily, the notifier could ask that only ranges of values are given

Summaries of toxicology and ecotoxicology studies. These summaries could be proposed by the notifier — they should include precise figures and interpretations

IX. CONCLUSION

Confidentiality is an undeniably fascinating subject. Nowhere else are the tension between public interest participation and the interests of government and industry brought into such sharp focus. It is the centerpiece for crystallization of what are usually characterized as conflicting interests. In fact, inherent in that visualization may be one of the most substantial problems — a conceptualization or, more accurately, a misconceptualization — for it is by no means clear that these differences are so starkly or irreconcilably divergent. There is an interest on the part of industry to have a degree of "openness" and in having government and public participation. Conversely, given existing economic and political realities, the "Public" and "Government" have an interest in protecting "Industry" and its economic viability and, not unimportantly, "Government" as the "Servant" of both has its intrinsic obligations to both. It is, as we have already seen in the final analysis, a question of balance, a balancing act which, however, is of increasingly critical importance and of ever-expanding dimensions, nationally and internationally.

While, in the short run, we must be concerned with understanding present and expected regulatory requirements, we must be equally mindful of long-term implications and the need for business, government, and public collaboration in the development and implementation of truly sound and beneficial public policy. Confidentiality, while perhaps only one example of where such collaborative efforts are required, remains an outstanding one and provides unique opportunities for participative policy development and implementation.

Chapter 8

ENVIRONMENTAL EFFECTS TESTING

Charles R. Ganz

TABLE OF CONTENTS

I. INTRODUCTION

In the original volume of the Guidebook (Ref. 1), the environmental effects section was structured to provide readers with an overview of the environmental factors which chemical manufacturers and processors would probably need to consider in formulating policies to comply with the Toxic Substances Control Act (TSCA). Since that writing, the U.S. Environmental Protection Agency (EPA) has taken a number of steps to clarify and begin to meet its statutory obligations under the Act.

In the present chapter, we will attempt to describe programs that have taken shape in response to the environmental testing requirements of TSCA. First those sections of the original legislation relating to the environmental safety and environmental testing of chemical substances will be summarized. In subsequent sections, proposed test protocols will be surveyed, anticipated good laboratory practice rules will be discussed, and more recent EPA actions and statements of intent in the environmental testing sector reviewed.

II. TESTING PROVISIONS OF THE TOXIC SUBSTANCES CONTROL ACT

The preamble to TSCA states that the objective of the law is to protect health and the environment by *"requiring testing* and necessary use restrictions on certain chemical substances . . . ." Several sections of the law refer to health, safety, or environmental testing. To assist in understanding the statute and its testing references, a flow chart highlighting some of the key testing sections follows. The complete text of these sections can be found in the full text section of this *Guidebook* or by reference to the Toxic Substances Control Act itself. Additional guidance can be found in Volume I of the *Guidebook*.

III. TEST PROCEDURES

A. EPA Approach to Determining Testing Requirements

In the March 16, 1979, *Federal Register* (Ref. 3) the EPA established, as its long-term goal, the publication of a series of premanufacture testing guidelines along with relatively detailed guidance concerning their use in a premanufacture testing program. Because of both the complexity of this task and the lack of adequate experience, EPA anticipates that the initial step toward this goal will be the publication of relatively general interim guidelines, with more detailed guidance being formulated as information and experience are gained. The final form of the premanufacture test guidelines will be based upon the proposed test procedures and will incorporate pertinent changes gleaned from a review of the public comments. The resultant test guidelines will also form the basis for testing requirements under Section 4(a) of the Act.

Section 5 of TSCA does not specifically authorize EPA to require premanufacture testing to develop health and environmental effects data except in cases where the substance is in a category subject to a Section 4 test rule. However, Section 5(e) does authorize the EPA to place limitations on the "use activities" (i.e., manufacture, distribution in commerce, use or disposal) of a new substance if:

1. The Administrator judges that there is not sufficient information to "permit a reasoned evaluation" of the potential risks
2. The substance in question may present an unreasonable risk; or
3. The substance is or will be manufactured in substantial quantities and result in substantial human and environmental exposure

Thus, although adherence to the proposed Section 5 premanufacture test guidelines will be voluntary (except for substances subject to a Section 4(a) ruling), the requirement that sufficient data be available to EPA to assess environmental and health effects is implicitly woven into the fabric of the prenotification section of the law.

The advantage of having EPA formulate explicit testing guidelines is that manufacturers would be able to anticipate, with some degree of certainty, what criteria EPA will use in its own risk assessments of new substances. This in turn would assist manufacturers in setting up testing programs which, at a minimum, would generate test results which the reviewing agency has identified as relevant. On the other hand, there are arguments that a rigid testing program would decrease the flexibility and innovation needed to assess the potential risks of as yet unknown chemical substances.

FLOWCHART OF ENVIRONMENTAL TESTING SECTIONS OF TSCA

	Definitions / Testing	Chemical Substance	Environment	Health and safety	Standards
Section 3	Definitions	(2) Chemical Substance -exclusions	(5) Environment	(6) Health and safety study	(12) Standards for the development of test data
Section 4	Testing of Chemical substances and mixtures	4 (a) Required testing rule "Use activities" may present unreasonable risk Substance will be produced in substantial quantities Insufficient experience or data to assess unreasonable risk Mixture for which individual substances don't predict effects 4 (e) Priority testing list Considerations of quantity of substance produced and potential human and environmental exposure Similarity to known hazardous substances Existing data and expected return from testing Publication of list Requirements for inclusion on list Timetable for revisions of list Makeup of Priority List Committee Rules for committee members	4 (b) Contents of rule Identity of substance Testing guide-lines (review every 12 months) Timetable for submitting data Some effects for which testing may be required Who must test Duration of testing rule Right to present data or arguments 4 (f) Required actions if data indicate substantial risk	4 (c) Exemptions Data already submitted Data in process of being developed Reimbursement period 4 (g) Petition for testing guidance for a substance	4 (d) Notice of receipt of test data
Section 5	Manufacturing and processing notices	5 (b) Submission of test data with premanufacture notification Compilation of current list of substances which may present an unreasonable risk	5 (c) Extension of premanufacture notification period	5 (d) Required content of premanufacture notification Publication of notice in *Federal Register*	5 (e) Regulating substances while information being developed Insufficient information to make judgment regarding health or environmental risk

FLOWCHART OF ENVIRONMENTAL TESTING SECTIONS OF TSCA
(continued)

Section 6 Regulation of hazardous chemicals	6 (a) Promulgation of Unreasonable Risk Ruling Publication in *Federal Register* of key reasons for making unreasonable risk assessment Right of manufacturer or processor to present data and arguments	5 (f) Protection against unreasonable risks Actions which Administrator may take	5 (g) Publication of reasons for not taking action	5 (h) Exemptions from prenotification rules Substantial quantities or human exposure expected Administrator may limit "use activities" and seek an injunction if order challenged
Section 8 Reporting and retention of information	8 (a) Required maintenance of environmental and health effects records Small manufacturer exemption Research quantity exemption	8 (c) What must be included in records Types of reports Length of time records must be retained	8 (d) Health and safety studies Studies which must be listed and reported	
Section 10 Research, development, collection, dissemination, and utilization of data	10 (a) Permission to cooperate with other agencies and to make grants and enter into contracts for research, development, and marketing	10 (b) Collection and dissemination of information	10 (c) Development of screening techniques	10 (d) Monitoring Projects

		10 (e) Basic Research projects	10 (f) Training projects	10 (g) Exchange of research and development results
Section 12	Exports	12 (a) Coverage of exports by the act	12 (b) When Notification of export required; Notice of availability of health and safety data to importing government	
Section 13	Entry into customs territory of the U.S. (imports)	13 (a) Required compliance with TSCA in order to import; Confiscation		
Section 14	Disclosure of data	14 (a) Conditions under which data will be disclosed	14 (b) Disclosure of data from health and safety studies; Protection of process and mixture information	
Section 18	Preemption	18 (a) Effect of TSCA on state and local right to regulate		
Section 21	Citizen's petitions	21 (a) Right of persons to petition for the issuance, amendment, or repeal of specific rulings	21 (b) Possible court actions in response to petitions	
Section 26	Administration of the act	26 (b) Payment of fees for reviewing submitted data	26 (c) Action with respect to categories of chemical substances or mixtures	
Section 27	Development and evaluation of test methods	27 (a) Authority to fund projects for development and evaluation of test methods	27 (b) Approval by secretary of HEW	27 (c) Publication of status reports in *Federal Register*

B. The EPA Approach to Risk Assessment

It is instructive to review EPA thinking with regard to both what constitutes an unreasonable risk and how testing programs should be designed to identify substances which will present varying degrees of risk. Although repeatedly referred to in TSCA, the expression "unreasonable risk" is never explicitly defined. The EPA approach to dealing with the unreasonable risk question will be to try to balance the magnitude of the potential risks with the social benefits associated with the chemical substance. Included in the EPA risk-benefit assessments will be consideration of the nature, extent, and reversibility of adverse effects; the availability of alternative substances; the potential risk which the alternatives may present; and the economic and social benefits. The magnitude of the potential risks will also determine what type of regulatory action will be needed (e.g., labeling requirements vs. a manufacturing ban).

Initially, in reviewing a premanufacture notification, EPA must answer the question of whether a risk exists. EPA views risk potential as being comprised of two contributing factors, hazard potential and exposure potential. The assessment of hazard potential ("hazard assessment") encompasses the inherent potential of a substance for causing injury to the health of humans and other living organisms and to the environment. Such injury may be in the form of both direct and indirect as well as acute (short-term) or chronic (long-term) toxic effects. "Exposure assessment" deals with the potential for individuals and the environment to be exposed to a hazardous substance. Clearly the risk of injury associated with a hazardous material will vary greatly depending upon its ability to reach target organisms and critical segments of the environment.

In making risk assessments, EPA expects to obtain, when available, certain insights from existing materials related structurally to the substance in question. In addition, certain information will be available from existing substances being introduced for new uses. In general, however, the following types of information are likely to be needed for the risk assessment:

1. Information for assessing potential human exposure and environmental release (includes existing or anticipated methods, locations, and volume of production, distribution, use, and disposal of the chemical substance)
2. Information to evaluate potential transport pathways, degradation, and bioaccumulation (includes the physical and chemical properties of the substance)
3. Information on potential human health effects to determine whether the substance is a potential health hazard
4. Information on potential effects on the environment to determine whether the substance is a biological or nonbiological environmental hazard.

Although the EPA strategies for risk assessment are, to a certain extent, clarified by the foregoing discussion, it must be remembered that these strategies will be put into action only after submission of the premanufacture notification. Therefore, a submitter, interested in obtaining speedy consideration and acceptance of his premanufacture notification, must be prepared to provide information with his submission which anticipates EPA risk assessment activities.

Except for substances subject to a Section 4(a) test ruling, decisions as to whether to test, to what extent to test, and how to test a new substance are left up to the individual submitter. Some factors which the submitter should keep in mind when arriving at these decisions include:

1. The anticipated costs of possible testing programs in relation to the potential profitability of the chemical substance
2. The ability or willingness of the submitter to bear the projected costs
3. Properties of the substance and expected manufacturing and use practices which will bear on the potential risks associated with the substance
4. The perception by the submitter of the amount and type of information which will be needed in order for EPA to judge that the substance will not present an unreasonable risk

If the decision to test is made, it is likely that a tiered approach will prove to be the most efficient both from a time and cost standpoint. The tiered approach consists of a preliminary battery of screening tests designed to weed out substances very likely to present an unreasonable risk and to identify chemicals for which further testing is warranted. Based upon both performance of the substance in the screening tests as well as the anticipated costs, projected manufacturing volumes and "use activities" of the substance, the material can then be subjected to higher test tiers. The latter are usually more costly and include more predictive and confirmative tests as well as tests for which there is no acceptable screening procedure. In building a tiered testing program it is important to establish prior decision criteria for determining when to stop and when to continue testing. A good discussion of the various criteria can be found in Reference 4. The stepped test program can be stopped when the submitter feels that sufficient information has been obtained to adequately assess the health and environmental risks of the substance.

EPA is considering, as one alternative, establishing a base set of tests, although it is aware of the inherent difficulties and the need for possible exceptions which are likely to arise. To ease the cost burden of such testing, EPA has suggested an interesting approach for new substances which it determines will present little potential risk at initial intended use and production levels. This approach would involve the application of the "significant new use rule" (SNUR) clause of Section 5(a)(2) of TSCA. Thus, premanufacture notices could be accepted for low-volume chemicals with an expected low risk after only a minimum of base set testing. There would be a provision, however, stating that if use or production exceeded preestablished levels this would constitute a significant new use and would consequently trigger a Section 5 review. The review would determine whether additional testing was required before production and use could be expanded. The SNUR approach would reduce the negative effects of TSCA on innovation by permitting low risk, new chemicals to get onto the market. Major testing and its concommittant costs would be deferred until such time as the potential exposure warranted it. A manufacturer would likely be better able to support additional testing costs at this time than at the inception of a product.

Another plan presently under study at EPA is one which would exempt from the normal premanufacturing notification procedure certain new chemicals which do not pose unreasonable risks. Included in the proposed exemption group are chemical intermediates which are not transported away from the manufacturing site, chemicals made in quantities of 10,000 kg or less, and certain polymers. Under the plan, manufacturers would be obligated to provide EPA with an "informational notice" containing the identity, use, and manufacturing site of the substance. In addition, an assessment of the toxicity and potential risk of the chemical would need to be submitted 14 days before the commencement of manufacturing.

C. Reference Tests and Methods

As the above discussion states, EPA, at the time of this writing, is still in the process of formulating its final approach to testing requirements. In the interim, the Agency has identified a number of reference tests which it feels are most appropriate for evaluating particular properties and effects. Use of such reference tests is voluntary, although tests carried out according to the recommended procedures "would assure acceptance of the test and method as valid". Note that the list of reference methods is a list from which a submitter can select tests for effects which he deems important. It is not a list of tests, all of which need to be applied to every new substance.

A key aspect of the reference test development process is an attempt by EPA to have other regulatory groups, which require testing, agree on standardized methods which will be acceptable to all of the participating groups. The advantages of this approach in terms of time and cost savings are obvious. To that end, EPA is working closely with sister organizations and has specified that methods developed under the following auspices "will be regarded as suitable for new chemical evaluation purposes":

- TSCA-Section 4
- Federal Insecticide, Fungicide and Rodenticide Act (FIFRA), as amended — Section 3
- Interagency Regulatory Liasion Group (includes the Occupational Safety and Health Administration [OSHA]; the Consumer Product Safety Commission [CPSC]; the Food and Drug Administration [FDA]; and the EPA)
- Organization for Economic Cooperation and Development (Ref. 13) (OECD) consists of 24, mainly European, countries; also includes U.S. and Japan

The European Economic Community (Common Market) recently issued its own toxic substances control guidelines, the so-called "6th Amendment." Dominguez (Ref. 5) has written an in-depth comparison of TSCA and the "6th Amendment". Although the main objectives of the two pieces of legislation are related, there are some important differences in approach and emphasis. The EEC rules appear to be more flexible and to cover a broader range than those of TSCA. For example, labeling and packaging guidelines are included in the "6th Amendment".

The reference testing procedures recommended by EPA are divided into three categories.

1.   Chemical fate tests — Tests for basic chemical and physical properties and other characteristics associated with environmental transport and persistence.
2.   Tests for health effects.
3.   Tests for ecological effects.

Since health effects testing will be the subject of another section of the Guidebook only the first and third categories will be discussed here.

1. Chemical Fate Tests

Chemical fate tests involve on the one hand determining properties of the chemical substance which bear on the likelihood of the substance being transported by environmental processes. Attention is then turned to the determination of the susceptibility of the substance to degradation through one or more of the many degradation pathways in the environment. Both of the aforementioned aspects will be influenced by the form in which the material is released (liquid, solid, or vapor) and by the environmental matrix into which it is released (air, water, or land). A detailed discussion of the distribution and interaction of chemical substances when they enter the environment can be found in Reference 1.

It should be emphasized that reference tests which deal with only single properties or degradation pathways in an isolated system can only be expected to give order of magnitude estimates of the ultimate fate of a substance. Such tests do not account for the complex and synergistic interactions which may occur when many different effects are operating in concert. The results from the reference tests, however, can be useful in identifying those areas which should be given greater or lesser emphasis in an overall testing program for the substance. The results may also induce the manufacturer to consider making changes in the manufacture or use practices of the substance in order to minimize what appears to be a possible adverse effect or to enhance a desirable degradation pathway.

Table 1 summarizes the chemical fate tests described by EPA in the March 16, 1979, *Federal Register* (Ref. 3). Sufficient detail has been provided here to assist the reader in making reasonable estimates of the level of effort which can be expected in obtaining certain data. The summaries are by no means complete protocols and the reader is urged to consult the March 16, *Federal Register;* the "Draft Technical Support Document" mentioned therein; the Pesticide Registration Guidelines (Ref. 6), and the references cited in each of these documents for more detailed background and instructions.

2. Ecological Effects Tests

Ecological effects tests are designed to assess the effects of substances directly on members of the hierarchical community of living organisms. Because of the varied interactions between and among lower and higher life forms in the environment, the most informative approach to assessing ecological effects is to allow a community composed of members of representative taxonomic groups to be exposed to the chemical over an appropriate period of time. The accumulation and distribution of the chemical within the community and its apparent effects on members of the community can then be periodically assessed by examining such parameters as relative tissue concentration of the substance, organism or plant growth, health, mortality, and behavior. At the present time, methods which would permit this type of examination are not sufficiently well developed to adopt them as standards. As a consequence, simpler tests using single species from various classes of organisms are being recommended until procedures for the more complex studies are refined. Extrapolation of the results of single species tests to ecosystems in general have often been found to yield good first approximations.

The need for ecological effects tests will have to be determined on a case-by-case basis as will the specific test organisms and conditions. The results of the chemical fate tests as well as the "use activities" of the substance in question will need to be closely studied in making such testing decisions.

Some criteria which EPA has taken into account in selecting the recommended procedures include:

● The use history and scientific acceptability of the test
● The usefulness of the test for risk assessment
● The applicability of the test to a wide range of substances
● The cost-effectiveness

Table 1
SUMMARY OF PROPOSED REFERENCE METHODS FOR CHEMICAL FATE TESTS

Property/test
Environmental importance
Summary of methods

Water solubility

Influences transport, likelihood of reactions in aquatic systems, bioaccumulation potential

Determine solubility at 10, 20, 30 ± 1°C. Prerinse transfer vessels with aqueous phase to minimize surface adsorption. Measure ionizable chemicals at pH 5.0, 7.0, 9.0. Use very pure, reagent grade (Type III) water or equivalent. Use analytically pure (>99% by weight) test substance (specify nature and % of impurities). Test in duplicate

For solid and liquid substances with solubility > 500 mg/ℓ — Approach equilibrium from above and below saturation. Determine amount of dissolved substance by specific chemical or physical analysis. True solubility reached when above and below values within 5% of each other. If solubility > 10 g/ℓ, must weigh known volumes of solution to determine density and use this to correct solubility value

Special precautions and exclusions

Usually not applicable to substances which are gases under ambient conditions; extreme care required for very poorly soluble substances

Federal Register page (Ref. 3)

16253-4

Property/test
Environmental importance
Summary of methods

Water solubility (continued)

As above

For solids with 1—500 mg/ℓ solubility — Use nephelometric (turbidimetric) method. Obtain estimate of solubility by repeatedly injecting a 1% acetone or ethanol solution of substance in 30 mℓ water and sonicating for 10 sec with an ultrasonic disrupter. If resultant suspensions quickly coalesce to larger aggregates, method will not work. If stable suspension formed, prepare series of suspensions using stop watch to time sonication. Immediately after sonication, measure temperature and turbidity. Plot turbidity vs. concentration. Extrapolate line to concentration at zero turbidity. This point will be the estimated water solubility

Method gives extrapolated value and is therefore an approximation

Special precautions and exclusions

Federal Register page (Ref. 3)

16253-4

Property/test
Environmental importance
Summary of methods

Water solubility (continued)

As above

For solids with very poor solubility (< 1 mg/ℓ) — Minimize colloid formation. Dissolve substance in suitable volatile solvent. Coat solvent solution on walls of test vessel (avoid bottom). Allow solvent to evaporate. Stir slowly with Teflon®-coated magnetic stirrer. Add water. Analyze concentration of chemical vs. time. When plateau is reached, equilibrium solubility is attained. Check absence of colloid formation by centrifugation at several G values until apparent concentration differences small

For liquids with very poor solubility (< 1 mg/ℓ) — Stir substance (10—20 mℓ) with high purity water (200 mℓ) in a 250-mℓ glass bottle fitted with a septum port near the bottom of the bottle. Shake vigorously for 1 hr or stir for 24 hr with a magnetic stirrer, using a Teflon®-coated stir bar. Make sure that vortex does not exceed 1/4 of solution depth. Withdraw a sample of the aqueous phase. To minimize emulsions, centrifuge in tubes rinsed with aqueous phase or allow to stand several days. Check solution with turbidimeter to ensure that no emulsions persist. Centrifuge at high speed if necessary to break emulsions. Make concentration-time studies until concentration plateau reached

Special precautions and exclusions

Promising liquid chromatographic procedures for very hydrophobic chemicals have been reported (Ref. 7). These may become preferred methods when more experimental work completed

Table 1 (continued)
SUMMARY OF PROPOSED REFERENCE METHODS FOR CHEMICAL FATE TESTS

Federal Register page (Ref. 3) 16253-4

Property/test Octanol/water partition coefficient

Environmental importance Indicator of bioconcentration potential in various segments of the environment and in living organisms (particularly in lipid portions)

Summary of methods A dilute solution of an organic solute distributes between octanol and water phases in contact with one another such that, at equilibrium, the ratio of the molar concentrations of the solute in each phase is a constant (P) at a given temperature. Thus:

$$\text{Partition coefficient} = P = \frac{C_{\text{octanol}}}{C_{\text{water}}}$$

Use dilute solutions (C < 0.01M). Temperature = 25 ± 1°C. Use very pure reagent grade (type III) water or equivalent. Use test substance (specify nature and % of impurities). Use high purity n-octanol. (washed with 0.1 N H$_2$SO$_4$, 0.1 N NaOH and high purity water until latter neutral, dried over CaCl$_2$ and distilled twice to give 99.9% pure n-octanol). Adjust ratio of solvent-water volumes to accomodate solubility of test substance. Shake 2 phase system gently in ground glass centrifuge tubes for 1 hr to ensure equilibrium. Centrifuge, if necessary, to break emulsions. Use large ground-glass-stoppered vessel for very hydrophobic substances. Transfer phases to centrifuge tubes prewashed with aqueous phase. Centrifuge and recombine aliquots of two phases. For volatile compounds fill tubes almost full to avoid losses to head space air. Analyze water and octanol for substance to determine P. Measure ionizable substances at pH 5, 7, 9.

Special precautions and exclusions Partition equation holds for molecules which remain unassociated or undissociated in solution; if test substance thought to associate or dissociate in solution make modifications in the P equation to account for reduced or increased number of unique molecular species

Federal Register page (Ref. 3) 16254-6

Property/test Vapor pressure

Environmental importance Provides information about the ability of a substance to be transported in the environment by vaporizing and becoming airborne; permits estimation of relative importance of interaction of the substance with air vs. water vs. soil

Summary of methods For all vapor pressure determinations — Test in triplicate at environmentally significant temperatures, 10°, 20°, and 30°C. Plot of log$_{10}$ vapor pressure (torr) vs. 1/K should be straight line if vapor pressure data correct. Note any changes of state which occur between 10—30°C along with temperature at which change occurs (at atmospheric pressure) and vapor pressure at 10° and 20° above and below change-of-state temperature. Determine vapor pressure at 10°, 20°, and 30°C for suitable standard reference material as verification of experimental conditions.

For liquids with vapor pressures from 1—760 torr — Use ASTM D-2879-70 isoteniscope procedure (Ref. 8). Place liquid sample in thermostated bulb connected to a mercury manometer and a vacuum pump. Evaporate and recondense liquid repeatedly to drive out lower boiling foreign gases until no further lowering in vapor pressure seen. This is equilibrium vapor pressure.

For liquids and solids with vapor pressures from 10^7—10 torr — Use gas saturation method. Pass current of inert gas over substance at slow enough rate to ensure saturation. Partial pressure of vapor from test substance in presence of other components can be computed from total gas volume and weight of material vaporized. Measure amount of substance (in g) evaporated per unit time at minimum of 3 gas flows. Plot of g vs. flow rate will be straight line if equilibrium achieved.

Special precautions and exclusions
Federal Register **page (Ref. 3)**

Vapor pressure determination not considered important for substances with standard boiling points 30°C or with vapor pressures 10^{-7} torr

16256-7

Property/test
Environmental importance

Adsorption

Gives information about the tendency of substances which have deposited on solid substrates in the environment to accumulate on or be transported from the solid substrate

Summary of methods

Adsorption isotherm method — Relationship of quantity adsorbed on substrate as a function of equilibrium concentration in aqueous phase. Applicable to soils and sediments. Use one soil and one sediment. Recommended soil types listed in *Federal Register*. Soil should have pH 4—8, 1—8% organic matter, >7 meq/g cation exchange capacity, < 70% sand. Report characteristics of soil such as soil order, series, classification, location, horizon, particle size analysis, % organic matter, pH, % CO_3 as $CaCO_3$, cation exchange capacity, extractable cations, clay mineralogy. Dry soil for 24 hr at 90°C and sieve through 100 mesh stainless steel or brass; air-dry sediments. Use pH 7, CO_2-free (by boiling) water in a ratio of 5 mℓ/g of solid. Equilibrate test vessels with solutions of test substance for 48 hr without soil or sediment. Measure adsorption. If water solubility permits, equilibrate 0, 0.05, 0.25, 1.25, 6.25, and 31.25 mg/ℓ concentrations with each substrate at 20°C. If water solubility of substance is too low, use lower range of concentrations covering at least one order of magnitude. Shake vigorously initially, then shake 5 min each hr for 48 hr at a rate which keeps substrate suspended. Centrifuge mixture at $\geq 20,000 \times$ G for 10 min. Remove aqueous solution and store at 5°C for analysis. To obtain a mass balance extract solid substrate portion three times with a volume of a suitable solvent equal to the volume of water used. Each time, shake vigorously for 10 min then centrifuge at $\geq 20,000 \times$ G. Mass balance calculated by adding test material in substrate to test material in equilibrium solution and comparing value with initial aqueous concentration. If difference $\geq 5\%$ report possible reasons. To obtain adsorption isotherm, plot the average μg test material adsorbed (initial μg/ℓ — final μg/ℓ) per g of substrate, vs. m, the final concentration of test material in solution (x/m = KC_e). Plot of log x/m vs. log C_e yields line of slope 1/n. Determine K (Freundlich constant) and 1/n.

Special precautions and exclusions
Federal Register **page (Ref. 3)**

—

16257-64

Property/test
Environmental importance
Summary of methods

Adsorption (continued)
As above

Soil thin layer chromatography — Perform in triplicate. Use one soil with pH 4—8, organic matter 1—8%, cation exchange capacity > 7 meq/100 g, $\leq 70\%$ sand. Sieve at 250 μm to remove large pieces but do not grind so hard that more fines are created. Add high purity, pH 7 (boiled to remove CO_2) water at approximately 0.75 mℓ/g of substrate. Mix slurry and quickly apply to a clean glass plate. Use a mechanical spreader or a glass rod with masking tape on either side of plate to create a uniform, 500—750 μm thick layer on the plate. Air dry minimum of 24 hr. Use purest grade of chemical (0.5—5 μg containing 0.01—0.03 μCi of ^{14}C-labeled compound is appropriate). Score a line across layer, 11.5 cm from bottom of plate. Apply spot of test chemical 1.5 cm from bottom of plate. Develop by placing bottom of plate in 0.5-cm deep water reservoir in closed chromatographic chamber. Allow to develop to score line. Determine distance test spot has eluted. Calculate R_f = average distance spot travels/distance solvent front travels (= 10 cm). R_f = 0.95 = very mobile, 0.25 = mobile, 0.10 = low mobility, 0.00 = immobile

Table 1 (continued)

SUMMARY OF PROPOSED REFERENCE METHODS FOR CHEMICAL FATE TESTS

Special precautions and exclusions	May be difficult if nonradiolabeled test substance used.
Federal Register **page (Ref. 3)**	16257–64
Property/test	Boiling/melting/sublimation points
Environmental importance	Help determine the likely physical state of a chemical at ambient temperatures in the environment
Summary of methods	Several methods acceptable (ASTM R324, D1078-70, D3451; CIPAC [Ref. 9] MT_1). For differential thermal analysis (DTA)/differential scanning colorimetry (DSC) use methods ASTM E487, E472, E537
Special precautions and exclusions	In virtually all cases slow heating is important for accurate results
Federal Register **page (Ref. 3)**	16264
Property/test	Density/specific gravity
Environmental importance	Helps determine whether substance will float or sink in water and, if a gas, whether its vapor will concentrate close to the ground or readily disperse in the atmosphere
Summary of methods	Required accuracy ± 0.1 g/cm³. Solids — Determine density at 20° ± 3°C by CIPAC MT_3 (for solids) or ASTM (35) D792066. Liquids — Determine specific gravity at 20° ± 3°C relative to water at 20°C. Use CIPAC MT_3 (for liquids), ASTM (23) D941-55, (23) D1480-62, or (23) D1481-62. The air (helium) comparison pycnometer (Beckman 930 or equivalent) can be used. Gases — Density will be calculated by EPA. Actual gas density measurements in the possession of the submitter should be submitted
Special precautions and exclusions	—
Federal Register **page (Ref. 3)**	16264
Property/test	Dissociation constant
Environmental importance	Gives information on the likelihood of ionizable substances being ionized under environmental conditions. Ionization will influence mobility in water and solid substrates and will affect accumulation in the lipid portions of living organisms.
Summary of methods	Test solids and liquids likely to have pK values from 3—11. Use pure samples. If solubility < $10^{-5}M$, testing not necessary. Use potentiometric titration or UV absorbance method for Brønsted acids and bases. Use a conductivity cell for salts, solvolyzable organic compounds and poorly soluble acids and bases. Determine at 25 ± 1°C to 0.1 pK unit. Concentrations < 0.01 M should be tested. Measure pK of one of the following test compounds as a positive control; acetic acid; benzoic acid; d-, m-, or p-nitrophenol; phosphoric acid ($H_2PO_4^- \rightleftharpoons H^+ + HPO_4^{-2}$); ammonium hydroxide; cyclohexylamine. Choose compound pK closest to test compound.
Special precautions and exclusions	
Federal Register **page (Ref. 3)**	16264-5

Property/test
Environmental importance
Flammability/explodability
Helps assess acute risk of substance to human health and safety. Gives information on conditions needed to ensure safe handling and disposal

Summary of methods
Flash point — Temperature at which liquid or volatile solid gives off sufficient vapor to form an ignitable mixture with air near the surface of the test substance or within the test vessel. Use for chemicals that are liquid below 60°C. Closed cup test should be used except in rare instances when open cup more appropriate. Use ASTM D93-73, D56-75, D3278-73, D1437-64, or E502-74 or CIPAC MT 12.

Gas flammability — Maximum and minimum concentrations of gas and air mixtures that will ignite. Use for chemicals which are gases under conditions of intended use unless chemical has totally oxidized structure. Use method described in (DOT, Bureau of Explosives) Code of Federal Regulations 49 Section 173.300.

Autoignition — Temperature at which a substance will spontaneously ignite in the absence of a spark or flame source. Use for chemicals that are likely to come in contact with hot (> 100°C) surfaces. Use ASTM D2155-66.

Shock and thermal explodability — Capability of a substance to undergo an uncontrolled, fast, violent chemical reaction resulting in a sudden increase in pressure. Use for all substances. For thermal explodability use DTA/DSC method ASTM E487 plus ASTM E474 for calibration. For impact explodability of solids use Bureau of Explosives impact apparatus (C.F.R. 49, 173.53, Note 4). For liquids use Bureau of Mines or Naval Ordinance Laboratory impact testers.

Special precautions and exclusions
—

Federal Register
page (Ref. 3)
16265

Property/test
Environmental importance
Particle size
Provides information on the respirability of crystalline or fibrous substances as well as their expected mobility in the atmosphere.

Summary of methods
Substances likely to form particles ≤ 500 μm in length should be examined. Use ASTM C690, E20-68, D1705, or D422 for particle size measurement. Dry sieves, light microscopy, Coulter counter, sedimentation, and electron microscopy acceptable. Sedigraph 5000D analyzer also acceptable.

Special precautions and exclusions
—

Federal Register
page (Ref. 3)
16266

Property test
Environmental importance
pH
Useful for estimating acute skin and eye irritation potential and effects on water quality which might affect aquatic organisms

Summary of methods
Measure pH of 1% (v/v) solution with a calibrated glass electrode pH meter. Use CO_2-free distilled water. If solubility < 1% use up to 80% dioxane/water (v/v) (Caution: dioxane is a suspect human carcinogen). If dioxane used, apply correction factor (Ref. 10).

Special precautions and exclusions
—

Federal Register
page (Ref. 3)
16266-7

Table 1 (continued)
SUMMARY OF PROPOSED REFERENCE METHODS FOR CHEMICAL FATE TESTS

Property/test
Environmental importance Chemical incompatibility
Predicts probability of creating hazardous conditions when test substance comes in contact with other substances

Summary of methods Observe significant temperature increases ($\geq 5°C$), evolution of gases, noxious fumes, splatter and evolution of flame when substance contacted with representative materials likely to be exposed to test chemical. Judgement of person familiar with substance should be used. Use high ratio of test chemical to substrate. Check, for example, contact (24 hr) with: water; carbon dioxide and/or ammonium phosphate (fire extinguishing agents); zinc and iron (moderately strong reducing agents) — standard corrosion testing may suffice; dilute, neutral, aqueous permanganate (moderately strong oxidizing agent); household solvents (kerosene, turpentine, gasoline) if for household use.

Special precautions and exclusions —

Federal Register page (Ref. 3) 16267

Property/test
Environmental importance Vapor-phase UV spectrum for halocarbons
Gives information regarding the potential of the substance to cause stratospheric ozone depletion

Summary of methods Applicable to substances containing chlorine and bromine which have a vapor pressure ≥ 0.5 torr at 25°C. Measure vapor phase absorption spectra (275—700 nm) at 3 different gas pressures against a blank cell. Use calibrated 10-cm gas cells evacuated before use. Use compound of known molar absorptivity for calibration. Calculate absorption cross sections at different wavelengths from absorbance spectrum, cell path and gas density in cell. Calculate tropospheric half-life assuming photoreactions proceed with a quantum yield of 1. If tropospheric half-life is greater then several years, then substance can be expected to reach the stratosphere where it may interact with sunlight and ozone causing ozone depletion.

Special precautions and exclusions —

Federal Register page (Ref. 3) 16267

Property/test
Environmental importance UV and visible absorption spectra in aqueous solution
Important in predicting likely interactions of substance with sunlight such as photoreactions, screening effects, photosensitization

Summary of methods Record spectra of analytically pure substance in reagent grade water using a minimum of methanol or acetonitrile as cosolvents, if necessary. Region above approximately 290 nm is of major importance since sunlight of lower wavelengths does not penetrate the upper atmosphere. Spectra at pH 5, 7, and 9 should be recorded for ionizable substances or substances which protonate.

Special precautions and exclusions —

Federal Register page (Ref. 3) 16267-8

Property/test
Environmental importance Chemical transformation: hydrolysis
Provides information about the rate and extent of degradation of the substance by water. Also helps to determine whether possible hydrolysis products are of potential environmental concern

Summary of methods

Use sterilized glassware to preclude microbial activity. Prepare solution of analytically pure test substance in reagent grade, sterile water at a concentration < 1/2 solubility in water and < 10^{-3} *M*. Add aliquot to buffers of pH 5.0 (0.01 *M* sodium acetate adjusted with 0.1 *M* acetic acid); pH 7.0 (30 ml 0.067 *M* sodium dihydrogen-phosphate mixed with 60 ml 0.067 *M* potassium monohydrogen phosphate and diluted 10-fold); and pH 9.0 (0.025 *M* sodium borate adjusted with 0.1 *M* acetic acid). Keep buffer concentration low to minimize side reactions between substance and buffer. If substance not sufficiently soluble, dissolve in acetonitrile and add buffer solution to an aliquot of acetonitrile solution. Maintain ≤ 1 volume % acetonitrile in solution. Perform hydrolysis in greaseless, stoppered flasks protected from light (if substance absorbs light). If substance volatile use almost completely filled, sealed tubes. Immerse reaction vessels, in duplicate, in 25° ± 1 °C bath. Determine initial concentration of test substance (C_o) then determine concentrations at regular time intervals (hr) to provide a minimum of 6 time points with hydrolysis between 20—70%. Rate should yield 60—70% hydrolysis in 4 weeks (672 hr). If hydrolysis only proceeds to 20—30% in 4 weeks, take more data points between 10—30%. If hydrolysis < 20% in 4 weeks determine the final concentration (C) after 4 weeks. When samples removed for analysis, analyze immediately or quench the hydrolysis reaction before sample storage. Assuming pseudo first order kinetics, calculate the hydrolysis rate constant (k_h) from the linear regression line (plot $\log_{10}C$ vs. t) of the equation:

$$\log_{10}C = \frac{k_h}{2.3}\, t + \log_{10} C_o$$

Calculate the hydrolysis half-life $t_{1/2}$ from: $t_{1/2} = 0.693/k_h$ where k_h is the mean vlue of the rate constant from the pH 5.0, pH 7.0, and pH 9.0 runs. If, after 672 hr, C_o = C then the substance would be considered persistent to hydrolysis. Substances which ionize or protonate at pH 5—9 may yield unusual pH-rate profiles.

Special precautions and exclusions

Interim procedure; improved test under development.

Federal Register page (Ref. 3)

16268-70

Property/test

Chemical degradation by oxidation

Environmental importance

Helps assess the potential for environmental degradation of the substance through atmospheric and aquatic oxidation

Summary of methods

Air oxidation tests — Applicable to organics with vapor pressure ≥ 10 torr at 30°C. Enclose reactants (hydroxy radicals or ozone appear to be likely atmospheric reactants) and substance in a suitable container at environmentally realistic concentrations. Follow change in concentration of substance or reactant.

Water oxidation tests — Only photochemically generated alkylperoxy radicals and singlet oxygen seem to be important. Alkylperoxy radical may be generated in water in the presence of test compound and the change in concentration of test compound followed.

Special precautions and exclusions

No simple procedure developed and validated. Only basic approaches given.

Federal Register page (Ref. 3)

16270-71

Property/test

Aqueous photochemical reactions

Environmental importance

Helps predict the effect of sunlight on the degradation of the test substance

Summary of methods

Use analytically pure substance (specify impurities). Use sterile, air-saturated, reagent-grade water. Sterilize glassware in an autoclave.

Table 1 (continued)

SUMMARY OF PROPOSED REFERENCE METHODS FOR CHEMICAL FATE TESTS

Use concentration < 1/2 water solubility, < 10^{-5} M and yielding < 0.02 absorbance at wavelengths > 290 nm. Prepare solutions of insoluble chemicals using aliquots of acetonitrile solutions. Keep acetonitrile < 1% of total volume. If chemical absorbs light below 340 nm, use ground-glass-stoppered quartz reaction cells. If chemical absorbs light above 340 nm, borosilicate glass can be used. Fill reaction cell almost completely and cap tightly. Run in duplicate along with two positive control samples wrapped in aluminum foil. Run photolysis between June and August. Place samples at 30° angle in direct, bright sunlight. Include samples containing a reference compound (being developed by EPA) from which the overall intensity of sunlight can be measured (i.e., an actinometer). If chemical photolyzes 60—80% in 28 days, measure concentration (C) at t = 0 and at periodic intervals at 12:00 noon. If chemical photolyzes 20—50% in 28 days, determine concentration after 28 days. If chemical photolyzes < 20% in 28 days, half-life is > 3 months. If chemical photolyzes 60—80% within 7 days, determine C_o and C at 12:00 noon each day. For chemicals which photolyze 60—80% in 1—2 days, determine C_o before sunrise and C after sunset the 1st day and after sunset the 2nd day. Repeat experiment 3 times on cloudless days. Calculate the photolysis rate constant (k_p) in days^{-1} in sunlight from:

$$k_p = k_a\phi$$

where ϕ = quantum yield of chemical reaction in sunlight in dilute aqueous solution (fraction of substance reacting per quantum of light absorbed) and k_a = sum of $k_a\lambda_a$ values for all wavelengths of sunlight absorbed by the chemical (a measure of the total quanta of light absorbed by the sample). Calculate photolysis half-life from:

$$\log_{10} C_o/C = \frac{k_p t}{2.30}$$

$$\text{where half-life} = \frac{0.693}{k_p}$$

Report latitude, site of photolysis and dates of photolysis. If chemical reversibly ionizes or protonates carry out photolysis at pH 5, 7, and 9.

Special precautions and exclusions	—
***Federal Register* page (Ref. 3)**	16271-2
Property/test	Biodegradability
Environmental importance	Provides information on the ability or inability of microorganisms in nature or in waste treatment plants to degrade substances and the rate and extent to which this degradation may occur in the environment. Biodegradation is generally the most important degradative mechanism for organic compounds in nature
Summary of methods	Shake flask method — Perform test and analysis in duplicate. Prepare microbial growth medium (basal medium) by first preparing dilution water from reagent-grade water (< 1 mg/ℓ total organic carbon), phosphate buffer, magnesium, calcium and iron salts and

trace elements needed for cell growth and metabolism. Add yeast extract solution prepared immediately before use to 0.5—1 ℓ of freshly prepared or sterile dilution water in an erlenmeyer flask stoppered with a cotton plug. For inoculum, use microbial culture derived either from fresh, filtered, secondary effluent from a treatment plant handling mainly domestic sewage; from garden soil suspended in chlorine-free tap water (100 g to 1 ℓ), allowed to settle 30 min and filtered through glasswool; or from a mixture of 100 mℓ of sewage culture and 50 mℓ of soil filtrate. (Note: cultures from other sources may be used as appropriate.) Add test material (as aqueous solution) at equivalent of 10 mg organic carbon per liter of basal medium; add 1 mℓ of microbial inoculum. Set up a positive control of 10 mg organic carbon equivalent of linear alkane sulfonate (obtained from EPA) and a blank control (no test compound) in the same manner as that for the test substance. Shake flask in the dark at 22 ± 3°C using a reciprocal or rotary shaker to ensure good aeration (dissolved oxygen 2 mg/ℓ or higher; liquid should be well agitated but should not spill or wet cotton plug). Acclimatize respective microbial cultures by transferring, at 48—72-hour intervals, 1 mℓ of shaken mixture into fresh 1-ℓ aliquots of the same mixtures as initially prepared for each. After the 4th transfer (on day 10) observe the flasks, especially the blank control, for evidence of microbial viability as indicated by cloudiness, film formation, or deposits of cellular material in the test flasks. Absence of these signs in the blank control may indicate a defective inoculum. Absence in the test solutions may indicate material is toxic or inhibitory to the microbial population. Three days (72 hr) after the last transfer, remove small samples from each of the well-mixed solutions for dissolved organic carbon analyses. This will be the day 0 sample. At this time set up 2 additional flasks containing basal medium and test compound but no inoculum and carry them through the rest of the procedure with the other samples. These samples, which should be kept sterile, will monitor nonmicrobial changes in dissolved organic carbon. Transfer 1 mℓ of the flask mixture to 1 ℓ of fresh basal medium as before and continue shaking. Repeat sampling (without any further transfers) so that samples are obtained on days 0, 1, 2, 3, 4, 7, 10, 15, 21 (or other appropriate 21-day schedule). Mark the flasks to designate the volume. Maintain volume by appropriate additions of reagent grade water between samplings. Within 3 hr of sampling, filter all samples for DOC analysis through a 0.2-μm filter which has been boiled in water for 1 hr and stored for less than a week at room temperature in reagent grade water (TOC = DOC after filtration). If samples are not analyzed immediately, acidify to pH ≤ 2 with H_2SO_4, HCl, or HNO_3, as appropriate for instrument used. If desired, follow disappearance of test material if method available. To determine rate of removal, calculate % removal for each sampling time from:

$$\% \ Removal_t = \frac{(C_0 - B_0) - (C_t - B_t)}{(C_0 - B_0)} \times 100$$

where: C_0 = mg/ℓ dissolved organic carbon in sample flask at t = 0. B_0 = mg/ℓ dissolved organic carbon in blank control flask at t = 0. C_t and B_t = the concentration of the same solutions at time t. Plot % Removal, vs. time (days) to obtain removal curve.

Activated sludge test — Degradation of test compound monitored in simulated activated sludge system. Construct cylindrical polyethylene or glass test vessels of approximately 3-ℓ capacity with cone-shaped bottom. The vessels should be fitted with aeration inlet tubes tipped with an aeration stone, reaching to the lower end of the cone, and a draw-off tube at the 500-mℓ volume mark on the cylinder. A mark should be made at the 2 ℓ volume level of the cylinder. Add 2000 mℓ activated sludge (adjusted to 2500 mg/ℓ suspended solids) from a sewage treatment plant treating mainly domestic sewage. Aerate with oil-free compressed air at about 500 mℓ/min. After each 24-hr aeration period, scrape down the inner surface of the chamber, determine suspended solids (mixed liquor suspended solids — MLSS) and adjust to 2000—3000 mg/ℓ if necessary, stop aeration, allow 30 min of settling, drain off 1500 mℓ supernatant, resume aeration, add 15 mℓ of a synthetic sewage solution (contains glucose, nutrient broth, phosphate buffer, ammonium sulfate, and mineral-nutrient solution) and bring to 2000 mℓ mark with tap water. Maintain pH 6.5—8.0 daily by adjusting with H_2SO_4 and NaOH. Maintain dissolved oxygen at ≥ 2 mg/ℓ. Increase aeration rate if necessary. Acclimatize 2 sets of duplicate units of the sludge to the test substance by daily feedings of the test material in increasing increments until amount of test compound added is equivalent to 100 and 200 mg/ℓ organic carbon. Maintain 1 test unit fed only with synthetic sewage as

Table 1 (continued)
SUMMARY OF PROPOSED REFERENCE METHODS FOR CHEMICAL FATE TESTS

a blank control and 1 test unit both fed with synthetic sewage and acclimatized and fed with 200 mg/ℓ organic carbon equivalent of diethylene glycol (DEG) as a positive control. Maintain temperature at 22 ± 3°C. On the 14th day or later, after acclimatization concentrations have been at their maxima for at least 2 days, retain a portion of the supernatant, after settling, for DOC analysis. Filter through 0.2 μm filter as for the shake-flask method. If necessary, centrifuge before filtering. Repeat procedure for at least 3 days continuing daily feedings of 100 and 200 mg/ℓ carbon equivalent of test compound. Compare DOC to DOC of filtered, simulated blank. DEG and test compound influents using equation: % Removal = $(1 - S)/1 \times 100$ where 1 = DOC of simulated influent and S = DOC of supernatant sample. Plot acclimatization data and continue feedings until 3 consecutive days' points on the plot show 70 ± 10% removal. If this level not achieved within 30 days, conclude the test. If level achieved within 30 days, then, on the 4th day of consistent removal, remove sufficient mixed liquor to yield 1000 mg suspended solids in each unit. Shut off aeration, allow sludge to settle for 30 min, discard all of supernatant above settled solids and resume aeration. Fill units to 2 ℓ mark as follows: blank unit-mineral-nutrient solution only; DEG-unit-442 mg DEG (200 mg organic carbon equivalent) plus mineral-nutrient solution; low concentration test unit — 100 mg organic carbon equivalent of test compound plus mineral-nutrient solution; high concentration test unit — 200 mg organic carbon equivalent of test compound plus mineral-nutrient solution. Remove samples for DOC analysis after 30 min (to detect adsorption), 3, 6, 24 hr then 2, 3, 4, 6, 8, 10, 13, 16, 19 days. Maintain 2-ℓ volume by additions of diluted nutrient solution. Calculate % removal of DOC making corrections for amount removed at each sampling. Use equation

$$\% \text{ Removal} = \frac{[DOC_A - (DOC_T - DOC_K)] \times 100}{DOC_A}$$

where DOC_A = initial mg/ℓ DOC based on amount added; DOC_T = mg/ℓ DOC at time of sampling; and DOC_K = mg/ℓ of blank at time of sampling. If ≥ 90% removal found on 2 consecutive days, test can end. Otherwise plot % removal vs. time over the 19-day test period.
Indirect test of ultimate fate of compound. Radiolabelled materials should be used to obtain more detailed information

Special precautions and exclusions
Federal Register page (Ref. 3)
16272-80

Property/test
Environmental importance
Summary of methods

Biodegradability (continued)
Gives information about degradation in anaerobic sewage digesters and in anaerobic portions of soil and sediments
Methane and carbon dioxide production in anaerobic digestion. Put anaerobic digester sludge from municipal sewage treatment plant in tube or flask at 25 mℓ/100 mℓ capacity of container. Displace air in container with O_2-free CO_2 and cap with butyl rubber stoppers or alternatively trap evolved gases in displacement-type gasometer. Maintain units at 35—37°C and collect and measure volume until daily gas production volume constant (± 10%) over 3-day period. Add test material to duplicate test vessels at 0, 10, 20, 40, 100, 200 mg/ℓ sludge. Add NYG medium to 2 additional vessels at 1 mℓ/100 mℓ sludge to be used as a positive control. Incubate 28 days. Measure gas volume periodically and analyze evolved gas for methane and CO_2. At end of 28 days convert any bicarbonate to CO_2 by adding 10 mℓ 5 N HCl per 100 mℓ sludge. Measure evolved gas for methane and CO_2. Calculate % of theoretical maximum of methane and CO_2 produced by:

$$\% \text{ Theoretical} = \frac{(G_T - G_M)\,100}{G_T}$$

where: G_T = total mg organic carbon in sample and G_M = mg of carbon in $CH_4 + CO_2$ formed in excess of that in the units with no test compound added.

Special precautions and exclusions

Federal Register page (Ref. 3)

As above

16272–80

Property/test

Environmental importance

Summary of methods

Biodegradability (continued)

Gives information about biodegradation under more dilute conditions than shake-flask and activated sludge tests

Carbon dioxide evolution test. Monitors complete conversion of organic carbon to CO_2 by microorganisms. Same 13-day acclimation as shake-flask test is used except that positive control is glucose vs. LAS. Several test materials can be acclimatized simultaneously in separate shake flasks. Acclimatized cultures from separate test materials can be combined on 13th day (500 mℓ of each) to give a composite inoculum. This allows use of a common blank control and glucose positive control for several different test materials. The head space of 4-ℓ test vessels containing inoculum (250 mℓ), and minerals (2750 mℓ) are continuously purged with air freed of CO_2 before the test vessel by passage through ascarite and a 0.05 N Ba(OH)$_2$ test trap. The CO_2 generated in the test vessel is purged by the treated air through 3 0.05 N Ba(OH)$_2$ absorber traps in series. After purging residual CO_2 from the basal-inoculum solution for 24 hr, the test material is added to 2 flasks at concentrations of 5 and 10 mgC/ℓ. The mixtures are incubated in the dark at 22 $\pm$ 3°C. The Ba(OH)$_2$ trap nearest the test flask is removed and titrated with 0.1 N HCl to a phenolphthalein endpoint usually on days 1, 2, 3, 4, 5, 8, 10, 15, 21, and 28. Each time a trap is removed, the next two in series are moved forward and a new trap placed in the third position. The titration determines the quantity of Ba(OH)$_2$ not converted to BaCO$_3$. Thus, to calculate CO_2 formed, subtract 0.1 N HCl required for sample from 0.1 N HCl required for blank. Net value is related to the amount of CO_2 generated in the sample as follows: net milliliter 0.1 N HCl $\times$ normality of HCl (0.1) = meq HCl required = meq CO_2 Note: 1 meq CO_2 = 22 mg. Calculate % theoretical CO_2 produced from:

$$\% \text{ Theoretical } CO_2 = \frac{\Sigma[CO_2]_1^n\,(100)}{\text{Theor. } CO_2}$$

where $\Sigma[CO_2]_1^n$ = sum of CO_2 formed from day 1 to day n and theoretical CO_2 = mg test compound added $\times$ 44/12. Plot % theoretical CO_2 vs. time(days) to obtain rate curve of CO_2 evolution.

Special precautions and exclusions

Federal Register page (Ref. 4)

As above

16272–80

Property/test

Environmental importance

Summary of methods

Biodegradability (continued)

Gives similar but less extensive information than CO_2 test

Biochemical oxygen demand (BOD). Use acclimated cultures prepared as for CO_2 evolution test. Dilution water same as mineral salts solution used in CO_2 test but stored in the dark for at least 1 day at 20°C before use. Inoculate dilution water, 1—2 hr before use, with 0.1 to 0.5 mℓ of mixed inoculum per liter. Aerate gently with oil-free compressed air for 15 min. Prepare 0.3 mg/mℓ

Table 1 (continued)

SUMMARY OF PROPOSED REFERENCE METHODS FOR CHEMICAL FATE TESTS

aqueous solution of test compound. If necessary, adjust pH to 6.5—8.0 range with H_2SO_4 or NaOH. Make duplicate additions of test solution (or directly weighed portions, if insoluble) to BOD bottles sufficient to yield 0.3, 0.6, 1.2, 2.4, and 4.8 mg/300 mℓ of dilution water. Also prepare blanks and a standard reference set (5 mℓ and 10 mℓ of 150 mg/ℓ each of glucose and glutamic acid). Immediately measure dissolved oxygen (DO) in each bottle in known order measuring two blanks first and another two blanks last. Incubate bottles in the dark at 20°C. Remeasure DO on days 1, 2, 3, 4, 5, 8, 10, 15, 21, 28. Maintain a water seal around bottle caps to prevent infiltration of air into the bottles. Calculate BOD at each sampling time from:

$$BOD_n = [(DO_o - DO_n) - DO_B] \times 1/DF$$

where BOD_n = BOD on day n in mg O_2 utilized per mg of test material; DO_0 = dissolved oxygen (mg/ℓ) on day 0; DO_n = dissolved oxygen (mg/ℓ) on day n; DO_B = average dissolved oxygen (mg/ℓ) in blanks on day n; and DF = dilution factor = mg added/0.3 1. For calculations use dilutions which yield at least 1 mg/ℓ residual DO and have DO depletions of at least 2 mg/ℓ. Also calculate % of theoretical BOD attained on each sampling day from

$$\% \text{ of Theoretical BOD} = \frac{(BOD_T - BOD_M)\ 100}{BOD_T}$$

where BOD_T = Calculated theoretical amount of oxygen required to completely oxidize the test material to CO_2, water and inorganic molecules (e.g., S → SO_4, N → NO_3, etc.) and BOD_M = measured BOD. Plot % theoretical BOD vs. time to obtain biodegradation rate curve.

Special precautions and exclusions —

Federal Register **page (Ref. 3)**   16279-80

- The ability of the test to adequately detect the subject effect
- The ecological importance of the class containing the test organisms
- The position which the test organism occupies in the food chains leading to man or other important species
- The availability of data and experience regarding the normal physiology and behavior of the test organism

A summary of the recommended ecological testing procedures is given in Table 2. Note that, in addition to the standard information which should be included in test reports (see Part IV) the following supplementary information should be provided in ecological test reports:

a. Description of organisms including
    - community type, species, strain, subline
    - rationale for selection of organism if different from those recommended by EPA
    - source
    - breeding history
    - pretest conditioning or acclimation (include diet, quarantine, etc.)
    - number of each sex, size, and weight in each test group
    - age and condition
    - method of randomizing test population in test groups
b. Environmental conditions
    - source, chemical characteristics, pretreatment of water, if applicable
    - photoperiod, including type and intensity of light used, and off-on periods
    - source, composition, and amount of feed or nutrients made available to test organism
    - temperature and humidity of test environment, as applicable
    - description of test chambers including dimensions and volumes
c. Preparation of the test mixture
    - identity and concentration of test chemicals and carriers
    - method of preparation
    - frequency and dates of preparation
    - physical state
    - pH
    - stability, decomposition, and storage conditions (if preparation unstable prepare fresh before each application)
d. Exposure
    - dosage or concentration in the test system
    - method, route, frequency of exposure (give rationale for selections)
    - volume of preparation added at each exposure
    - assays of test substance in the preparation to determine homogeneity and stability of the preparation
    - assays of test substance in the environment
    - methods of random sampling for assay purposes
    - mean total amount of test substance administered per organism, test population, or community
    - reasons for selection of the carrier
e. Treatment for infectious disease
    - describe any treatment to prevent or control disease before or during the test
f. Observations
    - describe method and frequency of observation
    - describe deviations from control data which may be indicative of ecological effects

IV. QUALITY ASSURANCE AND REPORTING — GOOD LABORATORY PRACTICE

As of this writing, efforts to promulgate so-called Good Laboratory Practice Rules (GLP) dealing with quality assurance (QA) procedures have been concentrated on QA procedures for health-related studies

Table 2
SUMMARY OF PROPOSED REFERENCE METHODS FOR ECOLOGICAL TESTING

Test parameter	Microbial effects tests
Environmental importance	Determines potential adverse effects on microbes which are primarily responsible for the cycling of key elements and nutrients in ecosystems
Summary of methods	Cellulose decomposition — Associated with carbon cycling in the environment via humification. Use cellulase-producing organism *Trichoderma longibrachiatum* (ATCC 26921) and monitor CO_2 production in the presence of cellulose as the sole carbon source. Dispense aseptically, 100 mℓ of a liquid mineral salts-growth medium containing ball-milled cellulose along with appropriate concentration of test material into special incubation-CO_2 collection flasks. Add 1 mℓ of *Trichoderma* spore inoculum 1 hr before start of run. Set up controls of only *Trichoderma* spore inoculum in liquid growth medium and liquid growth medium only (latter monitors nonbiological CO_2 evolution). Test minimum of 5 concentrations of test material in triplicate. If at least 3 of the test concentrations yield > 10% but < 90% of the control response then results can probably be extrapolated to any expected environmental concentration. Incubate at 22 ± 3°C in the dark. At 1, 3, and 7 days determine CO_2 collected in the KOH absorption trap by titration. Compare CO_2 production in controls with that in flasks containing test chemical. If latter show significantly reduced CO_2 production levels then inhibition of the degradation of cellulose by *Trichoderma* is indicated.
Special precautions and exclusions	Water insoluble gases (< 1 mg/ℓ) need not be tested. Recommended tests are adequate for preliminary assessment. Tests with natural soil samples should be considered if further testing is deemed necessary
Federal Register **page (Ref. 3)**	16282
Test parameter	Microbial effects tests (continued)
Environmental importance	Tests ability of organisms to convert organically bound nitrogen to ammonia. Important to the nitrogen cycle in the environment
Summary of methods	Nitrogen transformation — Test uses urea as an organic nitrogen source. Microbial degradation of urea to ammonia correlates closely with the degradation of protein, a major environmental nitrogen source. Add urea to sieved soil or sediment along with test chemical. Use 3 replicates of at least 5 concentrations, 3 of which yield > 10% but < 90% inhibition of ammonia formation compared to a control containing no test chemicals. Incubate at 22 ± 3°C for 2—3 weeks. Analyze ammonia formed by standard ammonia analysis techniques. Report in milligrams ammonia-*N* per 100 g of oven-dried soil. If the test substance yields no inhibition at or near expected environmental concentrations, no detriment to ammonification process likely
Special precautions and exclusions	As above
Federal Register **page (Ref. 3)**	16282
Test parameter	Microbial effects tests (continued)
Environmental importance	Tests ability of microorganisms to reduce sulfate. Important to sulfur cycling in the environment. Sulfur needed for sulfur-containing amino acids
Summary of methods	Sulfur transformation — Sulfate-reducing organisms contacted with test chemical in a growth medium. Organisms then transferred to sulfate reducing medium and growth measured. Use sulfate-reducing organism *Desulfovibrio desulfuricans* strain Hildenborough (NCIB8303) as test organism. Manitain stock culture of the organisms by weekly subculture into a sulfate reducer growth medium containing selected mineral salts plus sodium lactate, yeast extract, sodium thioglycollate, ferrous ammonium sulfate, and ferrous sulfate. Add 15 mℓ of 4—6-day stock culture to 1485 mℓ of fresh medium, invert to distribute organisms then dispense 99 mℓ aliquots into screw-cap milk dilution bottles. Add test chemical (1, 10, 50, 100, 500, 1000 mg/ℓ) to 1 set of bottles. Add chlorine

(1, 10, 50 mg/ℓ to a 2nd set. For a control add nothing to a third set of duplicate bottles. Transfer 1 mℓ of the latter to screw cap culture tubes and fill to capacity (to maintain anaerobic conditions) with reducer medium. Cap bottles and tubes, mix and incubate at 22 ± 3°C in the dark. After 3 and 24 hr, mix each bottle, transfer 1 mℓ aliquot into culture tubes, and fill the tubes with reducing medium. Incubate tubes at 30°C for 21 days. Determine the presence of sulfate reducing bacteria by observing blackening in the tubes. If sulfate-reducer growth appears in the control but not in the test samples, repeat the experiment at 1, 0.1, 0.01, and 0.001 mg/ℓ. Report minimum inhibitory concentration as greater than highest test concentration in which growth observed

Special precautions and exclusions
As above

Federal Register page (Ref. 3)
16282

Test parameter

Environmental importance
Photosynthetic processes in plants provide virtually all atmospheric oxygen. Plants also provide a major food source for aquatic and terrestrial organisms

Plant effects tests

Summary of methods
Algal inhibition test — Microscopic algae provide much of the photosynthetic oxygen of the earth and are a primary food and energy source for aquatic organisms. Inhibition of algae growth may therefore reduce population, reduce diversity, or alter in an undesirable manner the relative population of members of the aquatic food chain whose food source depends ultimately on algae. Use algae representative of various aquatic environments. *Anabaena flos-aquae* (blue-green) and *Selenastrum capricornutum* (green) should be grown in fresh-water medium at 24°C for 7 days. *Skeletonema costatum* (diatom) should be grown in artifical sea-water with Na$_2$SiO$_3$ at 24°C. *Navicula seminulum* variation *hustedii* (diatom) should be grown in fresh water medium with 20 mg/ℓ Si added as Na$_2$SiO$_3$·5H$_2$O. Expose algae to at least 5 concentrations of test chemical in the range of expected environmental concentrations. At least one concentration should yield no effect. If 3 concentrations yield > 20% but < 95% of normal growth (as determined by controls), then results can be extrapolated to virtually any expected environmental concentration. Changes in acetylene reduction by *Anabaena flos-aquae* can be followed to obtain information about effects on nitrogen fixation. If test chemical causes pH shift determine effect, if any, on inorganic carbon. Control light conditions carefully. Determine growth by dry weight increase (or a technique which correlates with dry weight increase) and by cell size. If using automatic particle counter, adequately disrupt aggregates and construct calibration curve.

Special precautions and exclusions
—

Federal Register page (Ref. 3)
16285-6

Plant effects tests (continued)

Test parameter

Environmental importance

Summary of methods
Results serve as an indicator of the effects of a chemical substance on higher aquatic plants

Lemna inhibition — *Lemna* are grown in a suitable growth medium along with a test chemical. Inhibition of *Lemna* growth is determined from root growth, frond number, and dry weight. Grow *Lemna minor* or *Lemna gibba* on nutrient medium in simulated sunlight (3000 lux) at 24°C for 7 days. Add test chemical at minimum of 5 concentrations at least 3 of which inhibit growth > 10% but < 90% compared to a control containing no test chemical. Use concentrations near the reasonably expected environmental range. Report 50% inhibitory concentration and sublethal effects such as discoloration, abnormal growth, etc.

Special precautions and exclusions
—

Federal Register page (Ref. 3)
16286-7

Table 2 (continued)
SUMMARY OF PROPOSED REFERENCE METHODS FOR ECOLOGICAL TESTING

Test parameter

Environmental importance Plant effects tests (continued)

Reveals adverse effects on the critical early stages in the life cycle of flowering plants (angiosperms). Angiosperms are the dominant vegetation in many areas of the U.S. and are the source of all major food crops

Summary of methods Seed germination and early growth — Early growth stages used because rapid growth, high metabolic rates, and immaturity of defense systems make plant more vulnerable to chemical inhibition. Test by exposure routes most likely to be important for the test chemical (e.g., feed through roots; dust or spray on foliage; fumigate for volatile chemicals). Effects on seedling growth measured following exposure of seedlings after seed germination. Use monocotyledons (oats, ryegrass, corn) and dicotyledons (cucumber, bean, tomato) as test species. Measure seed germination, seedling height, shoot dry weight and root dry weight as a % of untreated controls. Report any abnormalities noted.

Special precautions and exclusions Chemicals which are subject to high environmental release or show significant biological activity in other tests may require testing at later life stages to fully assess effects on angiosperms

Federal Register page (Ref. 3) 16287-8

Test parameter

Environmental importance Animal effects tests

Tests potential adverse effects on representative terrestrial and aquatic species as an indication of effects at higher community levels of the environment

Summary of methods Acute toxicity to aquatic invertebrates — Most appropriate test species to use depends on the physical-chemical characteristics and "use activities" of the test substance. Substances expected to concentrate in sediments should be tested on species which feed on or live in sediments while substances expected to be present in the water should be tested on invertebrates indigenous to these regions. Recommended marine and estuarine invertebrates include copepods, shrimp, crab, oysters, and polychaetes. For freshwater tests, daphnids, amphipods, crayfish, stoneflies, mayflies, midges, snails, and planaria are recommended. Young or immature specimens should be exposed to various levels of test chemical over a 48—96-hr period. Specific effects (e.g., lethality, immobility) should be monitored to determine 50% effect doses of the test substance at 24, 48, and 96 hour intervals. Report any other abnormalities noted as well as any effects on the controls. Include pH, oxygen concentration, loading (grams of organisms per liter of water), oxygen concentration during the test.

Special precautions and exclusions —

Federal Register page (Ref. 3) 16288-9

Test parameter

Environmental importance Animal effects tests (continued)

Useful for determining safe concentrations for fish and test concentrations for chronic and bioconcentration tests

Summary of methods Acute toxicity to fish — Rainbow trout (Salmo gairdneri) and bluegill sunfish (Lepomis macrochirus) are the cold and warm water species of choice, respectively, due to their high general sensitivity to most chemical substances. Goldfish (Carassius auratus) are generally not recommended due to their relatively low sensitivity. Specific geographical locations in which the test chemical is expected to be released may dictate that other species would be more appropriate to test. Young fish are exposed to various concentrations of test chemical usually spaced on a log concentration scale. The LC_{50} (concentration lethal to 50% of the test population) is then statistically determined at 24, 48, and 96 hours along with 95% confidence intervals. Abnormal behavior, pH, oxygen concentrations, and water hardness should be reported.

Special precautions and exclusions
Federal Register page (Ref. 3)   16289-90

Test parameter   Animal effects tests (continued)
Environmental importance   Gives information on possible effects of substance on terrestrial birds
Summary of methods   Quail dietary test — Use for chemicals which may become incorporated into terrestrial bird food. Bobwhite quail (*Colinus virginianus*) are given feed fortified with test chemical at various levels. A minimum of 10 birds are used at each level. Test concentrations should include at least one level which yields no-effect. The birds are fed for 5 days. Surviving birds are fed unfortified feed for an additional 3 days. The concentration at which 50% of the birds are killed (LC_{50}) is estimated from the dose-response relationship. If $LC_{50} > 5000$ ppm, then LC_{50} calculation and confidence limits are not needed. The nature, time of occurrence, severity, and duration of any abnormal effects are reported. Quantity of test substance consumed in the feed over 5-day period should be recorded.

Special precautions and exclusions   —
Federal Register page (Ref. 3)   16290

Test parameter   Animal effects tests (continued)
Environmental importance   —
Summary of methods   Terrestrial mammal tests — Testing omitted since virtually the same information is obtained during health effects testing.
Special precautions and exclusions   —
Federal Register page (Ref. 3)   16290

Test parameter   Animal effects tests (continued)
Environmental importance   Effects of chemical substances on aquatic invertebrates give information on the possible disruption of the food chain linking microscopic phytoplankton to desirable species of fish and shellfish
Summary of methods   *Daphnia* life cycle test — Use *Daphnia pulex* or *Daphnia magna*. Applicable to soluble or suspendable substances which may enter freshwater systems. Newborn, healthy *Daphnia* are exposed to various concentrations (within 10-fold of each other) of test substance for 21—28 days. Controls, no-effect level, as well as statistically significant effect levels on reproduction should be included. Chronic lethal effects on the original test population are noted as are the number of surviving offspring compared to the control population. About 80% of a healthy control population should survive for 28 days and should produce more than 100 young each during the same period. Report grams of organisms per liter, % survival, daily production of young, water quality parameters. Describe flow-through systems, if used.

Special precautions and exclusions   —
Federal Register page (Ref. 3)   16290-1

Test parameter   Animal effects tests (continued)
Environmental importance   Gives information about important food link between marine plankton and fish

Table 2 (continued)

SUMMARY OF PROPOSED REFERENCE METHODS FOR ECOLOGICAL TESTING

Summary of methods
Mysidopsis bahia (mysid shrimp) life cycle — Expose young (< 24-hr-old) shrimp to various concentrations of test chemical. Monitor survival and reproduction daily for 28 days. Compare to controls. Record grams of shrimp per liter of water, water quality parameters, survival, reproduction, and any abnormal effects.

Special precautions and exclusions
—

Federal Register page (Ref. 3)
16291

Test parameter Environmental importance
Animal effects tests (continued)
Gives a preliminary indication of the potential effects of a chemical on fish when long-term exposure probable

Summary of methods
Fish embryo juvenile test — Use for any chemical which will continuously reach natural water either in dissolved or suspended form. Tests effects of chemical at very sensitive life stages (fish eggs and the resultant fry exposed to various concentrations of chemical substance for several weeks). Hatchability, growth, and survival are determined as % of controls. Abnormalities and water quality are reported. Minnows are widely used as test fish because they can be readily reared and hatched at various times of year.

Special precautions and exclusions
—

Federal Register page (Ref. 3)
16291

Test parameter Environmental importance
Animal effects tests (continued)
Determines ability of a substance to be concentrated in living tissue and passed on to higher members of the food chain

Summary of methods
Fish bioconcentration — Especially important for highly lipid soluble materials (octanol-water partition coefficient > 1000), poorly soluble materials (< $0.9 \times 10^{-3}M$) and relatively stable, i.e., potentially persistent substances. Fathead minnow (*Pimephales promelas*) and bluegill sunfish (*Lepomis macrochirus*) are most frequently used species of fish. Expose fish to test chemical in > 60 ℓ aquaria fitted with continuous flow-through diluter system. Incoming water should contain test chemical below the toxic effect level (0.1—0.01 × 96 hr LC_{50}). Water should be at 22 ± 1°C and contain ≥ 60% of oxygen saturation. Photoperiod should be 12-hr light — 12-hr dark. Introduce chemicals without co-solvent, if possible. If co-solvent needed, use minimum of dimethylformamide, ethanol, methanol, or triethylene glycol. Use same solvent in control tanks. Use sexually immature fish since mature females tend to accumulate more than males. Expose fish to a constant level of chemical for 28 days. On days 1, 2, 3, 4, 7, 14, 21, and 28 remove 5 fish and a water sample from each tank and analyze for test chemical. If possible note the sex of each fish. Use analytical methods giving > 90% recovery, if possible. If the average concentration of test chemical reaches a steady-state before 28 days of exposure, terminate the exposure phase and begin the depuration phase. Otherwise, begin the depuration phase after 28 days by transferring the fish to aquaria free of test chemical. Remove 5 fish after 1, 2, 4, and 7 days and analyze to determine release, if any, of accumulated chemical back into clean water. Calculate bioconcentration as concentration in fish tissue divided by concentration in water. Plot bioconcentration factor vs. time to determine if steady-state was reached. If not, use curve fitting to estimate steady-state concentration. Determine rates of bioconcentration and depuration

Special precautions and exclusions
—

Federal Register page (Ref. 3)
16291-2

(Ref. 11, 12). Some preliminary reporting guidance for environmental studies was provided by EPA in the March 16, 1979, *Federal Register* (Ref. 3) and more detailed rules should be forthcoming. In addition, OECD Guidelines for Good Laboratory Practice (Ref. 13) have been developed and adopted.

In the present section, some general GLP guidelines gleaned from the proposed health effects guidelines (Ref. 12), from the March 16, 1979, *Federal Register* (Ref. 3), and from the author's experience will be outlined. This summary should be helpful in anticipating EPA QA requirements for environmental tests so that GLP can be given proper consideration in establishing in-house testing programs or choosing outside contract laboratories to perform testing.

A good QA program for monitoring environmental tests should include among its main ingredients, procedures which ensure that:

1. The study protocol is adequate to provide a valid measure of the response of the test material to the parameter in question (in actuality this is the function of the sponsor; however, it is the author's experience that a prestudy conference between the person supervising the performance of the study (i.e., the study director) and the quality assurance officer is extremely useful in uncovering possible deficiencies in the protocol which may have been overlooked by the sponsor)
2. The protocol is followed during the study
3. Facilities, equipment, and personnel assigned to the project are sufficient in number and capability to expeditiously and accurately complete the required tasks
4. Adequate supervision is provided to coordinate and oversee the studies
5. The required individual tasks are performed accurately and according to prescribed methods
6. Results and procedures are documented and records maintained in such a manner that all facets of the study can be reconstructed in sufficient detail to confirm the quality and integrity of the study
7. QA activities comply with applicable guidelines promulgated by appropriate regulatory authorities

The formation of a QA function should begin with the formulation of a standard operating procedure which specifies the objectives, requirements, and procedures to be used by those given the responsibility for performing the QA activities (i.e., the QA unit). The QA unit should be composed of an individual or group of individuals with the requisite training and experience to recognize deficiencies in the above-listed areas. However, to maintain objectivity, the members of the QA unit should not be directly involved in the conduct of the study under scrutiny.

V. CONCLUSION

From the foregoing it is apparent that environmental effects testing has not achieved the degree of uniformity that has evolved in the health effects testing area. However, developments are underway and we should be seeing far more refinement and consistency in the tests required, the nature of the protocols employed, and the application of good laboratory practices to environmental effects testing.

REFERENCES

1. **Ganz, C. R.,** Environmental effects monitoring, in *Guidebook: Toxic Substances Control Act,* Dominguez, G. S., Ed., CRC Press, Boca Raton, Fla., 1977, chap. 8.
2. Summary and analysis of the Toxic Substances Control Act, in *Guidebook: Toxic Substances Control Act,* Dominguez, G. S., Ed., CRC Press, Boca Raton, Fla., 1977, chap. 5.
3. Toxic Substances Control Act premanufacture testing of new chemical substances, *Fed. Regist.,* 44 (53), 16240, March 16, 1979.
4. **Duthie, J. R.,** The importance of sequential assessment in test programs for estimating hazard to aquatic life, in *Aquatic Toxicology and Hazard Evaluation,* Mayer, F. L. and Hamelink, J. L., Eds., ASTM Spec. Tech. Publ. 634, American Society for Testing and Materials, Philadelphia, Pa., 1977, 17.
5. **Dominguez, G. S.,** The 6th Amendment and TOSCA — contrasts in toxic substance identification and control, *Toxic Substances J.,* 1(4), 349, 1980.

6. Proposed guidelines for registering pesticides in the United States, *Fed. Regist.*, 43(132), 29696, July 10, 1978.

7. **May, W. E., Wasik, S. P., and Freeman, D. H.,** Determination of the Aqueous solubility of polynuclear aromatic hydrocarbons by a coupled column liquid chromatographic technique, *Anal. Chem.*, 50, 175, 1978.

8. **Part 24,** *ASTM Annual Book of Standards,* American Society for Testing and Materials, Philadelphia, Pa., 1974.

9. **Ashworth, R. De B., Henriet, J., and J. F. Lovett,** *CIPAC Handbook, Vol. 1: Analysis of Technical and Formulated Pesticides,* G. R. Row, Ed., CIPAC Ltd., Hertfordshire, England, 1970.

10. **Irving, H. H. and Mahnot, U. S.,** pH Meter corrections for titrations in mixtures of water and dioxane, *J. Inorg. Nucl. Chem.*, 30, 1215, 1968.

11. Food and Drug Administration, Non-clinical laboratory studies. Good laboratory practice regulations, *Fed. Regist.*, 43(247), 59986 — 60025, December 22, 1978.

12. Environmental Protection Agency, Good laboratory practice standards for health effects, *Fed. Regist.*, 44(91), 27362-27375, May 9, 1979.

13. OECD Guidelines for Good Laboratory Practices, Organization for Economic Cooperation and Development, Paris, France, 1981.

Chapter 9

INTERNATIONAL ASPECTS OF TOXIC SUBSTANCES CONTROL

Irving L. Fuller and Breck Milroy

TABLE OF CONTENTS

I. INTRODUCTION AND BACKGROUND

Since the end of World War II, chemical manufacturers have developed and marketed hundreds of new chemical substances and discovered new uses for existing substances, annually. Currently, about 5 million separate compounds are known to exist, with 55,000 in commercial production in the U.S. alone.

The benefits of these developments have not been insignificant. Our lives have been lengthened and made more convenient in countless ways by chemicals and chemical products. However, these benefits have not accrued without costs to human health and the environment. With increasing frequency, it seems, we have witnessed chemical accidents causing intense, high-level exposure to hazardous substances and both acute and chronic health effects.

The town of Seveso, Italy, became known world-wide because an explosion in a chemical plant resulted in the release of dioxin — one of the most potent carcinogens known. In the U.S., the names of Hopewell, Virginia, and the Love Canal in New York gained similar renown.

Increasingly, national and international attention has turned to formulating control measures aimed at preventing such effects, rather than at clean-up after the fact. The obstacles to implementing such approaches have become clear and include inadequate knowledge of the health and environmental effects of chemicals, lack of agreed standards for evaluating hazards, and insufficient number of facilities and trained personnel to conduct necessary evaluations.

In attempting to formulate the most effective legislative remedies, lawmakers in the U.S. and in other concerned countries face the same issue: "to provide society maximum protection from the adverse effects while at the same time not denying it access to beneficial products because of testing procedures that are prohibitive in terms of economic, scientific, or other resources."[1]

National decision makers have begun to recognize that the problems will not be solved by a patchwork of individual national laws. A certain amount of consistency in national approaches must exist for economic, as well as environmental, reasons. Some chemicals have effects that are global in nature, such as chlorofluorocarbons and PCBs, and require concerted international action to address the effects. Differing national testing or assessment requirements potentially could result in the misallocation of scarce technical and scientific resources, as well as the creation of international trade barriers. As a result, countries have begun to recognize the need for agreed standards for laboratories and lab practices, as well as testing standards and protocols.

With this background, the sections that follow will examine the ongoing international efforts of the U.S. as well as chemical control activities of other countries and several international organizations.

II. U.S. INTERNATIONAL ACTIVITIES RELATING TO THE CONTROL OF CHEMICALS

A. Introduction

For some time, the U.S. has acknowledged the need for international action to address the potential hazards of the widespread and growing use of chemicals. As a signatory nation to the Stockholm Declaration,[2] the U.S. government is committed to pursue activities for the protection of man and the environment from such hazards. Within the United Nations framework, as well as in the Organization for Economic Cooperation and Development (OECD), and in bilateral relationships with other concerned nations and the European Communities, the U.S. has been involved actively in international programs designed to bring about a better understanding of, and more effective ways to assess and control, the potential dangers of toxic substances.

The level and emphasis of U.S. effort has been an evolving one, varying from organization to organization. Therefore, this discussion will deal in more detail with certain of the international efforts in which the U.S. is involved than with others.

B. Organization for Economic Cooperation and Development (OECD)
1. General

The Organization for Economic Cooperation and Development (OECD), headquartered in Paris, was

---

[1]   National Academy of Sciences, *Principles for Evaluating Chemicals in the Environment*, 1975, p. 34.
[2]   Final Declaration of the Stockholm Conference on the Human Environment, 1972.

established in 1960 as an outgrowth of the postwar Organization for European Economic Cooperation.[3] Originally consisting of 20 Member States, the major objectives of the OECD were to promote economic growth among members, to aid less-developed countries (OECD members and nonmembers alike), and to encourage the expansion of trade.[4]

The major decision-making body of the OECD is its Council, which is comprised of representatives from all Member States. The Secretariat performs the administrative functions required to carry out directions given by the Council. The OECD Council acts in basically two ways, each requiring unanimous vote by Member States. It may issue Decisions, which are considered "binding on all Members and implemented by them in accordance with appropriate national procedures."[5] Council Recommendations are "mutually agreed Acts which are submitted to the members in order that they may...provide for their implementation."[6] While not binding, recommendations are felt to have a certain amount of influence upon national decision-makers.

Currently, the 24 Member States of OECD include most of the major industrialized nations of the world. The OECD therefore is of particular importance when discussing toxic substances control, since its members account for over two thirds of total world chemical trade.[7] The OECD activities in this sphere take place within the Environment Committee, particularly in the Chemicals Group (formerly the Sector Group on Unintended Occurrences of Chemicals in the Environment) and, as discussed later, in the Management Committee for the Part II Special Chemicals Program.[8] The major chemicals-producing countries of the world (also OECD members) were, in the early 1970s, either enacting or considering environmental legislation aimed at preventing or controlling the hazards of toxic substances.[9] The likelihood of differing national approaches and the potential thereby for the creation of nontariff trade barriers was clear. It was a logical area for the attention of the OECD Environment Committee.

At the first major meeting of the Council of Ministers responsible for environmental policies, in November 1974, the Ministers laid the foundation for what has since become known as the OECD program of "harmonization" of chemicals control policies. The Council adopted a recommendation calling on Member States to develop procedures for assessing the effects of chemicals on the environment and agreeing that assessments of the health and environmental effects of chemicals should be carried out prior to marketing. The Council further noted the desirability of greater collaboration and harmonization among Member States concerning the development of assessment procedures.[10]

Work within the OECD began on the development of a set of guidelines for "Anticipating the Effects of Chemicals on the Environment" to try to develop agreed procedures to generate data that would be internationally acceptable for assessing the effects of chemicals. The Report of this work was completed in April 1977,[11] and led to the establishment of the OECD Chemicals Testing Program with the adoption of a recommendation later that year.[12] The Report had several significant aspects. It recognized: the importance of testing chemicals for environmental as well as health effects, the responsibility of industry for generating the necessary data, and the potential need for additional premarket testing in addition to an initial screening.

---

[3]   Treaty of Paris, signed 14 December 1960.

[4]   The original OECD member states were: Austria, Belgium, Canada, Denmark, France, Germany, Greece, Iceland, Ireland, Italy, Luxembourg, Netherlands, Norway, Portugal, Spain, Sweden, Switzerland, Turkey, U.K., and the U.S. The following countries joined later: Japan (1964), Finland (1969), Australia (1971), and New Zealand. Yugoslavia is a special status country.

[5]   Organization for Economic Cooperation and Development, *OECD and the Environment*, OECD, Paris, 1979, p. 9.

[6]   Id.

[7]   OECD, *The Control of Chemicals within the OECD Context, ENV/CHEM*, 79, 22, 1980.

[8]   The Environment Committee was established in 1970 by the OECD Council for an initial period of 5 years, in recognition of the need for growth and development to be carried out in a way that is compatible with environmental protection. Its mandate was renewed in 1975 to examine common environmental problems, to review and consult on environmental actions taken by member states, to provide member states with policy options and guidelines, and to encourage, where appropriate, the harmonization of environmental policies among member states. Several specialized subgroups have been established to deal with specific aspects of environmental policy, such as air, water, chemicals, waste management, urban affairs, and land use, noise, energy, and transportation.

[9]   See discussion of national legislation, Sections III and IV, *supra*.

[10]   *The Assessment of the Potential Environmental Effects of Chemicals*. Recommendation adopted 14th Nov. 1974. C(74)215.

[11]   *Anticipating the Effects of Chemicals on the Environment*, Organization for Economic Cooperation and Development, Paris, 1 April, 1977.

[12]   *Guidelines in Respect of Procedures and Requirements for Anticipating the Effects of Chemicals on Man and the Environment*. C(77)97 Final.

Finally, the Council directed the Environment Committee to "pursue a programme of work designed to facilitate the practical implementation of this recommendation, with particular attention to the need for further development and improvement in respect of experimental techniques and for the validation of the capability of laboratories for performing tests."[13]

The OECD Chemicals Group assumed responsibility for the Chemicals Testing Program. Six Expert Groups were established, under a lead country approach: five on particular health and environmental effects testing and one to examine stepwise approaches to testing. The Groups, except for the last, were directed to submit final reports of their activities within 2 years, or by the end of 1979. They were:

- Physical/chemical properties: lead country, Federal Republic of Germany
- Ecotoxicology: lead country, Netherlands
- Long-term toxicology: lead country, U.S.
- Short-term toxicology: lead country, U.K.
- Degradation/accumulation: lead countries, Japan, Federal Republic of Germany
- Step-systems: lead country, Sweden

As the OECD Chemicals Testing Program got underway, momentum for even more far-reaching international action was building. In April 1978, primarily as a result of a U.S. initiative, the Government of Sweden hosted a meeting of 16 nations concerned about regulation of toxic substances in Stockholm.[14] Representatives at the Stockholm meeting agreed on the need for concerted international action and discussed prerequisites for such action. They identified several areas in which immediate work was necessary, including the development of consistent test methods; procedures for good laboratory practice (as well as a means to achieve compliance); mechanisms to facilitate information exchange; international exchange of data on health and safety and provisions for protecting the proprietary value of data; the development of consistent methodologies for analyzing the economic impacts of regulation; and an agreed glossary of key terms and definitions.[15]

Participants acknowledged that some of these activities were being carried out already within the OECD. They agreed that "the OECD seems to be the logical forum in which expedient action could be taken on the near term job of harmonizing national regulatory schemes.[16,17] They therefore recommended that the OECD expand the Chemicals Program to include these elements.

The OECD Environment Committee and Chemicals Group gave careful consideration to the recommendations resulting from the Stockholm meeting at subsequent meetings during the spring. While certain of the recommended activities were either underway or could be accommodated within the existing Chemicals Testing Program, it was clear that additional efforts would have to be initiated, and that special funding would be required for the remaining activities that had been proposed.

The proposals formulated as a result of the Stockholm meeting were endorsed in principle by both the Chemicals Group and the Environment Committee in April 1978, and in the fall, the OECD Council formally took a Decision establishing an additional program to deal with proposals.[18]

This additional program is formally referred to as the Special Program on the Control of Chemicals and also is called the Part II Program (to differentiate it from the Part I, Chemicals Testing Program). The OECD Council provided the Special Program a mandate for expert groups organized under the lead country approach, under the guidance of a Management Committee, to undertake work in the areas of good laboratory practices, confidentiality of data, a glossary of key terms and definitions, and an expanded Information Exchange Program.[19] Using the example of the Chemicals Testing Program, Expert Working Groups were established for the first three areas, with U.S. serving as lead country for the Good Laboratory Practices Group, France for the Group on Confidentiality, and the Federal Republic of Germany for the Glossary

---

[13] *Id.*

[14] Australia, Austria, Belgium, Canada, Denmark, Federal Republic of Germany, Finland, France, Italy, Japan, the Netherlands, Norway, Sweden, Switzerland, the U.K., and the U.S.

[15] *The Control of Toxic Substances,* Proceedings from an International Meeting arranged by the Swedish National Products Control Board, Hasselby Castle, Stockholm, 11 to 13 April 1978, p. 12.

[16] Bureau of National Affairs, *International Environment Reporter* (IER), 10 February 1978, p. 27; 10 May, p. 121.

[17] *IER,* 10 May 1978, p. 123, from remarks made by EPA Administrator Douglas M. Costle.

[18] *Decision of the Council Concerning a Special Programme on the Control of Chemicals,* C(78) (Final), 21 Sept., 1978.

[19] *Id.*

Group. Because the Special Program would involve only those countries interested and would be funded outside the OECD Environment Committees and Chemicals Group mechanism, by discrete contributions from interested Members, the Management Committee was established in 1978 and the U.S. was elected to the Chair for the initial 2 years. The Management Committee mandate was to deal with resource allocation questions, and to review and give direction to the progress of the Expert Groups.[20] Member States also decided that the Chemicals Group would meet from time to time at High Level to give direction to the work of the Chemicals Program.

OECD Environmental Ministers, meeting in Paris in May 1979, called the first of these meetings for May 1980, to review the work of the Chemicals Program to date, to determine areas where agreement was currently possible, and to provide guidance for the future activities of the Chemicals Group.[21] The High Level Meeting was scheduled for 19—21 May, 1980, in Paris, and a Steering Committee was formed to begin careful preparations.

The efforts of the Steering Committee, combined with those of the OECD Secretariat, resulted in a remarkably successful enterprise.[22] The High Level Meeting Participants endorsed the work of the five Expert Testing Groups, agreeing in principle to adopt their Test Guidelines, insofar as completed and validated, and to support an editing panel to finalize and polish the Guidelines. They agreed on the need for an updating mechanism to incorporate advances in the state of the art into the Test Guidelines. The Expert Group on Good Laboratory Practices (GLPs), though a Part II Group and underway only since late 1978, produced a set of General Principles of Good Laboratory Practice, which the High Level Meeting delegates agreed to recommend for adoption. This recommendation was made with the understanding that the Group would continue working to investigate issues of compliance with the GLPs and would examine the GLPs with reference to Test Guidelines to identify and correct gaps and overlaps.

The Step Systems Group had developed the components of a minimum premarketing set of data (MPD) that should be available to regulators to assess the effects of new chemicals. The High Level Meeting delegates agreed that this "MPD", with provisions for its flexible application, should be generated for new chemicals, and endorsed the continuing work of the Step Systems Group to formulate a logical framework for moving beyond the MPD — to identify criteria for utilizing the MPD most effectively and for judging when it is necessary to undertake testing beyond the MPD.

Delegates also agreed to adopt the principle that data generated in one country in accordance with OECD Test Guidelines and GLPs would be accepted in OECD Member States for purposes of assessment. Although this principle, called Mutual Acceptance of Data, was not a source of controversy in the U.S., some other OECD Member States traditionally had been opposed to accepting data that had not been generated indigenously.

Finally, because of the magnitude and importance of the work that remains to be done as well as the fact that certain activities, such as the updating mechanism, will require a secure and continuing level of OECD resource allocation, the participants agreed to recommend that the Council extend the Special Program for an additional 3 years. The Council adopted the principle of Mutual Acceptance of Data and extended the Special Program in May 1981. They also called for another High Level Meeting, which will occur in October 1982.

Since 1973, the OECD has consistently demonstrated its capability to deal with complex and technical issues related to the control of toxic substances.[23] The success of the 1980 High Level Meeting impressed even critics of the OECD.

The head of the U.S. Delegation, the Administrator of the Environmental Protection Agency, spoke of a new era in international cooperation on toxic substances issues, terming the result of the High Level Meeting a "significant breakthrough" because of the consensus that was reached on the necessity for premarket screening of new chemicals and on a common set of tools for national regulators.[24]

[20] *Id.*

[21] *Id.*

[22] Organization for Economic Cooperation and Development, *OECD High Level Meeting on Chemicals, 19 to 21 May 1980 Chairman's Statement*, Press/A(80)34, Paris, 21 May 1980.

[23] *See*, for example, OECD Council Decision on PCBs, *Protection of the Environment by Control of Polychlorinated Biphenyls* (Decision adopted on 13 February 1973) C(73)1(Final), and two OECD assessments of the CFC/ozone issue: *First Interim Report on Fluorocarbons: An Assessment of Worldwide Production, Use and Environmental Issues* (1976), and *The Economic Impact of Restrictions on the Use of Fluorocarbons* (1978).

[24] Remarks by EPA Administrator Douglas M. Costle at the close of the High Level Meeting on Chemicals, 21 May 1980.

2. Outlook

The OECD is a major forum for U.S. efforts in the area of international harmonization of the control of chemicals. It is by no means the only one, as will become clear. However, the OECD provides a promising avenue for progress in areas that are crucial to achieving a workable international chemical control scheme.

3. OECD

The Second High Level Meeting of the Chemicals Group was scheduled for October 26 to 28, 1982 in Paris. Most of the Part II expert working groups completed their work and presented final reports to the Management Committee in late 1981 or early 1982. Participants will be reviewing progress made in implementing the conclusions of the first High Level Meeting and providing guidance for on-going Chemicals Program activities. Participants also will be asked to recommend action in three areas: mutual recognition of GLP compliance schemes, information transfer on chemicals, and data interpretation guidance developed under the Hazard Assessment Project of the Step Systems Group. Certainly, difficult problems remain to be faced. It is, of course, still unclear the extent to which problems will arise in national implementation of the agreements reached already. The issue of confidentiality of data and the continued work of the step systems group are not only technical issues but also difficult policy considerations as well.

C. United Nations

Within the framework of the United Nations, several international programs have been undertaken dealing with aspects of toxic substances control. UN attention was first drawn to the issue largely as a result of the Stockholm Conference on the Human Environment in 1972. The Stockholm Declaration called for improved international acceptability of testing procedures through "development of international schedules of tests for evaluation of the environmental impact potential of specific contaminants or products. Such a schedule of tests should include consideration of both short-term and long-term effects of all kinds, and should be reviewed and brought up to date from time to time to take into account new knowledge and techniques."[25]

1. United Nations Environment Programme (UNEP)/International Register of Potentially Toxic Chemicals (IRPTC)

The 1972 Stockholm Conference resulted *inter alia* in the establishment of the United Nations Environment Programme (UNEP), whose headquarters are in Nairobi, Kenya. The UNEP cooperates with several other UN bodies on chemicals activities and "should be understood as a coordinating, catalyzing and financing body, rather than as an organization which plans activities internally (though it *has* internal projects)."[26]

Prior to the Stockholm Conference, the Scientific Committee on Problems of the Environment (SCOPE), a subsidiary body of the International Council of Scientific Unions, recommended that data concerning the environmental effects of chemicals be gathered and assembled into a register. This concept was endorsed at the Stockholm Conference, which called on the Secretary-General "drawing on the resources of the entire United Nations system and with active support of governments and appropriate scientific and other international bodies, to develop plans for an International Register of Data on Chemicals in the Environment, based on a a collection of available scientific data on the environmental behavior of the most important man-made chemicals, and containing production figures of the potentially most harmful chemicals together with their pathways from factory via utilization to ultimate disposal or recirculation."[27]

In 1975, UNEP assumed responsibility for this register, which has become known as the International Register of Potentially Toxic Chemicals (IRPTC). The IRPTC itself began work in 1976, in cooperation with the World Health Organization and, when fully operational, should facilitate access to many types of scientific and regulatory information on chemicals. Basic objectives of the IRPTC are: to promote efficient use of national and international resources available for the evaluation of the effects of chemicals and their control by providing access to existing data on chemicals; the identification of gaps in existing knowledge

---

[25] Stockholm Declaration, Recommendation 74.

[26] DeReeder, P. L., *Environmental Programs of InterGovernmental Organizations* (UNEP-6-1), The Hague, Martinus Nijhoff, 1977.

[27] Stockholm Declaration, Recommendation 74(e).

on the effects of chemicals and the research needed to fill the gaps; identification of the potential hazards from chemicals; and the distribution of information about national, regional, and global policies, regulatory measures, and standards.[28]

Immediate efforts of the IRPTC were directed at gathering information about short-term effects of chemicals. However, ultimately the IRPTC will be designed to include data on physical and chemical properties; environmental fate; statistics on production, transportation, and use; and information about disposal.[29]

Basic information categories have been developed and IRPTC files contain information on about 34,000 chemicals. Data are contributed by other UN organizations, such as the World Health Organization, International Agency for Research on Cancer (IARC), and International Labor Organization (ILO), as well as other international organizations and national governments.

Due to resource constraints, IRPTC efforts in the near term are likely to concentrate on building contacts with other information networks and data banks. The U.S. has provided its support to date through EPA, which acts as the national contact point for the IRPTC, the National Institute of Occupational Safety and Health (NIOSH), and the National Library of Medicine (NLM). Contacts have also been made with the Commission of the European Communities Environmental Chemicals Data Information Network (ECDIN), the International Council of Scientific Unions, the OECD, and the International Center for Industry and the Environment, among others. National correspondents for IRPTC have been designated in 83 countries to date.[30]

The IRPTC clearly could become an invaluable tool for international cooperation on toxic substances issues. A principal problem has been the lack of up-to-date, easily accessible information about the effects of chemicals. The UNEP Governing Council recognized the potential of IRPTC and at its eighth Session in 1978 endorsed the IRPTC objectives and called upon Member States to take necessary measures to strengthen the Register, by improving support at the national level. Ultimately, the usefulness and effectiveness of the IRPTC depends upon the contributions of individual national governments, industry, the scientific community, and other international organizations.

2. International Agency for Research on Cancer (IARC)

The International Agency for Research on Cancer (IARC) was established in 1965 as an autonomous body within the framework of the World Health Organization to "promote international collaboration on cancer research,"[31] particularly with respect to the role that may be played by environmental factors in inducing cancer. IARC has collected and computerized data on environmental carcinogens from 60 geographical areas and publishes bulletins on information about cancer research currently underway worldwide. By its furtherance of communication concerning chemical carcinogensis testing, it is hoped that duplication in research efforts can be avoided.

3. World Health Organization (WHO)

The World Health Organization (WHO) has been involved with several activities of importance for international efforts to control toxic substances. In addition to its initial cooperative efforts with the IRPTC, WHO has several projects designed to fulfill the objectives of its broad program to promote environmental health. The major components of the program are:

● Technical advisory services for providing Member States with basic sanitary services
● Establishment of environmental criteria on health effects of pollution in air, water, food, land, and in the work environment, and assistance in developing national standards
● Assistance in methodology for and implementation of programs of environmental health monitoring levels, trends, and effects of pollution exposures
● Assistance in establishing or strengthening national institutions required to manage national health programs[32]

[28] United Nations Environment Programme, *Experience of the United Nations Environment Programme (UNEP) in the Exchange of Information on Potentially Toxic Chemicals.* May 1980, p. 3.
[29] DeReeder, *supra,* n. 2 at UNEP 6.2-2.
[30] UNEP, *supra,* n. 4 at 6-8.
[31] DeReeder, *op. cit.*
[32] DeReeder, *supra,* n. 2 at WHO 6.1-2.

Some of these elements have been the responsibility of the Environmental Health Criteria Program, established in 1973 to promote a united approach to determining maximum permissible levels of exposure to different pollutants in several media and to "promote coordination of relevant national and international collaborative research (especially on comparability of development of methodologies for toxicological and epidemiological tests)."[33] The Program has produced a series of criteria documents, summarizing existing information on certain chemical substances.

In 1978, WHO, at the 31st World Health Assembly, agreed to strengthen its efforts in the area of evaluating the effects of chemicals on health.[34] A new International Program on Chemical Safety was established.[35]

The WHO International Program on Chemical Safety (IPSC) assumed the research objectives of the earlier criteria program and coordinates toxic substances control activities of other United Nations agencies. The Chemical Safety Program will continue selecting priority chemicals for testing (under international coordination) and evaluation, reviewing previously prepared Criteria Documents, and preparing new ones. Other objectives are to: promote appropriate methodology for toxicology and epidemiology testing; promote research and manpower development; promote development of information on chemical accidents and emergencies; and to undertake cooperative projects with Member States on problems of chemical safety.[36] WHO also has collaborated with the Food and Agriculture Organization (FAO) in evaluating the risks of food additives and pesticide residues and in recommending safe levels. This joint enterprise is known as the *Codex Alimentarius*.

III. EUROPEAN ECONOMIC COMMUNITY (EEC)

The activities underway in the OECD and in the UN to bring about a unified approach to the evaluation and control of toxic substances are not the only multilateral efforts that have been undertaken with that aim. The European Economic Community (EEC) recently adopted a Directive on toxic substances that mandates "harmonized" legislation in its Member States.[37] In order to understand more fully the implications of this Directive, it will be useful to clarify the organization, composition, and operations of the EEC.

The European Economic Community, one of three European Communities, was established in 1957 to promote common economic policies and to ensure that national actions did not result in barriers to trade among the Member States.[38] The EEC Member States are Belgium, Denmark, Federal Republic of Germany, France, Greece, Ireland, Italy, Luxembourg, the Netherlands, and the U.K.[39]

The major policymaking body of the EEC is the Council of Ministers (EC Council), consisting of representatives of the Member States. Council representation may change, depending upon the issues under consideration. (For example, the toxic substances legislation was considered by environmental ministers of the Member States.) The administrative functions of the EEC are handled by the Commission (EC Commission), which is staffed by civil servants selected from the Member States, though they operate independently of their national governments.[40]

Responsibility for applying and interpreting the Community treaties rests with the Court of Justice. The European Parliament, which votes on the budget and advises on proposed actions, originally consisted of

---

[33] *Id.*

[34] *WHO Official Records,* 247, 18, 1978.

[35] *Ibid.*

[36] World Health Organization, *International Program on Chemical Safety: Proposed Activities 1980/81.* No date.

[37] Council Directive of 18 September 1979, Amending for the Sixth Time Directive 67/548/EEC on the Approximation of Laws, Regulations, and Administrative Provisions Relating to the Classification, Packaging, and Labeling of Dangerous Substances. (79/831/EEC) O.J.L. 259/10, 15.10.79.

[38] Treaty Establishing the European Community signed in Rome, 25 March, 1957. The other two Communities are: the European Coal and Steel Community, established in 1951 and the European Atomic Energy Community, established in 1957.

[39] The original Treaty of Rome establishing the EEC was signed by the governments of six nations: Belgium, France, Germany, Italy, Luxembourg, and the Netherlands. Denmark, Ireland and the U.K. joined the Community on 1 January 1973, under the terms of the Treaty of Accession, Paris, 1 January 1972. The Community will soon grow to 12 members with the addition of Spain, Portugal, and Greece.

[40] The Council and Commission serve not just the EEC, but the other Communities as well. Originally, each Community had its own Council and Commission, but this situation changed as a result of the Convention relating to certain Institutions common to the European Communities (an Annex to the Rome Treaties) and the Treaty Establishing a Single Council and a Single Commission of the European Communities, signed in Brussels 8 April 1965.

198 appointed representatives from the Member States. In mid-1979, the first direct elections were held throughout Europe for 5-year terms in the 410-member Parliament.[41]

The EC Council can take several kinds of regulatory action — Regulations, Directives, Decisions, Recommendations, and Opinions — all requiring unanimous vote of the Members. The first three types of action bind the Members to their provisions, but only Directives specifically require Member States to implement them by means of national legislation within 2 years time. Regulations are binding without additional national regulatory measures; Decisions require formal legislation or regulations to implement them.[42] Recommendations and Opinions are not binding to the Member States.

The beginnings of the EEC approach to toxic substances control can be found in 1973 Community efforts to formulate an environmental policy. As in other parts of the world, increasing attention was being given in Europe to environmental problems in the early 1970s. As a result, the EC Council in late 1973 adopted a set of environmental principles and guidance for their implementation. Among the principles adopted were those advocating the preventive approach to pollution control and the consideration of environmental impacts of policy actions at an early stage in the policymaking process.[43] In addition, Member States agreed on the need to develop environmental policy "in harmony among Member States, rather than in each country in isolation." This principle reflects earlier agreement and community policy on harmonization of environmental legislation.[44]

The 1973 Action Program was divided into 3 broad areas: prevention of pollution and nuisances, actions to improve the environment, and international actions. Specific concern was voiced about the potential threats from chemicals in the Program: "The protection of man and his environment calls for special attention to be paid to products, the use of which may lead to harmful consequences for man and his environment."[45] The Council recognized that existing community measures might be inadequate to protect man and the environment from the hazards of chemicals and that prohibitions or restrictions of use and marketing might be required. It specifically directed the Commission to investigate measures for premarket assessment of chemicals for safety.

It must be kept in mind that such measures were already in force in some countries and under consideration in several others.[46] One of the EEC countries considering introducing a chemicals control law was France. In accordance with agreed information exchange procedures,[47] the government of France notified the Commission in June 1975 that it intended to introduce such legislation. The Commission, in July, requested the French government to delay the introduction of its legislation because the Commission was preparing its own proposal, in the form of a Sixth Amendment to a 1967 Directive on Classification, Packaging, and Labeling of Dangerous Substances and Preparations.[48]

As proposed, the Amendment would have required the mandatory notification of new chemicals, with certain exceptions, by manufacturers. The notification process would consist of notifying the responsible national authorities of the marketing of a new chemical and submitting a dossier of technical information indicating the results of a base set of tests done to evaluate the risks of the substance for man and the environment.

The proposed Sixth Amendment was debated for some 2 1/2 years, passing the Council on 19 June 1979. It was formally adopted on 18 September 1979.[49] Consideration of the measure involved some controversy, and political compromises were required in order to secure its passage. The extent of disagreement may not be known because debates were not public. However, probable areas of debate were: the need for, and duration of, an interval between notification and marketing; exemptions; contents of base

[41] For background see Direct Elections Act of 20 September 1976. The Parliamentary seats are divided as follows: U.K. (81); Ireland (15); France (81); Federal Republic of Germany (81); Italy (81); Netherlands (25); Belgium (24); Denmark (16); Luxembourg (6). For a general discussion of the elections and the new Parliament, see EC, *Europe Goes to the Polls,* Luxembourg, 1979.
[42] A recent Council Decision, for example, required Member States to take such steps as were necessary to ensure that aerosol use of CFCs was reduced by at least 30% by the end of 1981. This Decision did not require any formal action in FRG — government had negotiated a voluntary agreement with industry to bring about the required reductions.
[43] Declaration of the Council of the European Communities. . .on the Program of Action on the Environment. Official Journal of the European Communities (O.J.C.), 112, 20.12.73.
[44] Agreement on Information Exchange and Harmonization. O.J.C. 9 of 15.3. 1973.
[45] Action Plan, *supra* n. 7.
[46] See discussion of national initiatives, Sections III and IV, *supra.*
[47] O.J.C. 9, 15.3. 1973. *Op. cit.*
[48] Council Directive 67/548/EEC of 27 June 1967.
[49] Council Directive of 18 Sept. 1979, *supra,* n. 1.

set of tests; triggers for additional testing beyond initial screening, and indicators for that testing; public access to notification information; and provisions to govern imports.

The Sixth Amendment (hereinafter called simply the Amendment), is designed to "protect man and the environment against the potential risks which could arise from the placing on the market of new substances,"[50] and to eliminate barriers to trade. The Amendment establishes a system of mandatory notification for new chemicals 45 days prior to marketing.[51] "New" chemicals are those not included on the "inventory of substances on the Community market," which the Commission is to complete by 1984.[52]

A. Inventory

The Commission has published a core inventory (ECOIN) and compendium of known substances (similar to the TSCA candidate list). Manufacturers had until December 31, 1982, to report for the final inventory (EINECS), which is expected in 1984. With certain exceptions,[53] new substances may not be placed on the market within the community, unless notified, packaged, and labeled in accordance with the Amendment.[54] A proper notification includes a dossier of technical information stipulated in the Annex, consisting of physical-chemical properties of the substance and the results of toxicological and ecotoxicological investigations.[55] The notification must also contain information about unfavorable effects, proposed classification and labeling, and recommended precautions for safe use.[56]

Subsequent notifications are required if production volume reaches certain levels, if new knowledge becomes available about health or environmental effects, if new uses are intended, and if there are any changes in the chemical properties of the substance as a result of a modification.[57] Annex VIII to the Amendment sets forth the production levels that require additional notification and procedures for determining test requirements. Thus, it is not just one notification but a series of up to three, depending on production volume.

A dossier submitted in one country is forwarded, either completely, or in summary form, to the EC Commission, and from the Commission, may be forwarded to other Member States.[58] Member States may question information or suggest that additional testing be carried out.[59] At the end of the 45-day premarketing period, the substance may enter the market in the country of notification and in all Member States of the Community.

While the EC action appears fairly comprehensive in scope, a number of issues remain unclear, and raise questions about how it will function when fully implemented. Of course, it must be kept in mind that it is a framework, filled out by national legislation. This raises the first question — the extent to which national legislation can vary from the 6th Amendment and still be considered in compliance.

All but one of the member states, Belgium, have begun to implement the Sixth Amendment, by adopting either new legislation or administrative procedures.

All of the approaches parallel the Sixth Amendment fairly closely, though some go beyond its provisions. For example, under the provisions of the Danish law, a chemical substance must have been introduced onto the *Danish* market by October 1, 1980 to be considered an existing chemical (and therefore not subject to premarketing notification requirements). The German law covers existing, as well as new chemicals. Divergences from the Sixth Amendment that violate the Treaty of Rome will be brought to the European Court of Justice for resolution.

The inventory will be static; anything not included on this inventory is considered "new" and a premarket notification must be submitted. This means that the same new substance must be notified by each individual manufacturer, and that the same information and data will be required of each notifier. The 6th Amendment states that manufacturers can satisfy the notification requirements by referring to the results of studies carried out by other notifiers, "provided the latter have given their agreement in writing."[60]

---

[50]   Directive (79/831/EEC), findings.

[51]   *Id.*, Article 6.

[52]   *Id.*, Article 13(1).

[53]   Outlined in *Id.*, Article 8(1).

[54]   *Id.*, Article 5.

[55]   *Id.*, Annex VII.

[56]   *Id.*, Article 2(2) (a)-(h).

[57]   *Id.*, Article 6(4).

[58]   *Id.*, Arts. 9, 10(1).

[59]   *Id.*, 10(2).

[60]   Directive 79/831/EEC, Article 6(2).

The Commission will maintain a list of notified substances; but it is not clear if this list will be published. Claims of confidentiality should be anticipated. How can a manufacturer ascertain whether a substance he intends to market has actually been notified, or how can he contact the first notifier if the identity of the first notifier has been claimed confidential? Even assuming contact, there is no requirement that the first notifier *must* negotiate rights to use the data.

Policies have not been enunciated with respect to the requirements for multinational corporations. Technically, notification in one Member State ensures that a substance can be marketed in all Member States, and a multinational corporation should simply be required to notify in one of the countries in which it is operating. However, there appears to be a great sensitivity to the possibility of selective notification, for importers as well as for domestic manufacturers, i.e., avoiding notification to authorities that are assumed to be more rigorous, or choosing to notify in a country with an obvious lack of sufficient resources or motivations to process carefully a large number of notifications. Currently, therefore, it appears that multiple notifications may be required in such instances.[61]

In any event, the EEC measure and its implementation cannot be viewed in isolation. The Commission and individual Member States are represented in the OECD, and, for example, the OECD Test Guidelines will be incorporated directly into the Directive as a technical annex that will be supplied by the Commission.

In addition, the U.S. and the EC Commission have held regular bilateral discussions on toxic substances issues since mid-1978. The Commission initially sought to begin formal bilateral negotiations. It requested, and received, a mandate from the Council to negotiate with the U.S. on technical questions related to TSCA, such as recognition of test data.[62] The meetings have been discussions only, not negotiations, primarily because of ongoing activities concerning the major issues within the OECD. They have however, resulted in productive dialogues about various regulatory aspects of implementation of TSCA and 6th Amendment,[63] and it is anticipated that they will continue to play an important role.

IV. OTHER APPROACHES TO THE CONTROL OF TOXIC SUBSTANCES

Because the U.S. and the Member States of the European Community constitute the majority of major chemicals producers in the world,[64] the U.S. Toxic Substances Control Act (TSCA) and the EEC Directive on Classification, Packaging, and Labelling of Dangerous Substances have received most of the attention in discussions of chemicals control legislation and its implications. However, as noted in the Introduction of this chapter, a number of other laws have been enacted in the past decade to deal with hazards posed by dangerous chemical substances. While sharing a common purpose, legislators in different countries have chosen several different ways to accomplish the purpose. To illustrate some of the differences, a discussion of two of the earliest pieces of legislation, both enacted in 1973, is presented below.

A. Japan

Japan is a chemical producer of increasing importance in the world market and a net exporter of chemicals.[65] The rapid industrialization of Japan, particularly in the post-World War II period, largely in the absence of the constraints of environmental (pollution control) legislation, began to have painful consequences by the late 1950s. Residents of the town of Minimata, suffering from a painful, incurable disease, were found to have mercury poisoning, contracted from eating fish contaminated by discharges from a neighboring chemical plant. Over 1400 people had been affected by 1975.[66] Another painful, sometimes fatal, illness was diagnosed among residents of the Jintsu River Basin as cadmium poisoning from mining and smelting operations.

However, it was a 1968 "unfortunate" polychlorinated biphenyl (PCB) experience,[67] that prompted efforts to bring about a comprehensive chemicals control law. (PCBs are directly toxic to humans when ingested and pose long-term problems because they are not degradable, but persist in the environment and

[61] Statement by Herr Stuffmann at Seminar on Toxics Legislation, Bonn, West Germany, 13 June 1980.
[62] *IER*, 10 June 1978, p. 121.
[63] See U.S./EC Discussions, Transcripts of Meetings held in September 1979, January 1980. Washington, D.C.
[64] Germany alone has the three largest chemical producing companies in the world — Bayer, BASF, and Hoechst.
[65] Organization for Economic Cooperation and Development, *L'industrie Chimique*, Paris, OECD (1978), 20.
[66] OECD, *Environmental Policies in Japan*, Paris, 1977, 14.
[67] Philosophy of Bio-Accumulation Test in "Law Concerning the Examination and Regulation of Manufacture, etc., of Chemical Substances" in Japan, Chemical Products Safety Division, Basic Industries Bureau, MITI, Tokyo, September 1978.

thereby accumulate as they move up the food chain.) In this instance, PCBs leaked from a pipe in a factory making rice bran oil. It was mixed with the rice bran, causing acute poisoning and some deaths among those people who ate the contaminated rice bran. As an OECD study points out, "lethal cases of environmental pollution, although not unknown elsewhere (people died from air pollution in Donora, Pennsylvania, in 1948, and in London, in 1952), have been particularly numerous in Japan."[68] Most instances of chemical poisoning could be addressed under existing water pollution, air pollution, or pesticide laws.[69] However, the source of damage in the PCB incident was not emissions or discharges for which standards could be set.

The Law Concerning the Examination and Regulation of Manufacture, Etc., of Chemical Substances (hereinafter called the Chemical Substances Control Act)[70] was enacted to address cases such as that presented by PCBs, before marketing and widespread use of a chemical substance. The introductory portion of the law makes this clear: "This law, in order to prevent pollution of the environment by chemical substances which have persistence or other such properties and which may possibly be harmful to human health, has as its purpose the establishment of a system of examination to determine, before the manufacture or import of new chemical substances, whether such substances have these properties or not, and the implementation of necessary regulations in the manufacture, import, use, etc., of chemical substances having these properties."[71]

1. New Chemical Substances

In order to carry out this purpose, the law first provides for the compilation of a List of Existing Chemical Substances.[72] Persons wishing to manufacture or import any chemical not on the list, a "new chemical substance," must provide notification in advance to the Minister of Health and Welfare and the Minister of International Trade and Industry, (MITI)[73] of the name of the new chemical substance and other information as prescribed by Ministerial Order.

Required additional information that has been already prescribed,[74] includes the substance name, structural formula, physical-chemical properties, and composition, use, amount to be produced or imported annually for 3 years, and the name and location of manufacturer.

2. Classification

On the basis of all available information, the Ministers have 3 months to determine how to categorize the substance. The law establishes the category of "specified chemical substance," which:

1. (a) Does lend itself easily to chemical changes caused by natural effects and is also easily accumulated in biological organisms;
   (b) When ingested continuously, there is a fear that it may be harmful to human health;
2. Lends itself easily to changes caused by natural effects, but produces substances that correspond to (a) or (b) above[75]

If the new substance falls into neither of the categories, the notifier is told that he may proceed with manufacture or import, and the name of the substance is published by MITI.

A substance that can be categorized as a specified chemical substance is placed on the List of Specified Chemical Substances and its manufacture or importation requires special authorization from the government.[76]

However, if the substance cannot, with certainty, be classified as either "safe" or "unsafe"[77] the Ministry must determine promptly, "on the basis of the results of tests carried out by the manufacturer or importer for the new chemical substance," how to categorize the substance and inform the notifier. The

[68] *Environmental Policies, supra,* n. 4.
[69] In addition, a law was passed in 1973 to provide compensation to the victims of pollution-related damage. Law No. 111, 1973.
[70] Law No. 117 (18 September 1973), in force 16 April 1974.
[71] *Id.,* Article 1.
[72] *Id.,* Art. 3(3) .
[73] *Id.,* Art. 3.
[74] Ministerial order concerning the Notification, etc. of Manufacture or Import of New Chemical Substances, Art. 2, (1) — (6) promulgated 15 April 1974.
[75] Art. 2(2) (1) and (2).
[76] Art. 6, 7, 8.
[77] See Japan Chapter, in Stoel, T. S., Miller, A. S., and Milroy, B., *Fluorocarbon Regulation,* Lex Books, Lexington, Mass., 1980.

test requirements are specified in a later Ministerial Order,[78] and must include tests to determine the degradability, bio-accumulation, and toxicity of the substance.

MITI reported[79] that 960 substances were notified between 1974 and 1979: 602 for manufacture domestically and 358 for import. Of these, 629 had been designated as safe by the end of 1979.

3. Regulation

The manufacture, import, or uses of chemical substances can be regulated through the authorization procedures outlined in Articles 6 to 10 of the law. Authorization must be requested from MITI to begin manufacture, import, or use of a specified chemical substance. In order to grant the authorization the MITI must determine that:

1. The capacity for manufacture must not be excessively great in view of the demand
2. The equipment for manufacture must conform to the technical standards prescribed by MITI and the Ministry of Health and Welfare
3. Adequate fiscal basis and technical capacity exist for performing business efficiently

With respect to imports of a specified substance, authorization may be granted only if the importation is necessary to meet domestic demand.[80] Authorizations are considered on a case-by-case basis, with each specified substance considered separately, so that restrictions or conditions can be individually specified. Article 22 provides authority for emergency interventions, as necessary, to "prevent the progress of environmental pollution" by a chemical substance.

4. Existing Chemicals

A second feature of the Law is its coverage of existing chemicals. Recognizing that existing chemicals also present threats to health and the environment, but that their number makes comprehensive examination impossible, a system for review of existing chemicals was established to consider about 800 chemicals.[81] The chemicals are chosen for review primarily on the basis of three indicators:

1. PCB substitutes and chemicals with structural similarities to PCBs
2. Chemicals used in Japan at a level greater than 100 tons yearly
3. Chemicals used in volumes of less than 100 tons a year but with structure deemed "very significant to environment and health"

Toxicity testing is carried out not by industry but by government — by the Ministry of Health and Welfare. As of the end of 1978 a total of 448 existing chemicals had been examined with the result that 209 were designated as safe and 3 thus far designated as specified substances.[82]

Based on these activities, screening of new and existing chemicals, the Ministry of International Trade and Industry of Japan concludes that the "aim of preventing a recurrence of PCB-type problems has been well and truly satisfied."[83]

B. Sweden

Unlike Japan, Sweden is not counted among the major chemical producers of the world, and, in fact, is a net chemical importer.[84] In 1978, the Swedish chemical industry employed approximately 39,700 persons in a population of somewhat over 8 million people.[85]

---

[78] Order prescribing the Items of the Test Relating to New Chemical Substances (Order of the Prime Minister, the Minister of Health and Welfare and the Minister of International Trade and Industry, No. 1 (Promulgated July 13, 1974).

[79] Kobayashi, K., (MITI) Review of the Implementation of the Chemical Control Law of Japan, paper presented to the OECD Chemicals Forum, Paris, 24 October, 1980.

[80] Art. 12.

[81] Kawasaki paper (Review of Current Status of the Chemical Control Law of 1973, Kawasaki, M., Director, Chemical Products Safety Division, MITI, Tokyo, 1979).

[82] *Id.*, p. 5.

[83] Philosophy Regarding Japan's Chemical Substance Screening System and Evaluation Steps, MITI, Chem. Prod. Safety Div., Basic Industries Bureau, Tokyo, September 1978, 7.

[84] OECD, *L'industrie Chimique*, 1978, Paris, 1980, 21.

[85] *Id.*, 15.

Sweden, however, was similarly motivated to evaluate existing controls on chemical substances and acknowledge that they were inadequate to deal with the problems presented by a widely used, nonagricultural chemical with both direct and indirect, acute and chronic, toxic effects.

While Sweden had several pieces of chemical control legislation at the time, with premarketing and licensing and registration requirements for several types of chemicals, coverage extended only to pesticides and herbicides, and to poisons hazardous to man.[86] (For example, a special law was required to deal with the PCB problem.)

To investigate the gaps and overlaps in the legislative situation, the Minister of Agriculture created a special Commission on the Control of the Environment in 1969, which was also to make recommendations on the most effective ways to achieve comprehensive products control legislation. The Commission delivered its report in 1972, and included a comprehensive draft of products control legislation. This measure was endorsed, with few exceptions, by the Government, and became its Bill on Products Hazardous to Man and the Environment.[87]

Of particular interest is the broad scope of application and administrative authority provided by the Act and its Ordinance.[88] All products are covered that, by reason of their chemical or similar properties, are *liable* to cause harm to health or to the environment.[89] Certain duties and responsibilities fall to manufacturers and to government authorities, which become effective upon "the mere suspicion that a chemical may be harmful to man or the environment."[90]

The Act creates a Products Control Board, which is reponsible for implementing the provisions of the Act and Ordinance governing the manufacture, sale, handling, and importing of hazardous products. A National Environment Protection Board and National Board of Industrial Safety have joint supervisory responsibility for ensuring compliance with the provisions of the law.

1. Provisions of the Act

Central to this legislation is the so-called "rule of prudence," reflected in the provision of the Act that "anyone handling or importing a product hazardous to health and the environment shall take such steps and otherwise observe such precautions as are necessary to prevent and minimize harm to human beings or the environment."[91] Manufacturers and importers have the duty to "inquire carefully" into health and environmental effects of the products with which they deal. Authority is provided to prohibit the handling of certain hazardous products, to limit or make it subject to special permissions or conditions.[92] Authority is further given to the responsible authorities to require persons handling hazardous products to provide notification of the activity, as well as all information on the product necessary to assess the health and environmental risks.[93]

2. Ordinance

Against the broad framework of this Act, the Ordinance on Products Hazardous to Man and the Environment[94] provides specifics. The term "hazardous products" is defined to include poisons and dangerous substances, and their manufacture requires a permit from the Industrial Safety Inspectorate or, with respect to dangerous substances, a notification to the Inspectorate.[95] The Ordinance requires clear labeling of poisons and dangerous substances, in Swedish, showing the name of the product and character, hazard warning, precautions, and the name and location of the manufacturer.

Pesticides and allied products are also considered hazardous substances.[96] Their handling is subject to

---

[86]   *OECD Environmental Policy in Sweden*, Paris, 1977, Discussion pp. 26—28 is helpful on evolution of legislation.

[87]   Swedish Code of Statutes, 1973:329 and 1973:334. (Proclamation on Products Hazardous to Man or the Environment), April 27, 1973.

[88]   The Swedish Government has provided a detailed discussion of the Act's provisions and evolution for interested readers. *See* the Act on Products Hazardous to Health and to the Environment: the New Legislation with a Commentary, Minister of Agriculture (no date).

[89]   Commentary, 8.

[90]   *Id.*

[91]   Act, §5.

[92]   Act, §6 and 7.

[93]   Act, §8 and 9.

[94]   Swedish Code of Statutes 1973:334.

[95]   Ordinance, §2-8.

[96]   *Id.* 22-23.

the same requirements as poisons and dangerous substances, and their sale requires advance registration with and permission from the Products Control Board.[97]

The Ordinance deals, in Sections 39 to 45, with another category of hazardous products — PCBs and PCB products. Handling without a permit from the Products Control Board is prohibited. Transformers and capacitors may not be offered for sale unless they conform to specifications; likewise prohibited from sale are paints, inks, caulking compounds, hydraulic and lubricating oils, heat-transfer media, and separate capacitators.

Other products are specified that may not be offered for sale if containing, or treated with, a hazardous substance as defined in the Act. Although the Act, as a framework law, is broadly constructed, it offers a powerful tool for control of harmful chemical substances. This is due to the underlying rule of prudence and the burden of proof that is explicit throughout the legislation.

The Product Control Board is bound to consider the mere risk of harm and does not have to prove that harm will, in fact, result, in order to take restrictive action.[98]

V. CONCLUSIONS AND OUTLOOK

Due primarily to the economic implications of varying national chemical control measures, bilateral and multilateral efforts toward harmonizing control approaches have increased and gathered momentum during the 1970s. The efforts have been motivated as well by the recognition that consistency in approach offers the most promising means to prevent health and environmental dangers that could be posed by hazardous substances.

To date, these discussions and negotiations have been characterized by the participation of only the most industrialized countries of the world. The most important technical work has been conducted by the major chemical producers, those with largest stake in the outcome of any negotiations. However, change is indicated in two recent trends.

First, less industrialized, chemical-importing countries are beginning to express their concerns about health and environmental effects of chemicals. They desire to have a voice in determining aspects of any international control scheme that is developed. Within the OECD framework, attention has been directed to identifying the information needs of chemical importing countries.

In addition, the movement of chemical manufacturing capabilities to the nonindustrialized world is beginning. This is noticeable particularly since the price for petrochemical feedstocks has increased in recent years. Petroleum-exporting countries, with easy access to cheap feedstocks, also have the means to import manufacturing technology.

Both of these factors indicate an increase in the number of nations with a stake in the results of international discussions of chemical control schemes. A broadening of the discussion forum, even an international treaty, open to accession by any interested country, cannot be overlooked.

[97] *Id.* 24-25-26.
[98] Staffan Westerlund, The Swedish Legislation on Toxic Substances, paper presented at the International Seminar on Chemical Control, Bonn, West Germany, 11 to 14 June, 1980.

INDEX

A

Accumulation, chemicals, see also Bioaccumulation, 97, 176
Acetonitrile, 26, 44
Acrylamide, 25, 44
Activated sludge test, 185—186
Actual knowledge that Administrator has been informed, terminology explained, substantial risk reporting, 147
Acute toxicity testing, 24, 29, 34, 45, 96, 98, 142, 174, 192—193
Additional production and use information, 109
Administrative costs, government agencies, 18
Adsorption test, 179—180
Advance substantiation, confidentiality claims, 121—124
Adverse reactions, significant, 34—35
Algal inhibition test, 191
Alkyl epoxides, 25, 44
Alkyl phthalates, 25
Alkyltin compounds, 27
Aluminum Company of America v. DuBois, 16
Amendment, TSCA, 164
Anaerobic digestion, methane and carbon dioxide production test, 186—187
Aniline, 26, 44
Animal effects tests, 192—194
Annual production quantity, chemicals, effect of, 44—45
Antimony, 26, 44
Antimony sulfide, 44
Antimony sulfoxide, 26
Antimony trioxide, 26, 44
Aquatic invertebrates, acute toxicity test, 192
Aqueous photochemical reaction test, 183—184
Aqueous solution, UV and visible absorption spectra test in, 182
Articles, manufacture of, rection products created during, 82—83
Aryl phosphates, 25, 44
Asbestos, 5, 8, 34—35, 85

B

Base set data, 29, 96—98, 175
Behavioral disorders, testing for, 24
Benefits, regulatory action, 19—20, 33, 38—39
  societal, 38, 97, 138, 141, 174, 198
Benzidine dyes, 26
Benzyl butyl phthalate, 27
Bilateral confidentiality agreements, 166
Bill on Products Hazardous to Man and the Environment (Sweden), 210—211
Bioaccumulation, chemicals, see also Accumulation, 45, 137, 174
  pyridine, 58
Biochemical oxygen demand test, 187—188
Biodegradability testing, see also Degradation, 184—188
Biphenyl, 27, 44
Block diagram, premanufacturing notification, 111

BOD, see Biochemical oxygen demand
Boiling/melting/sublimation point test, 180
Bromo-anilines, 26, 44
Business information, confidential, see also Confidentiality, 162—163
Butyl glycolyl butyl phthalate, 27
By-product, 81—84, 103—104
  definition, 81
  exemption from PMN, 81—82
  premanufacturing notification studies, 78, 81—84, 103—104, 108—109, 111—113, 116, 134
  pyridine, 59
  substantial risk reporting studies, 151

C

Candidate list, chemical inventory, 74
Capable of appreciating significance, terminology defined, 144—145
Capital spending, 36
Carbon dioxide evolution test, 187
Carbon dioxide production in anaerobic digestion test, 186—187
Carcinogens, carcinogenicity, and carcinogenesis, 24—27, 38, 42, 45, 59, 97, 137—138, 142
  carcinogenicity/chronic toxicity studies, 29
  dioxin, 198
  pyridine, 59
Cartridges, firearm, 78
Cash flow, economic impact studies, 28, 37
  curves, 22—23, 31—32, 34—35
Categories, see Chemical substances, categories
Causative agent, true, identification of, 150—151
CBI, see Confidential business information
Cellulose decomposition test, 190
Chemical Controls Program, OECD, 200—202
Chemical degradation by oxidation test, 183
Chemical fate tests, summary of methods, 176—188
Chemical identity, chemical substances
  confidentiality studies, 119—129
  disclosure as part of health and safety study, 119—121
    exemption, 119
  environmental testing studies, 175
  litigation, 16
  manufacturer identity linked to, 126
  premanufacturing notification studies, 69—70, 75—76, 104, 106—107, 114, 118—129
  specific substantiation questions of EPA, 124—126
Chemical incompatibility test, 182
Chemical industry
  cost impact and economic impact on, see also Cost impact; Economic impact, 18—39
  management response to ITC, 47
  multinational corporations, 207
  operation of, 2—3
  overall industrial framework, confidentiality issues, 159—161

M

N

O